Perspectives in Animal Ecology and Reproduction

Perspectives in Animal Ecology and Reproduction

— *Volume 7* —

Editor-in Chief
Dr. V.K. Gupta

Editors
Dr. Anil K. Verma
Dr. G.D. Singh

2011
DAYA PUBLISHING HOUSE®
Delhi - 110 002

ISBN 81-7035-746-2
ISBN 978-81-7035-746-9

Published by : **Daya Publishing House®**
4760-61/23, Ansari Road, Darya Ganj,
New Delhi - 110 002
Phone: 23245578, 23244987
Fax: (011) 23260116
e-mail : dayabooks@vsnl.com
website : www.dayabooks.com

Laser Typesetting : **Classic Computer Services**
Delhi - 110 035

Printed at : **Chawla Offset Printers**
Delhi - 110 052

PRINTED IN INDIA

Editorial Board

– Members –

Prof. Clement A. Tisdell

Professor Emeritus,
School of Economics,
The University of Queensland, Brisbane,
Australia
E-mail: c.tisdell@economics.uq.edu.au

Prof. Tej Kumar Sherstha

Head,
Central Department of Zoology,
Tribhuvan University,
Kathmandu, Nepal
E-mail: drtks@ccsl.com.np

Dr. Justin Gerlach

Chief Scientist,
The Nature Protection of Seychelles,
133 Cherry Hinton Road,
Cambridge CBI 7BX,UK
E-mail: justgerlach@aol.com

Prof. Vanitha Kumari

Head,
Department of Zoology,
Bharathiar University,
Coimbatore, India
E-mail: regr@bharathiaruni.org

Dr. Sunil Kumar

Deputy Director (Sr. Grade),
Division of Reproductive Toxicology & Histochemistry,
National Institute of Occupational Health,
Meghani Nagar, Ahmedabad – 380 016, India
E-mail: sunilnioh@yahoo.com

Dr. Deep N. Sahi
Professor in Zoology
Wildlife & Animal Behaviour Lab,
Department of Zoology
University of Jammu,
Jammu – 180 006

Tel 0191-2457136 (O)
Fax 0191-2435256
0191-2481830 (R)
094191-46190 (M)

ISO 9001-2000 Certified

Foreword

"We shall not desist from reconnoiter
And the end of all our exploring
Will be to arrive where we commenced
And know the place for the first time."

T.S. Eiiot

On the above context, the effort of the editors in bringing about this worthwhile publication of Volume 7 of "*Perspectives in Animal Ecology and Reproduction*" is laudable. All living forms have the characteristic to reproduce which separates them from nonliving entities but all living forms interact with their environment that is what we call as ecology. The editors have chosen very correct title of the book and have come out with a series of volume under the heading. The current volume of the series is an agglomeration of diverse research articles bestowed by research workers in the twin

Residence: Lane No. 5, Little Flower School Road, Greater Kailash, Jammu
E-mail: dnsahi@yahoo.com

fields of Animal Ecology and Reproduction which would really be an eye opener reflecting a wide spectrum of the various aspects of the two fields.

The present volume edited in a very lucid language will enliven the quest of the readers thereby leading to their enlightenment,

The earlier 6 volumes of the series have been overwhelmingly well endured. I hope and endorse that present volume will also be well accepted.

I congratulate the editors for this endeavor and for their salubrious contribution to the society at a large.

I wish them all the best.

Prof. D.N. Sahi

Preface

Current patterns of resource exploitation by human kinds do not bode well for the future of bio-diversity. Ecologists Normen Meyers recently estimated that some 600,000 species have vanished since 1950 and today, two of every three species are estimated to be in decline. No one knows the scope of bio-diversity and also that how many species of plants and animals share the planet with human beings. Whatever be the actual number of species, but the preservation of bio-diversity itself is vital to humanity.

India a mega-diversity nation with numerous hot-spots of bio-diversity represents a blend of variety of terrain, forests, physiography. Flora and faunal elements has lured the intelligentia since time immemorial. Ecology and reproduction have become the focus of attention of the scholars, students and scientist and efforts are underway for a better understanding of this higher complex interactions among various organisms and the environment.

Animal ecology and reproductive strategies forms an important part of curriculum of almost of all universities in India and the study of animal ecology is useful to understand the co-relation between the breeding techniques employed by various animals during diverse environmental conditions. The objective of this series has been to collect, compile and disseminate information on the subject of animal ecology and reproduction so as to make the reader updated with the latest in the fields where propagation of scientific knowledge has been enormous in the past two decades.

Every effort has been made to make the book of prime utility for vigorous students taking higher courses in animal sciences and the environmental sciences. The chapters incorporated herein represents only an extremely modest selection of several facts of the breeding ecology of the animals and shall hopefully fill the gaps in the present knowledge of the subject existing today.

This volume like other volumes of the series will be a valuable, and an important research manual, that will stimulate interest and satisfy the need for further knowledge of this existing and rapidly expanding discipline. The articles in this volume cover a wide array of current research topics dealing with both fundamental and applied researches in the fields of animal ecology and reproduction. The articles are highly authoritative and thought provoking and contributed by the scientists in their respective fields. The articles on study of *Vibrio* through ERIC- PCR and REP -PCR for development of fish and human health management; Phagocytosis of charcoal particulate as immunological marker of Azadirachtin exposure; Phenotypic plasticity of *Drosophila melanogaster* at different temperatures; Amelioration of anti-oxidant vitamins on ciprofloxacin induced toxic changes in blood; Adverse effect of ampicilin on brain lipids in albino rats; Estrogen induced standard cytotoxic model in the uterus of rat; Efficacy of ovaprim and ovatide in induced breeding of *Clarias batrachus*; and the photoperiod and biological rhythms in regulation of avian reproduction etc. are important contributions.

We hope that this volume of the book series *Perspectives in animal ecology and reproduction* like our previous volumes will be highly useful to students, scientists and research workers in the animal ecology and reproductive aspects. We are highly grateful to all the eminent scientists for taking troubles in writing chapters and for constant support and encouragement to bring our this book series. We are specially grateful to M/s Daya Publishing House, New Delhi for printing volume in stipulated time.

Dr. V.K. Gupta

Dr. Anil K. Verma

Dr. G.D. Singh

Contents

Section II: Animal Reproduction

Section I
Animal Ecology

Chapter 1

Occurrence of Amphizoic Amoebae in Domestic Water Supply from Lucknow City

Tabrez Ahmad, Kavyanjali Shukla, Newton Paul and A.K. Sharma*

Protozoology Research Laboratory, Department of Zoology, University of Lucknow, Lucknow – 226 007, U.P., India

ABSTRACT

A study was planned to screen the domestic water supply of municipal water from the old as well as new areas of Lucknow city. Samples were collected from both the areas as: (a) from municipal water supply without any water purifier attachment and (b) from municipal supply with water purifier attachment. The samples were screened for microbiological analysis with special reference to protozoans. The biological characterization of amoebic isolates was carried out and it was found that a total no. of 5 strains of amoebae (1 each of *Acanthamoeba culbertsoni, Acanthamoeba rhysodes* and *Naegleria gruberi* from water supply of old Lucknow without any water purifier attachment and one each of *Acanthamoeba rhysodes* and *Naegleria gruberi* from water supply of new Lucknow without any water purifier attachment) were isolated. Presence of

* Corresponding author: E-mail: tabrez.ahmad17@gmail.com

Amphizoic amoebae in domestic water supply devoid of any water purifier attachment is a signal alarming for proper preventive measures before various uses at home, because some of the species of Amphizoic amoebae (*A. culbertsoni, A. rhysodes and N. fowleri*) are known to cause Meningitis affecting CNS and Keratitis affecting eyes.

Keywords: *Amphizoic, Domestic water, Water purifier, Lucknow.*

Introduction

The importance of pure water is realize since ancient time but due to rapid industrialization, urbanization and indiscriminate use of water bodies, no source of water is left unpolluted. Aquatic pollution is mainly caused by sewage various pathogen and chemicals. Among all microorganisms present in water, protozoans are most commonly found organisms and amoebae are the most common protozoan present in the nature. Earlier interest in small free-living amoebae was largely academic, because they were known to play an important role in soil and aquatic ecosystem (Singh, 1975).

Medical interest in amoebae developed when Gross (1849) described a parasitic amoeba from oral mucosa which was later called *Entamoeba gingivalis* F. Losch (1875), a russian zoologist first discovered *E. histolytica* in the stools and ulcers of patients suffering from chronic dysentery. Till 1958, it was generally believed that *E. histolytica* was the only pathogenic amoeba causing amoebic dysentery and other forms of amoebiasis. But these preconceived notions became untenable with the discovery made within the last 5 decades that Small Free Living Amoebae belonging to the genera *Naegleria* and *Acanthamoeba* were pathogenic and causing fatal human diseases affecting central nervous system and eyes. These diseases are Amoebic Meningoencephalitis and Amoebic Keratitis respectively (John, 1993 and Singh, 1985). These discoveries have presented medical science with entirely new diseases in human being. Normally these amoebae do not cause disease, but live as a photograph in their natural habitat (Mainly soil and water) feeding on bacteria; however, as opportunist they produce serious infections in brain and eyes. Because these amoebae have an adaptability to live both as free-living organism and also as endoparasite, thus they are referred as Amphizoic (Page, 1974). Pathogenic strains of free-living amoebae, which are the etiological agents of Primary Amoebic Meningoencephalitis (PAM) (Curson and Brown, 1976) and Granulomatous Amoebic Encephalitis (GAE) (Martinez, 1980 and Visvesvara *et al.*, 2007) and Amoebic Keratitis, appear to be truly ubiquitous organisms. They have been recorded from a variety of environment such as air (Kingston and Warhurst,1968),fresh and brackish water (De Jonckheere *et al.*, 1975, Stevans *et al.*, 1977),domestic water (Cerva and Huldt,1974),from soil (Curson *et al.*, 1978) and from sewage (Singh, 1972). Amphizoic amoebae have been isolated from surface water sources like lakes, river, thermally polluted discharge (Kasprzak and Mazur, 1974; De Jonckheere, 1981), from artificially heated water (Sheehan *et al.*, 2003), domestic tap water (Kilvington *et al.*, 2004), Pandey and Sharma (2006) and from different environmental sources (Rezaeian *et al.*, 2008).

Keeping in view of the significance of these new diseases caused by amoebae in human being, a study has been planned to screen domestic water supply so as to asses their occurrence in water. For this plan water sampling has been done in two sets, one from domestic supply having water purifier attachment and the other without water purifier attachment.

Materials and Methods

To isolate Amphizoic amoebae from domestic water supply of Lucknow city, with or without water purifier attachment, a total 20 samples were collected from old and new areas of Lucknow city (See plan of Experimental Design). For the isolation of amoebae, 15-20ml of sterilized non-nutrient agar (2.5 per cent w/v) without NaCl, pH6.6 to 6.8, is poured into Petri dishes and allowed to set for 24 hours. *Escherichia coli* bacteria from, 24 hours old cultures were used as food of amoebae. These bacteria were grown on nutrient agar slants.The bacteria were spread as a thick suspension on the solidified non-nutrient agar surface in the form of circular patches of about 20-22 mm in diameter. Three to four such bacterial circles can be made in a petridish, which is now ready for inoculation of any substrate under study. About two liters of water from each source was collected aseptically in sterilized bottles and was then filtered with a help of a conical funnel through a sterile filter paper (Whattman No. 1). After the filtration, the inner narrow end of filter paper was cut aseptically into 4 small pieces and each piece was then placed in the center of each bacterial circle in a petridish. The

Experimental Design

The sampling plan and isolated amoebic strains were as follows:

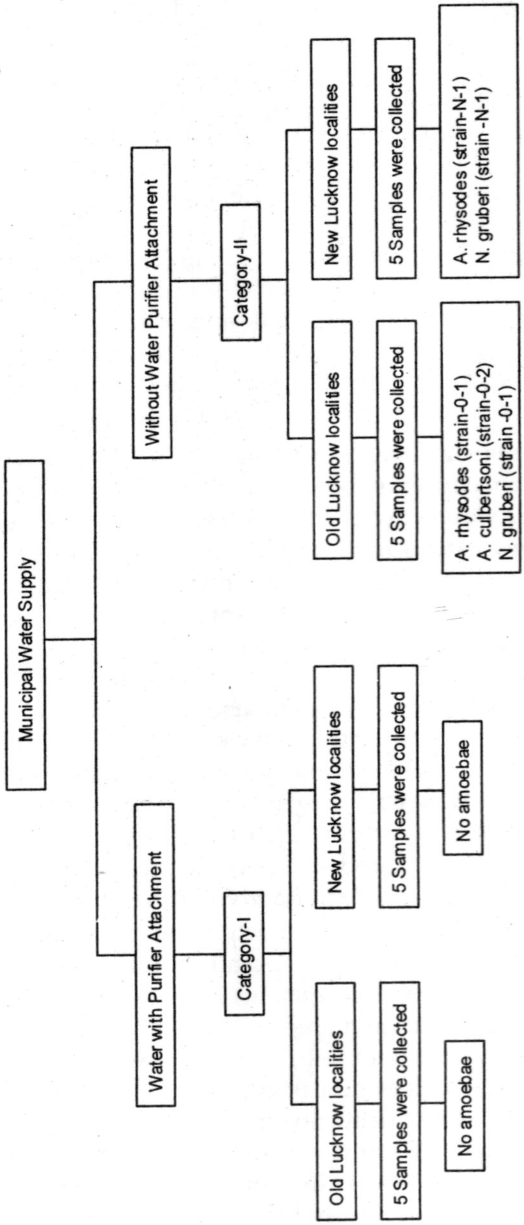

plates were incubated at 35°C-37°C for 5-10 days to allow for the growth of amoebae. These plates were examined daily under low power of a microscope for the presence of trophic and cystic stages amoebae. For clonal monobacterial culture of amoebae, the cysts from positive culture plates, were taken and put in sterile distilled water in a glass cavity slides. When cysts settled down on a glass surface and were well separated from each other, a single cyst was picked up with a help of a fine sterile micropipette, made of soft glass, under low power of microscope and finally transferred to non-nutrient agar plates pre-seeded with *Escherichia coli*. Plates were incubated at 37°C for 4-5 days for allowing amoebae to grow. Well grown amoebae were studied for their biological characterization. The morphology, locomotion and nuclear characteristic of amoebae were studied as per standard method of Singh and Dutta (1984).

Results and Discussion

The study was planned to screen domestic water supply of Lucknow city for the occurrence of Amphizoic amoebae. The water samples were collected from municipal supplies of old and new Lucknow city areas having with and without water purifier attachment. Out of 20 water samples collected from of different sites (See Experimental Design) a total five strains of small, aerobic amoebae were isolated. Our amoebic isolates were further studied for their biological characterization by performing studies on their morphology, physiology and pathogenicity etc. The amoebae were identified as *Acanthamoeba culbertsoni, Acanthamoeba rhysodes* and *Naegleria gruberi* (Volkonsky, 1931) (Table 1.1). While observing the trophozoites of strain O-1, isolated form old Lucknow area there was no temporary production of flagella. The amoebae of this strain produced broad anterior lobopodium with characteristic fine hyaline projection called acanthopodia (Plate 1.1, Figure 1) during active locomotion on glass surface. Cysts of this strain are characteristically polyhedral; cyst wall consisting of more or less polygonal endocyst and rippled ectocyst (Plate 1.1, Figure 2). The strain O-1 was thus identified as *Acanthamoeba rhysodes*. Other cloned strains like strain N-1 isolated from new Lucknow areas resembling and showing similar biological characteristics as of strain O-1, therefore the strain N-1 was also identified and designated as *A. rhysodes*. Where as cysts of strain O-2, isolated form old Lucknow area are characteristic rounded or oval and have two walls; in some cyst the outer wall is

Table 1.1: Biological Characterization

Sl.No.	*Trophozoite*	*Size*	*Trophozoite Characters*	*Cyst Characters*
1	*Acanthamoeba rhysodes* (N-1) ☆ Non pathogenic ☆ Nuclear division-Mesomitosis.	15-38µm	☆ During locomotion hyaline locopodium produces projection called acanthopodia.	☆ Double layered. Ectocyst irregular, wrinkled and has folds with ripples, 1 or more operculum and endocyst polygonal having truncated rays.
2.	*Acanthamoeba rhysodes* (O-1) ☆ Pathogenic ☆ Nuclear division-Mesomitosis.	17-40µm	☆ Granular endoplasm with a nucleus and a contractile vacuole.	
3.	*Acanthamoeba culbertsoni* (O-2) ☆ Pathogenic ☆ Nuclear division-Mesomitosis.	16-23µm	☆ No amoebo-flagellate stage.	☆ Double layered cyst, oval and rounded and wrinkled in outer appearance 1 or more opercula perforated cyst.
4.	*Naegleria gruberi* (O-3) ☆ Non pathogenic ☆ Nuclear division-Promitosis	14-26µm	☆ During locomotion eruptive pseudopodia formed consisting mainly of ectoplasm and with out hyaline filamentous projection characteristic amoebo-flagellates.	☆ Cyst were rounded or slightly oval in appearance and were double layered and 1 or more plugs or opercula was seem and outer surface smooth.
5.	*Naegleria gruberi* (N-2) ☆ Non pathogenic ☆ Nuclear division-Promitosis	16-27µm		

Plate 1.1

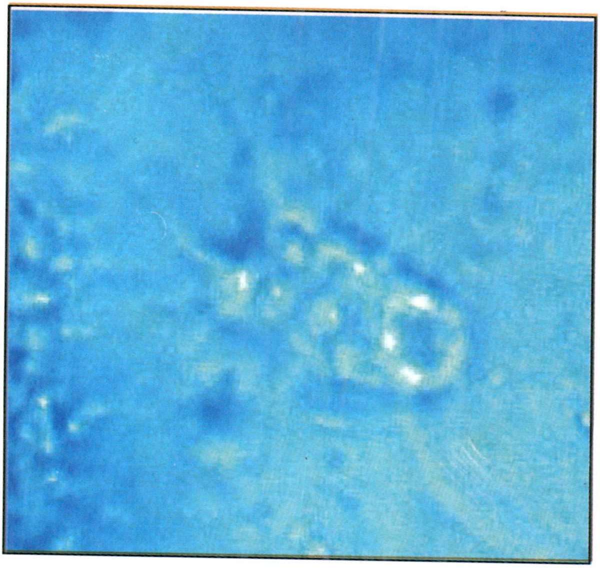

Figure 1: Trophozoites of *Acanthamoeba rhysodes* (10X40x)

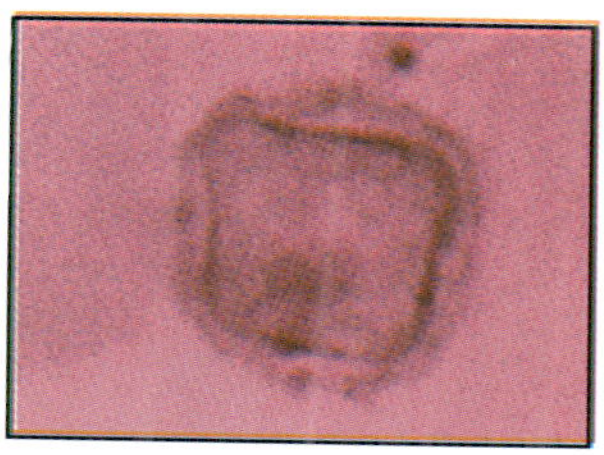

Figure 2: Cyst of *Acanthamoeba rhysodes* (10X40x)

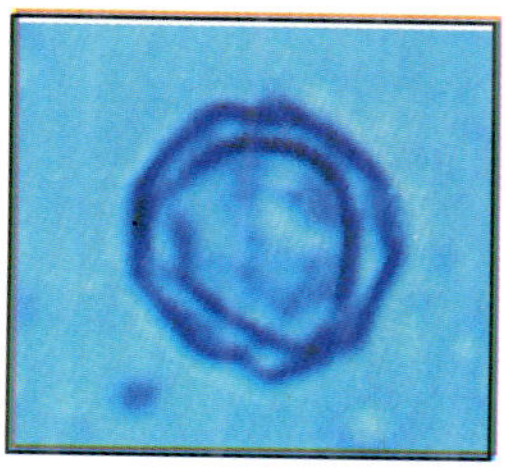

Figure 3: Cyst of *Acanthamoeba culbertsoni* (10X40x)

Contd...

Plate 1.1–Contd...

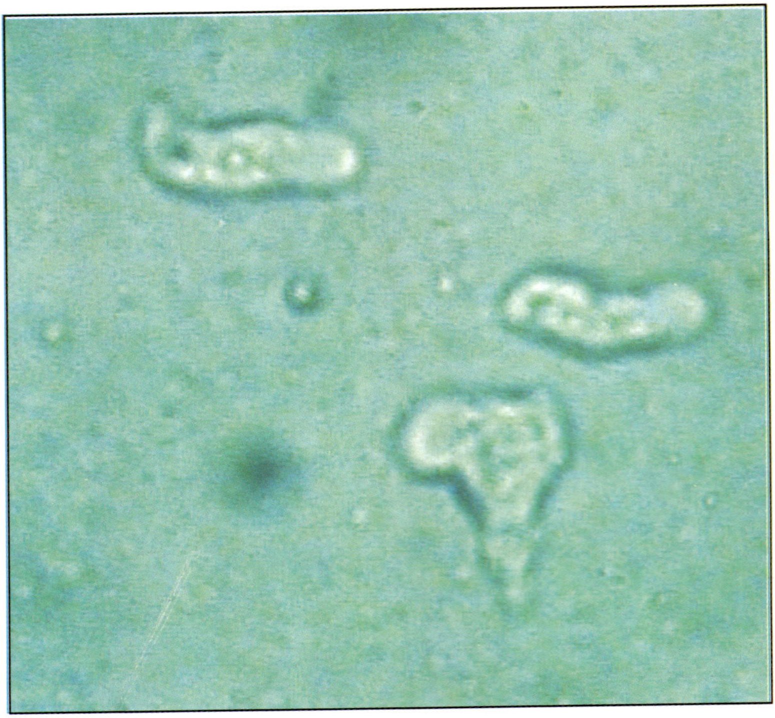

Figure 4: Trophozoites of *Naegleria gruberi* (10X40x)

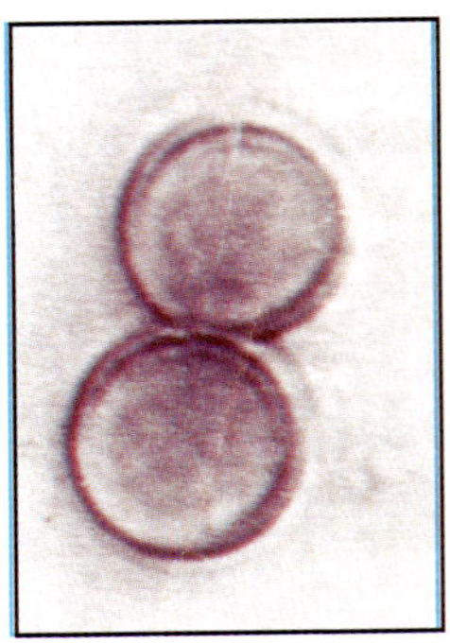

Figure 5: Cyst of *Naegleria gruberi* (10X40x)

nearly circular, in other it is irregular in out line (Plate 1.1, Figure 3), one or more pores or opercula perforated the majority of cysts, the inner wall is in contact with the outer wall at the point of operculum. The amoebae of this strain during active locomotion on glass surface produced typical character of *Acanthamoeba spp.* Thus the Strain O-2 isolated from old Lucknow area was identified and designated as *A. culbertsoni*. The occurrence of *Acanthamoeba* strains from water has also been reported by other workers (Sharma, 1991; Culbertson *et al.*, 1958; Lorenzo-Morales *et al.*, 2005).The cloned strain N-2 produced characteristic temporary amoebo-flagellate stage, when the trophozoites were kept in cavity slides filled with distilled water and incubated in a moist chamber at 28°C-37°C for 30 min. The amoebae got transformed into a transient, non-feeding and non-dividing amoebo-flagellate stage. The trophozoites of N-2 strain during active locomotion on glass surface formed eruptive pseudopodia and move like "limax" form. The ectoplasm was hyaline and endoplasm was granular in appearance. The pseudopodia chiefly constituted ectoplasm, produced toward the direction of movement (Plate 1.1, Figures 4 and 5). The cysts in living condition were rounded or slightly oval in appearance and variable in their size. The cysts were double layered and one or more plugs or operculum were visible and outer surface was smooth. Thus on the basis of locomotion, cystic and trophic morphology; the clone N-2 was identified and designated as *Naegleria gruberi* (Plate 1.1, Figure 5). Another strain O-3 isolated from old Lucknow showed all similar characteristic in locomotion, cystic and trophic morphology as of typical *N. gruberi*.

Findings pertaining to occurrence of *N. gruberi* in water are in conformity with other workers (Singh *et al.*, 1970, 1972 b and De Jonckheere, 2006).These results demonstrate that the domestic water supply is also not safe as it is also a significant source of such organisms (Jeong *et al.*, 2005). It is a general practice that people used to store water before using, though these storage containers also may get contaminated during prolonged use. Isolation of free-living amoebae from main water supply of municipal is a matter of concern. Since the water purification plant treat the water properly by chlorination and other method so as to render it safe and potable. It is also not clear whether the free-living amoebae are found directly from the main water supply or from within the home (Kilvington *et al.*, 2004).

In conclusion, we have shown for the first time, that domestic water supply is also not absolutely safe. So use of any type water purification practices neglates the chance of occurrence of these organisms. Further a greater awareness is also essential to educate people not only about these new diseases caused by free-living amoebae but also about safe sanitation practices while using water.

Acknowledgments

We are thankful to Head of Zoology Department, Prof. (Mrs.) M. Srivastava for providing necessary laboratory facilities to carry out this work. Thanks are also due to Prof. U.D. Sharma who has helped in photography and also in finalization of paper.

References

Cerva, L. and Huldt, G., 1974. Limax amoebae in five swimming pools in Stockholm. *Folia Parasitologica (Praha)*, 21: 71–75.

Culbertson, C.G., Smith, R.W. and Minner, J.R., 1958. *Acanthamoeba*: observation on animal pathogenicity. *Science*, 127: 1506.

Currons, R.T.M., Brown, T.J. and Keys, E.A., 1978. Primary Amoebic Meningoencephalitis (PAM) is New Zealand a etiological agent, distribution, occurrence and control. *Proceeding of the Ninth New Zealand Biotechnology Conference*, 1977: 96–110.

Cursons, R.T.M and Brown, T.J., 1976. Identification and classification of the a etiological agents of Primary meningoencephalitis. *New Zealand of Marine and Freshwater Research*, 10: 254–262.

De Jonckheere, J.F., 1981.*Naegleria australiensis* sp. nov., another pathogenic *Naegleria* from water. *Protistologica*, 17: 423–429.

De Jonckheere, J.F., 2006. Isolation and Molecular identification of Vahlkampfiid amoebae from an Island (Tenerife, spain). *Acta Protozool.*, 45: 96.

De Jonckheere, J.F., Van Dijck, P., Van de Voorde, H., 1975. The effect of the thermal pollution on the distribution of *Naegleria fowleri*. *J. Hyg.*, 75: 7–13.

Jeong, H.J. and Yu, H.S., 2005. The role of domestic tap water in Acanthamoeba containing in contact lens storage cases in Korea. *Korean. J. Parasitol.*, 43: 47–50.

Kasprzak, W. and Mazur, T., 1974. Small free-living amoebae isolated from "warm" lakes investigation on epidemiology and virulence of the strain Proc.III Int.Congs. *Parasitol.*, 1: 188.

Kilvington, S., Gray, T., Dart, J., Morlwt, N., Beeching, J.R., Frazer, D.G. and Matheson, M., 2004.*Acanthamoeba keratitis*. The role of domestic tap water contamination in United Kingdom. *Invest. Opthamol. Vis Sci.*, 45: 165–169

Kingston, D. and Warhurst, D.C., 1968. Isolation of amoeba from the air. *Journal of Medical Microbiology*, 1: 27–36.

Lorenzo-Morales, J., Mirand, M.A., Cardas, J., Concepcion, Tejedor, L.M., Valladares, B. and Ortega-Rivas, A., 2005. Evaluation of Acanthamoeba isolates from environmental sources in Tenerife, canary islands, Spain. *Ann. Agric. Environ. Med.*, (12): 233–236.

Martinez, A.J., 1980. Is *Acanthamoeba encephalitis* an opportunistic infection? *Neurology*, 30: 567–574.

Pandey, R. and Sharma, A.K., 2006. Isolation of pathogenic *Naegleria fowleri* and non cyst forming free–living amoebae *Vanella mira* and *Flabellula calkinsi* from domestic water supply U.P. India. *Lucknow Journal of Science*, 3(1): 27–31

Rezaeian, M., Farnia., Sh. and Haghi., M.A., 2008. Isolation of *Acanthamoeba spp.* from different environmental sources. *Iranian J. Parasitol.*, 3(1): 44–47.

Sharma, A.K., 1991. Virulence in mice of small free-living aerobic amoebae. *Biol.Mem.*, 18(1, 2): 1–5.

Sheehan, K.B., Michael, F.J., Henson, J.M.(2003). Detection of *Naegleria* sp. in a thermal acidic stream in Yellowstone National Park. *The Journal of Eukaryotic Microbiology*, 50(4): 263–265.

Singh, B.N. and Das, S.R., 1972a. Occurrence of pathogenic *Naegleria aerobica, Hartmannella culbertsoni* and *H. rhysodes* in sewage sludge sample of Lucknow. *Curr.Sci.*, 41: 277–281.

Singh, B.N. and Das, S.R., 1970. Studies on pathogenic and nonpathogenic small free living amoebae and the bearing of nuclear division on the classification on the order *Amoebida. Philos. Trans. R. Soc. London*, 259: 435–476.

Singh, B.N. and Das, S.R., 1972b. Intranasal infection of mice with flagellate stage of *Naegleria* aerobic and its bearing on the epidemiology of human meningoencephalitis. *Curr. Sci.*, 41: 625–628.

Singh, B.N. and Dutta, G.P., 1984. Small free-living aerobic amoebae: soil as a suitable habitat, isolation, culture, classification, pathogenicity, epidemiology and chemotherapy. *Indian Journal of Parasitology*, 8(1): 1–23.

Stevens, A.R. and Stein, S., 1977. Isolation of the etiological agent of primary amoebic meningoencephalitis from artificially heated waters. *Appl. Environ. Microbiol.*, 34: 701–705.

The Water Act, 1989. London: HMSO.

Visvesvara, G.S., Moura, H. and Schuster, F.L., 2007. Pathogenic and opportunistic free–living amoebae: *Acanthamoeba* spp., *Balamuthia mandrillaris, Naegleria fowleri* and *Sappinia diploidea*. *Immunol Med Microbiol.*, 50: 1–26.

Volkonsky, M., 1931. *Hartmannella castellani* Douglas et classification des *Hartmannelles*. *Arch. Zool. Exp. Gen.*, 72: 317–339.

Chapter 2

Study of *Vibrio parahaemoliticus* Through ERIC-PCR and REP-PCR for Development of Fish and Human Health Management

Agniswar Sarkar[1]*, Mousumi Saha[2] and Pranab Roy[1]

[1]Department of Biotechnology (DBT- Recognized), The University of Burdwan, W.B., India

[2]Department of Biotechnology, Oriental Institute of Science and Technology, Vidyasagar University, Burdwan, West Bengal, India

ABSTRACT

Development of suitable diagnostic tools to combat various human and fish disease is one of the most challenging aspects in modern days, as these are associated with different bacterial, viral, fungal and parasitic pathogens and also these pathogens has different virulence mechanism. Thus development of rapid and sensitive molecular techniques for detection and characterization of different human and fish pathogens is very important. The genus *Vibrio* includes several food-borne pathogens which cause a spectrum of clinical conditions including septicaemia, cholera, diarrhea and milder forms of gastroenteritis. The species most commonly associated

* Corresponding author: E-mail: ognish@gmail.com

with food-borne transmission include *V. cholerae*, *V. parahaemolyticus*, and *V. vulnificus*. *Vibrio parahaemolyticus* is a gram-negative halophilic bacillus found in marine environments worldwide. It colonizes filter feeding animals such as oysters, crabs, mussles and fish but can also be found free-living in seawater. This organism is the leading cause of seafood-associated bacterial gastroenteritis and it is one of most important food-borne pathogens in Asia and particularly in India causing approximately half of the food poisoning. Infections caused by *V. parahaemolyticus* specially have increased globally in the last few years. The main cause of infection is intake of raw or partially cooked seafood. The major virulence factors of *V. parahaemolyticus* have a major role for diagnosis and analysis the cause associated with the human and fish disease from *V. parahaemolyticu*. Different molecular typing methods, such as Pulsed field gel electrophoresis (PFGE), Ribotyping, Randomly Amplified Polymorphic DNA (RAPD) fingerprinting, Restricted Fragment Length polymorphism (RFLP), Polymerase Chain Reaction (PCR) based Fingerprinting methods etc. are making increasing inroads into clinical laboratories. Molecular methods have suppressed traditional methods of detection for many fasticidious organisms, both in terms of sensitivity and rapidly. Thus the aim of this study is to use conserved repetitive sequences for typing virulence factor and have wider scopes of application in molecular characterization, molecular development and molecular typing of this pathogen.

Keywords: *Vibrio parahaemolyticus, REP- PCR, ERIC- PCR.*

Introduction

Members of the genus *Vibrio* are Gram-negative; rod shaped with single polar flagellum, oxidase positive, facultatively aerobic bacteria and does not form spores. Like other members of the genus *Vibrio*, this species is motile, with a single, polar flagellum and are naturally occurring, free-living inhabitants of marine and estuarine environments throughout the world. Whereas the vast majority of *Vibrio* are nonpathogenic to humans, select strains from four species, *V. cholerae*, *V. parahaemolyticus*, *V. vulnificus*, and *V. mimicus*, are known to be important human pathogens that are predominantly associated with food and waterborne illness, humans seafood-borne gastroenteritis and traveler's diarrhea in humans after they consume contaminated raw or partially cooked fish or shellfish, particularly oysters (Blake *et al.*, 1980; Hlady *et al.*, 1995; Levine *et al.*, 1993; Honda *et al.*, 1993). *Vibrio parahaemolyticus* causes gastrointestinal illness, abdominal cramps, nausea, vomiting, headache, fever, and chills may be associated with infections caused by this organism. The illness is usually mild or moderate, although some cases may require hospitalization. Occasionally, this disease may manifest itself as a dysentery-like illness with bloody or mucoid stools, high fever and a high white blood cell count. Outbreaks of *V. parahaemolyticus* food poisoning are most common in worldwide and India also and are associated with diverse serovars.

A lethal toxin (Honda *et al.*, 1976; Sarkar *et al.*, 1987) and a vascular permeability factor (Honda *et al.*, 1976), as well as thermostable direct hemolysin and other related hemolysins (Honda *et al.*, 1991; Miyamoto *et al.*, 1969; Nishibuchi *et al.*, 1989), have been identified in *V. parahaemolyticus*. The incidence of *V. parahaemolyticus* in the environment and shellfish varies greatly, depending on the season, location, sample type, level of fecal pollution, and analytical method (Bartley *et al.*, 1971; DePaola *et al.*, 1988; DePaola *et al.*, 1990; Watkins *et al.*, 1985). Environmental strains of *V. parahaemolyticus* can produce thermostable direct hemolysin-related hemolysin which is physiochemically, biologically, and immunologically indistinguishable from the toxin from clinical isolates. The laboratory methods most often used to detect and identify the pathogenic *Vibrio* spp. rely on culture followed by conventional biochemical, serological, and susceptibility testing. However, these methods are time consuming, labor intensive, and reagent intensive and usually do not directly characterize the virulence factors associated with human illness. Molecular genetic identification methods have the ability to not only accurately identify these bacterial pathogens, but to also provide information pertaining to the pathogenic and biological potential of the organism. The disease mechanism of *V. parahaemolyticus* infections has not been fully elucidated.

A DNA-based typing technique, that is frequently used to generate strain-specific fingerprintings (Hulton *et al.*, 1991), relies on the polymerase chain reaction (PCR) and primers directed to specific nucleotide sequences, designated ERIC (Enterobacterial Repetitive Intergenic Consensus) sequences. Versalovic *et al.*, in 1991 demonstrated that complex amplification patterns could be generated in *Enterobacteriaceae* isolates by PCR amplification using PCR primers complementary to the inverted repeat. The ERIC PCR technique is now widely used for typing Gram-negative and Gram-positive isolates. We are interested in developing a PCR protocol specific to *V. parahaemolyticus* from food or environmental isolates, which could be useful for epidemiological surveys, determination of distribution and source, identification, and routine monitoring of this bacterium in contaminated shellfish and other seafood.

A significant increase in the number of cases of *V. parahaemolyticus* infections was reported in 1996. A unique clone of *V. parahaemolyticus* O3:K6 is responsible for many of the recent *V. parahaemolyticus* outbreaks, including epidemics in India, Russia, Southeast Asia, Japan, and North America (DePaola *et al.*, 1990; De Paola *et al.*, 2000; Matsumoto *et al.*, 2000; Smolikova *et al.*, 2001). Strains of the O3:K6 serovar emerged in Calcutta, India, in 1996 and have accounted for 50 to 80 per cent of *V. parahaemolyticus* infections annually since then.

Common Genotyping Methods

The advent of molecular biology has caused a significant shift in the types of approaches used to characterize and identify microbial pathogens and to devise disease management strategies. The shortcoming of phenotypically based typing methods have led to development of typing methods based on the microbial genotype or DNA sequence, which minimize problems with typeability and reproducibility and, in some cases, enable the establishment of large databases of characterized organisms (Olive and Bean, 1999). Genotyping methods are those that are based on an analysis of genetic structure of organisms and include polymorphism in DNA restriction patterns based on cleavage of the chromosome by the Res that cleave the DNA into hundreds of fragments (frequent culture) or into 10 to 30 fragments (infrequent cultures) and the presence or absence of extrachromosomal DNA. DNA based bacterial characterization methods can be divided into PCR amplification-dependent and independent genomic analysis approaches.

Basic Criteria of Molecular Typing Techniques

Subtyping or stain classification can be accomplished by a number of different approaches. All of these methods must meet several criteria in order to be broadly useful i) their typing capability, ii) differentiating power, iii) reproducibility. All typing system can be characterized in terms of its reproducibility, typeability, discriminatory power, ease of performance and ease of interpretation. A reproduce method is one that yields the same result upon real testing of bacterial strain. This is very important in the context of epidemiological study; this means the same strain recovered from epidemiologically linked patients will give the identical or similar typing results. This discriminatory power of a technique refers to its ability to differentiate among epidemiologically unrelated isolates. Traditional phenotyping methods such as antibiogram typing, serotyping and biotyping usually show lower discriminatory power than molecular method. All organisms within a species must be typeable by the method used. Along with considerations related to a particular method's ease of interpretation, its ease of use is also important. The technical difficulty, cost and time to obtain a result must also be evaluated in assessing the utility of a particular typing method (Olive and Bean, 1999).

Pulse Field Gel Electrophoresis (PFGE)

PFGE was first discovered in 1984 as a tool for examining the chromosomal DNA of eukaryotic organisms (Schwartz and Cantor, 1984). Subsequently PFGE has proven to be a highly effective molecular technique for many different bacterial species (Tenover *et al.*, 1997). In this method, the bacterial genome which typically is 2000 to 5000kb in sizes digested with a restriction enzyme that has relatively few recognition sites and thus generated approximately 10 to 30 restriction fragments ranging from 10 to 800kb. These fragments can be resolved as a part of distinct bands by PFGE, using specially digested chamber that positions the agarose gel between three sets of electrodes that form a hexagon around the gel. Instead of applying an electric current to the gel in a single direction, in PFGE, the current is applied first in one direction from one set of electrodes for a short time (a pulse) and then shifts to third sets of electrodes. Thus electric field that causes the DNA to migrate in the gels provided in pulses that alternate from three sets of electrodes.

This causes the DNA to wriggle through the gel and the back and the forth movements results in the higher level fragment resolution seen with the technique. Most species are thus typeable using this technique.

In some cases the bacterial genome spontaneously degrades during cell lysis, thus making this approach impractical. PFGE is often considered the "Gold Standard" of molecular typing methods and has been applied successfully to a wide range of bacterial species. PFGE has proven to be superior to most other methods for biochemical and molecular typing. It is highly discriminatory and superior to most methods for analysis. PFGE was also more discriminatory than repetitive element sequence- based PCR (REP-PCR) for differentiating strains of many bacteria (Olive and Bean, 1999). With the aid of the computerized gel scanning and analysis software, it is possible to create data banks of PFGE patterns for all organisms, enabling the creation of reference databases to which any new strain could be compared for identifying its phylogenetic relationship to other similar strains (Olive and Bean, 1999).

Amplified Fragment Length Polymorphisms (AFLP) Assay

The AFLP technology was developed by "KeyGene" in the early 1990's and has become one of the most popular genetic fingerprinting techniques which detect multiple DNA restriction fragments by means of PCR amplification. It is based on the selective amplification of a subset of genomic restriction fragments using PCR. AFLP is a technique for fingerprinting genomic DNA and the technology based on classical, hybridization- based fingerprinting and PCR- based fingerprinting. The AFLP comprises of the following steps: the restriction of the DNA with two restriction enzymes, preferably a hexa-culture and a tetra-culture; the ligation double-strand adapters to the ends of the restriction fragments; the amplification of a subset of the restriction fragments using two primers complimentary to the adapter and restriction site sequences and extended at their 3′ ends by "selective" nucleotides gel electrophoresis of the amplified restriction fragments on denaturating slab gels; the visualization of the DNA fingerprints by means of autoradiography, phosphor-imaging or other methods. The amplification primers, known as

AFLP-primers are generally 17 to 21 nucleotides in length and anneal perfectly to their target sequences; *i.e.*, the adapter and restriction sites and a small number of nucleotides adjacent to the restriction sites. The renders AFLP a very reliable and robust technique, which is unaffected by small variations in amplification parameters (*e.g.*, thermal cyclers, template concentration, PCR cycle profile) (Vos *et al.*, 1995).

A major advantage of the technology is the high marker density that can be obtained. A typical AFLP fingerprints contains between 50 and 100 amplified fragments. The frequency with which AFLP markers are detected depends on the level of sequence polymorphism between the tested DNA samples. The application of the AFLP technology requires no prior sequence information. If sequence information is available an *in-silico* analysis can be done to select the most informative enzyme and primer combinations. Studies to date have demonstrated that aflp is reproducible and has good ability to differentiate clonally derived strains (Olive and Bean, 1999). The differentiation power of AFLP appears to be greater than that of PCR based ribotyping (Olive and Bean, 1999). The reproducibility of the banding patterns for a given strains facilitates storing pattern in databases for use in identifying new bacterial strains. This combined with strong discriminatory power may make AFLP attractive to those laboratories performing frequent epidemiological studies such that a DNA sequencer becomes cost-effective.

Ribotyping

Rationales for the application of ribotype-based differentiation of independent isolates within a species have included taxonomic classification, epidemiological tracking, geographical distribution, population biology and phylogeny. Such wide interest has led to the development of various schemes from the initially described methods. The 16S rRNA gene is considered to be universally present in bacteria and shows a high degree of sequence conservation. The sequence homology analysis of these genes demonstrates some interesting results for the phylogenetic analysis of some genera. Typically, each ribosomal operon consists of three genes encoding the structural rRNA molecules, 16S, 23S and 5S, co-transcribed as a polycistronic operon. Among bacterial species, the average lengths of the structural rRNA genes are 1, 522 bp and 120 bp for 16S, 23S and 5S respectively. The copy numbers, overall ribosomal operon sizes, nucleotide sequences and secondary structures of the three rRNA genes are highly conserved within a bacterial species due to their fundamental role in polypeptide synthesis. Because the 16S rRNA gene is the most conserved of the three rRNA genes, 16S rRNA gene sequence has been established as the "Gold Standard" for identification and taxonomic classification of bacterial species. Knowledge of intraspecies conservation of the 16S rRNA gene sequence and basic 16S-23S-5S ribosomal operon structure led to the first insights into its usefulness in developing Ribotyping for bacterial classification. Different variants of ribotyping are used.

RFLP- Ribotyping

Ribotyping or RFLP of genes coding for rRNA uses a labeled probe containing 16S or 23S or both 16 and 23 ribosomal cDNA. Before hybridization the DNA is digested with *Bam*HI, *Eco*RV, *Puv*II or *Nar*I and transferred on a member by Southern Blot. The hybridization patterns or ribotypes produced by hybridization probe to different fragments of DNA digested allow to do the differentiation between the bacterial species. For the hybridization with a specific probe, the nucleic acid fixed on a solid support, nitrocellulose or nylon membrane, by dot blot or colony hybridization. The probe can be a single oligonucleotide or cloned and characterized DNA fragment labeled with biotin or dioxigenin to produce a colorimetric reaction or radiolabeled probes the amount of hybrid form is determined by autoradiography. Fluorescence *in-situ* hybridization (FISH) on a microscopic slide was also used to detect and to determine the population of *Bifodobacterium* spp. in different samples (Ward and Roy, 2005).

PCR-RFLP-Ribotyping

The PCR has enabled specific genetic loci to be routinely amplified and examined for differences indicative of strain variation and antimicrobial resistance. The specific locus to be examined is amplified with gene-specific primers and subjected to RFLP-analysis. The DNA fragments are separated on an agarose gel or small polyacrylamide gel, and the digestion patterns are visualized following ethidium bromide (EtBr) staining. The 16S, 23S and 16S to 23S spacer regions have also been used as targets for locus specific RFLP (Vila *et al.*, 1996). In this variation of Ribotyping, the ribosomal

DNA is amplified and subjected to digestion with a restriction enzyme and the DNA fragments are visualized by following separation by gel electrophoresis, alleviating the need for Southern blotting. In addition to length variation, many species demonstrate high degrees of sequence variability among multiple copies of ISR. Such variability is due, in part, to the fact that intergenic regions often encode one or two tRNAs. Ribotyping can be successfully applied for differentiation of bacterial strains that display a high degree of heterogeneity of within the rRNA operons. However, the discriminatory power of locus-specific PCR is generally not as good as that of other methods, due primary to the limited region of the genome which can be examined. It has been noted PCR- Ribotyping to have poor discriminatory power in comparison to PFGE and biochemical typing methods.

REP-PCR and ERIC-PCR

Repetitive DNA PCR-Based Genomic Fingerprinting (REP-PCR) analysis was developed based on the observed occurrence of specific conserved repetitive sequences, repetitive extragenic palindromic (REP) sequences, enterobacterial repetitive intergenic consensus (ERIC) sequences, and BOX elements distributed in the genomes of diverse bacteria. However, the term has been expanded to include the use of primers for PCR genomic fingerprinting that anneal to any repetitive DNA sequences. Three primer sets are commonly used for REP-PCR genomic fingerprinting analysis, corresponding to REP, ERIC and BOX sequences. The protocols are referred to as REP-PCR, ERIC-PCR and BOX-PCR, respectively, and REP-PCR, in general. The primers are designed to amplify intervening DNA between two adjacent repetitive elements (Louws *et al.*, 1999). ERIC sequences are 126-bp elements which contain a highly conserved central inverted repeats and are located in extragenic regions of the bacterial genome. They have been defined primarily based on sequenced data obtained from *E.coli* and *Salmonella typhimurium*. This PCR uses primers to match short consensus repetitive sequences. A complex array of 10 to 30 or more PCR fragments is generated per genome, ranging in size from less than 200 bp to more than 6kb. REP-PCR has been extensively used to identify pathogens, to differentiate strain, and to assess the genetic diversity of plant pathogens. Recent reviews have provided detailed protocols and applications of REP-PCR including application to medical and environmental microbiology. Each primer set (REP, ERIC and BOX) is useful to fingerprint diverse bacteria, including Gram-negative and Gram-positive bacteria, as well as plant associated actinomycetes (Louws *et al.*, 1999). Different techniques can be used, elaborate in the Figure 2.2.

DNA Sequence Analysis

Ultimately, all molecular genetic methods for distinguishing organism subtypes are based on differences in the DNA sequences. Logically, then, DNA sequencing would appear to be the best approach to differentiating subtypes. DNA sequencing was originally performed by using radioactive labels for detection of the reaction products. Current DNA sequencing protocols employs fluorescent nucleotides to label the DNA. The sequences are then read with an automated instrument (Olive and Bean, 1999). There are several considerations that must be evaluated before undertaking the use of DNA sequencing for subtyping. First, for pratical purposes, DNA sequencing must be directing at only a very small region of the chromosome of an organism. It is impractical to sequence multiple or large regions of the chromosome of an organism. Thus, in contrast to techniques like PFGE, REP-PCR, or RAPPD analysis which examines the entire chromosome, DNA sequencing examines only a very small portion of the sites which can potentially vary between bacterial or fungal strains. Furthermore the short region of DNA used must meet several criteria before it can be used for strain differentiation. The structure of the region of DNA selected must consist of a variable sequence flanked by highly conserved regions. This enables PCR amplification and typing of all members of a species. The variability within the selected sequence must be sufficient to differentiate different strains of a particular species. Unfortunately, for bacteria and fungi, few sequences meet these criteria. While the 16S rRNA genes have been used to identify new species of organisms, they show limited variability between strains of a bacterial species. The intergenic region between the 16S and 23S rRNA genes has also been used with variable results. In contrast DNA sequencing is considered the good standard for viral typing. The regions used for viral genotyping and detection of drug resistance mutations are in short, well-defined sequences that fulfill the criteria needed for the application of nucleic acid sequencing methodologies (Olive and Bean, 1999).

Sequencing Analysis of Specific Genes in Typing

Sequences of specific genes can be used for identification and characterization of microorganisms. It is known that the DNA sequences of protein coding genes are more effective than 16S rRNA gene sequencing for the characterization. Some genes other than the 16S rRNA are used for the differentiation of bifidobacteria. These genes are: L-lactate dehydrogenase gene (ldh), recA gene, heat shock protein (HSP60) gene, Pyruvate kinase (PK) gene. Before sequencing, a part of gene is selected and amplified by PCR. The PCR product is sequenced, analyzed and compared with other sequences. With a short region of the recA (231bp) and Idh (312 bp) gene is possible to distinguish between the bacterial species. Finally, analysis of partial hsp60 gene (538 bp) sequences is very effective for the differentiation between all human *Bifidobacterium* species. The hsp60 gene is a powerful tool for the phylogenetic study of *Bifidobacterium* species (Ward and Roy, 2005). However, other functional genes like bacterial toxin production genes can also be used marker genes for typing of different species of bacteria.

Multilocus Sequence Typing (MLST) Analysis

In long term epidemiological investigation, the relationship between the strains recovered over greater periods of time and often from a broader geographic range is studied. Several different kinds of intra-species variation are used for molecular typing using a variety of molecular tools like RAPD, RFLP, ERIC-PCR, BOX, REP-PCR etc. and the results vary between the techniques employed and between the laboratories. Again since highly variable genes that are targeted in these approaches are highly variable for a season and environmental pressure and thus their high rates of evolution will obscure the true relationship between isolates. Genes that are not subject to any unusual selective forces and which diversity slowly by the random accumulation of neutral variations should provide more reliable information about the relationship between the isolates. This is the very concept of MLST for characterization and typing of microbial isolates using housekeeping genes as the target genes. MLST is actually the modified extension method of an old but reliable method called Multilocus Enzyme Electrophoresis (MLEE) that has been used for genetic characterization of species. In MLST the alleles at multiple house keeping loci are assigned or characterized directly by nucleic acid sequencing than indirectly from the electrophoretic mobilities of their gene products. MLST makes use of rapid sequencing techniques to uncover allelic variants in conserved genes or house keeping genes for microbes for the purpose of characterization, subtyping and classifying them. Internal approximately 450bp fragments of house keeping genes are used in MLST. For each gene fragments every unique sequence is assigned as a different allele. Bacterial strains are therefore defined unambiguously by a string of integers or the allelic profiles or sequence types. It has been particularly useful in studying the population genetics of a variety of microbial pathogens including and their epidemiological prevalence study. MLST has been evaluated to characterize *Klebsiella* isolates. Amplification of seven housekeeping gene *e.g.* rpoB (beta-subunit of RNA polymerase), gapA (glyceraldehyde 3-phosphate dehydrogenase), pgi (phosphoglucose isomerase), tonB (periplasmic energy transducer), mdh (malate dehydrogenase), phoE (phosphorine E) and infB (translation initiation factor 2) was carried out using PCR (Mishra and Goyal, Unpublished data) and following the specified annealing temperature for primers given at the *K. pneumoniae* MLST data base (http;//pubmlst.org/kpneumoniae). Each distinct sequence within a locus was assigned a unique allele number. Each nucleotide sequence was analyzed by BLASTn algorithm against the *K.pneumoniae* MLST database and a matching allele number in the database was searched. Nucleotide sequences having no matching allele in the MLST database were considered new and got their respective allele number and sequence type (ST) assigned by the curator of the *K.pneumoniae* MLST database web site. After getting the allele numbers and sequence types, cluster analysis was performed by Bionumerics software version 4.0 (Applied Maths, Martens-Latem, Belgium) using categorical coefficient. MLST analysis has indicated that in many species, recombination replacements contribute more too clonal diversification than due to point mutation. A major advantage of this new technique is that the sequence data are unambiguous and electronically portable, allowing molecular typing of bacterial pathogens via internet. MLST was initially evaluated using *Neisseria meningitides* because it provides a good example of species in which the rate of recombination is high but in which distinct clones can clearly be distinguished by MLEE. MLST has also been used for characterizing

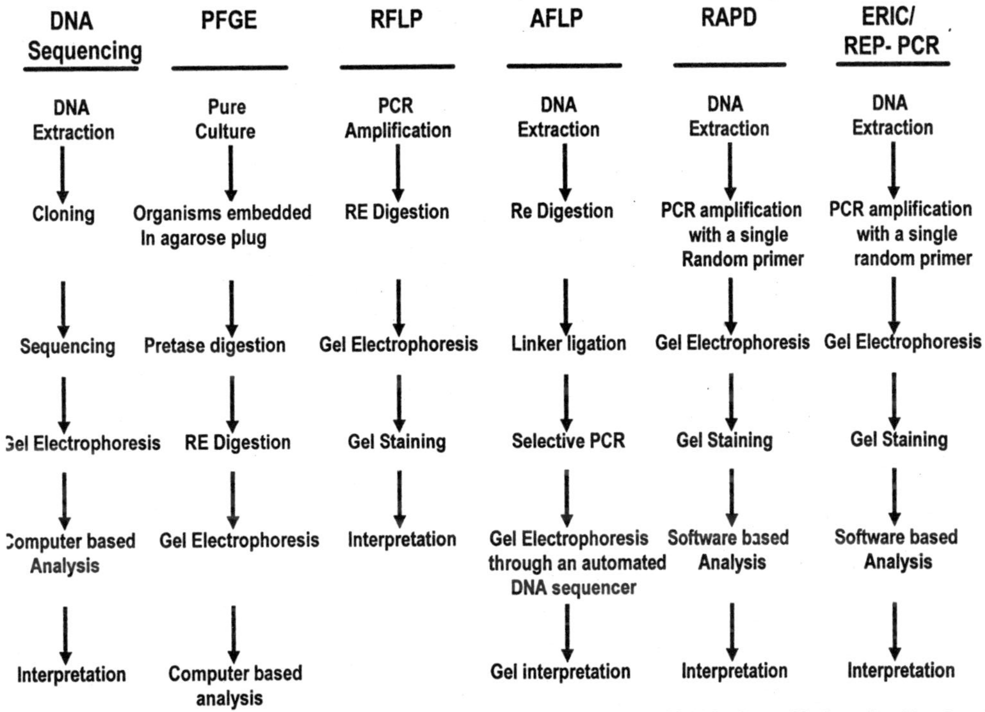

Figure 2.1: Comparison of the Different Procedural Steps Involved in Various Molecular Typing Methods

***Source*: Olive, D. M. and Bean, P., 1999. Principles and Applications of Methods for DNA-Based Typing of Microbial Organisms, *J Clin. Microbiol.*, 37 (6): 1661–1669.**

many other bacterial and fungal isolates. The biggest advantage of the technique over other molecular typing methods is that sequencing data are repeatable, reproducible and portable between laboratories. This has led to creation of global database for many microbes on the line of Genebank that allows the exchange and verification of molecular typing data via internet or web among various laboratories around the world.

Conclusions

The identification and differentiation of various microbial (*e.g., Vibrio*) isolates within a species and between isolates is central to may aspects of clinical microbiology and epidemiological investigation. The advent of molecular biology in general and the polymerase chain reaction in particular have greatly facilitated genomic analyses of microorganisms, provided enhanced capability to characterize and classify strain, and facilitate research to assess the genetic diversity of populations. The diversity of large populations can be assessed in a relatively efficient manner using PFGE, AFLP, Ribotying PCR, REP-PCR and AP-PCR/RAPD based genomic fingerprinting methods, especially when combined with computer assisted pattern analysis. Genetic diversity maps provide a framework to understand the taxonomy, population structure, and dynamics of microbes. Although the importance of phenotypic methods in identification and characterization of microbes can not be rules out, molecular based genotyping methods have been the area of intense application. PCR based protocols, combined with computer based analysis, have provided novel fundamental knowledge of the ecology and population dynamics of bacterial pathogens, and present exciting new opportunities for basic and applied studies in microbiology (Louws *et al.*, 1999; Olive and Bean, 1999). The choice of a molecular typing method, therefore, depends upon the needs, skill level, and resources of the laboratory. Modern epidemiologists are fortunate to have a variety of tools which provide good molecular differentiation and which can be tailored to fit the needs of both the laboratory and the clinical study. Several of these molecular methods could enable the creation of large references libraries of typed organisms to which new outbreak strains can be compared across laboratories in order to monitor changes in microbial populations. In conclusion, we believe that REP-PCR genomic fingerprinting coupled to computer assisted phylogenetic analysis and library search programs will constitute a useful method for the identification or diagnosis of various pathogens. This technique has already been successfully used to study the population structure of plant pathogens, to follow greenhouse infections, to examine nodule occupancy, to identify (brady)-rhizobial strains that could not be distinguished by any other method, and to carry out phylogenetic analyses of world-wide collections of rhizobia and plant pathogens. It therefore is another useful molecular approach in molecular microbial ecology.

References

Bartley, C.H. and Slanetz, L.W., 1971. Occurrence of *Vibrio parahaemoliticus* in estuarine waters and oysters of New Hampshire. *Appl. Microbiol.*, 21: 965–966.

Blake, P.A., Weaver, R.E. and Hollis, D.G., 1980. Diseases of human (other than cholera) caused by vibrios. *Annu. Rev. Microbiol.*, 34: 341–367.

Chiou, C.-S., Hsu, S.-Y., Chiu, S.-I., Wang, T.-K. and Chao, C.-S., 2000. *Vibrio parahaemolyticus* serovar O3:K6 as cause of unusually high incidence of food-borne disease outbreaks in Taiwan from 1996 to 1999. *J. Clin. Microbiol.*, 38: 4621–4625.

DePaola, A., Kaysner, C.A., Bowers, J. and Cook, D.W., 2000. Environmental investigations of *Vibrio parahaemolyticus* in oysters after outbreaks in Washington, Texas, and New York (1997 and 1998). *Appl. Environ. Microbiol.*, 66: 4649–4654.

DePaola, A., Hopkins, L.H., Peeler, J.T., Wentz, B. and McPhearson, R.M., 1990. Incidence of *Vibrio parahaemolyticus* in U.S. coastal waters and oysters. *Appl. Environ. Microbiol.*, 56: 2299–2302.

DePaola, A., Hopkins, L.H. and McPhearson, R.M., 1988. Evaluation of four methods for enumeration of *Vibrio parahaemolyticus*. *Appl. Environ. Microbiol.*, 54: 617–618.

Hlady, W.G. and Klontz, K.C., 1995. The epidemiology of *Vibrio* infections in Florida, 1981, 1993. *J. Infect. Dis.*, 173: 1176–1183.

Honda, T. and Lida, T., 1993. The pathogenicity of *Vibrio parahaemolyticus* and the role of the thermostable direct haemolysin and related haemolysins. *Rev. Med. Microbiol.*, 4: 106–113.

Honda, T., Abad-Lapuebla, M.A., Ni, Y., Yamamoto, K. and Miwatani, T., 1991. Characterization of a new thermostable direct haemolysin produced by a Kanagawa-phenomenon-negative clinical isolate of *Vibrio parahaemolyticus*. *J. Gen. Microbiol.*, 137: 253–259.

Honda, T., Shimizu, M., Takeda, Y. and Miwatani, T., 1976. Isolation of a factor causing morphological changes of Chinese hamster ovary cells from the culture filtrate of *Vibrio parahaemolyticus*. *Infect. Immun.*, 14: 1028–1033.

Honda, T., Taga, S., Takeda, T., Hasibuan, M.A., Takeda, Y. and Miwatani, T., 1976. Identification of lethal toxin with the thermostable direct hemolysin produced by *Vibrio parahaemolyticus*, and some physico-chemical properties of the purified toxin. *Infect. Immun.*, 13: 133–139.

Hulton, C.S.J., Higgins, C.P. and Sharp, P.M., 1991. ERIC sequences: a novel family of repetitive elements in the genomes of *Escherichia coli, Salmonella typhimurium* and other enterobacteria. *Mol. Microbiol.*, 5: 825–834.

Levine, W.C., Grion, P.M. and the Gulf Coast Vibrio Working Group, 1993. *Vibrio* infections on the Gulf Coast: results of ¢rst year of regional surveillance. *J. Infect. Dis.*, 167: 479–483.

Louws, F.J., Rademaker, J.L.W. and Brujin, F.J. de., 1999. The three Ds of PCR-based genomic analysis of Phytobacteria: Diversity, detection and disease diagnosis. *Annu. Rev. Phytopathol.*, 37: 81–125.

Matsumoto, C., Okuda, J., Ishibashi, M., Iwanaga, M., Garg, P., Rammamurthy, T., Wong, H.-C., Depaola, A., Kim, Y.B., Albert, M.J. and Nishibuchi, M., 2000. Pandemic spread of an O3:K6 clone of *Vibrio parahaemolyticus* and emergence of related strains evidenced by arbitrarily primed PCR and *toxRS* sequence analyses. *J. Clin. Microbiol.*, 38: 578–585.

Miyamoto, Y., Kato, T., Obara, Y., kiyama, S., Takizawa, A.K. and Yamai, S., 1969. *In vitro* hemolytic characteristic of *Vibrio parahaemolyticus*: its close correlation with human pathogenicity. *J. Bacteriol.*, 100: 1147–1149.

Nishibuchi, M., Taniguchi, T., Misawa, T., Khaeomanee-Iam, V., Honda, T. and Miwatani, T., 1989. Cloning and nucleotide sequence of the gene (*trh*) encoding the hemolysin related to the thermostable direct hemolysin of *Vibrio parahaemolyticus*. *Infect. Immun.*, 57: 2691–2697.

Olive, D.M. and Bean, P., 1999. Principles and applications of methods for DNA-based typing of microbial organisms. *J. Clin. Microbiol.*, 37(5): 1661–1669.

Sarkar, B.L., Kumar, R., De, S. P. and Pal, S.C., 1987. Hemolytic activity of and lethal toxin production by environmental strains of *Vibrio parahaemolyticus*. *Appl. Environ. Microbiol.*, 53: 2696–2698.

Smolikova, L.M., Lomov, I.M., Khomenko, T.V., Murnachev, G.P., Kudriakova, T.A., Fetsailova, O.P., Sanamiants, E.M., Makedonova, L.D., Kachkina, G.V., and Golenishcheva, E.N., 2001. Studies on halophilic vibrios causing a food poisoning outbreak in the city of Vladivostok. *Zh. Mikrobiol. Epidemiol. Immunobiol.*, 6: 3–7.

Versalovic, J., Koeuth, T. and Lupski, J.R., 1991. Distribution of repetitive DNA sequences in eubacteria and application to ¢ngerprinting of bacterial genomes. *Nucleic Acids Res.*, 19: 6823–6831.

Ward, P. and Roy, D., 2005. Review of molecular methods for identification, characterization and detection of bifidobacteria. *Lait.*, 85: 23–32

Watkins, W.D. and Cabelli, V.J., 1985. Effect of fecal pollution on *Vibrio parahaemolyticus* densities in an estuarine environment. *Appl. Environ. Microbiol.*, 49: 1307–1313.

Chapter 3

Light Microscopic as well as Ultrastructural Changes in Thyroid Gland of *Mus norvegicus* Induced by Sublethal Heroin Administration

Kaminidevi K. Bhoir[1], *S.A. Suryawanshi*[1] *and A.K. Pandey*[2]*

[1]*Department of Zoology, Institute of Science, 15 Madam Cama Road, Mumbai – 400 032, India*

[2]*National Bureau of Fish Genetic Resources, Canal Ring Road, Dilkusha, Lucknow – 226 002, U. P., India*

ABSTRACT

Thyroid gland of *Mus norvegicus* consisted of spherical or round follicles lined with low cuboidal as well as columnar epithelial cells and lumina filled with eosinophilic colloid. Ultrastructurally, the thyroid follicular cells showed the presence of round nucleus, polymorphic mitochondria, Golgi complex as well as lysosomes located on the apical side of the nucleus and cytoplasm with different size of lipid droplets and smooth as well as rough endoplasmic reticulum (RER). Basal lamina of the follicular cells was often seen in association with the endothelium of the capillaries. Sublethal heroin

* Corresponding author: E-mail: akpandey_cifa@yahoo.co.in

administration (0.50 LD_{50}; 13.5 mg/kg/day) for 30 days elicited degenerative changes in the follicular epithelial cells as evident by the vacuolization of cytoplasm, pycnotic nuclei and reduced colloidal content. Ultrastructurally, the thyroid follicular cells showed indented nucleus with heavy deposition of chromatin material on inner membrane of nucleus and dilated rough RER. Along with RBC infiltration, vesiculated mitochondria owing to the loss of cristae were also seen. Diffused electron dense material was seen at periphery of cell body. Heroin treatment caused cellular necrosis as revealed by the fragmentation of cytoplasmic materials in follicular epithelial cells of the gland.

Keywords: *Heroin, Thyroid gland, Light Microscopic, Ultrastructure, Mus norvegicus.*

Introduction

Chronic abuse of heroin has diverse effects on various body systems due to widespread distribution of specific receptors in many tissues and organs (Martin, 1984; Sawynok, 1986; Cami and Farre, 2003). The drug (diacetylmorphine) is initially metabolized to 6-acetylmorphine and subsequently to morphine in human body (Sawynok, 1986; Goldberger *et al.*, 1994; Jenkines *et al.*, 1994; Sporer, 1999). Despite of its long history of clinical therapeutic use and protracted abuse by addicted subjects (Sawynok, 1986; Cami and Farre, 2003), little is known regarding possible influences of these drugs on the endocrine system (George *et al.*, 2005; Brown *et al.*, 2006; Al-Gommer *et al.*, 2007). Establishment of the functional integrity of the hypothalamic-pituitary axis in man receiving narcotic agents is of great clinical as well as pharmacological importance (Cushman *et al.*, 1970; Brambilla *et al.*, 1984; Pechnick, 1993; Koob and Le Moal, 1997; Blesener *et al.*, 2005). It is well known that pituitary hormone secretion is under control of brain and pituitary, in turn, control all the other endocrine glands located in different parts of the body (Becu and Libertun, 1982). When any toxic substance or chemical enters the vascular system, it reaches the brain and pituitary in a short span of time which, in turn, affects the target or concerned organ of the body. The adversely affected pituitary secretes less hormone resulting in suppression of secretion of the target endocrine gland. Acute or repeated administrations of psychostimulants like amphetamine to male rat do not make any changes in the level of thyroid hormone (Budziszewska *et al.*, 1996). Benzodiazepines, a group of pharmacologically active compounds used as minor tranquillizers and hypnotics, have potent inhibitory effects on circulating TSH secretion in response to stressful and pharmacological stimuli after seven days of the treatment (Humbert, 1994).

Nicotinic acid, one of the principal components of chewing tobacco, tends to affect T_3 and T_4 levels showing the symptoms of hypothyroidism. The measurement of thyroid function returned to the pre-treatment levels after discontinuation of nicotinic acid therapy. The treatment revealed significant decrease in the circulating T_3 (13 per cent) and T_4 (21 per cent) levels whereas no change was noted in TSH value suggesting that nicotinic acid decreases serum thyroid hormone concentration without involving pituitary and maintaining euthyroid state. This effect may be mediated through reduction in thyroxine binding globulin (Shakir *et al.*, 1995). Chronic administration of methadone (narcotic analgesic) in animal for a larger period suppresses TSH secretion resulting into hypothyroidism due to which T_3 and T_4 decreased significantly during the course of treatment (Kuhn and Bartolome, 1985). Morphine, the main alkaloid of opium, slightly increases TSH while synthetic allergic and sedative drugs like pentazocine analgesic drug and nalorphine an antagonist to morphine failed to produce any change in TSH level. In case of man, morphine administration resulted into significant increase in TSH from basal value (Grossman and Jones, 1983).

Opium reduced secretion of various hormones of pituitary gland including thyrotrophic hormone (TSH), thus inducing decreased secretion of thyroid hormones (Kalant, 1997). Acute administration of morphine to male wistar rat depresses the release of TSH and thyroid function resulting in depletion of T_3 and T_4. However, in case of man, morphine did not alter the serum TSH level (Tolis *et al.*, 1975). In contrast to this observation on morphine, Kotani *et al.* (1973) examined the extrathyroidal colloid droplets and found that pentazoane did not abolish an increase in TSH secretion. This may be related to the fact that pentazocine is an opioid-derived anatogistic analgesic. Morphine and heroin elicit structural changes in the endocrine glands. In case of fetal intravenous administration of both the drugs, thyroid gland exhibits nodular stroma in the follicles

(Kringsholm and Christoffersen, 1989). An attempt has been made to record the effects of sublethal dose of heroin administration on light microscopic as well as ultrastructural alterations in thyroid gland of *Mus norvegicus*.

Materials and Methods

Healthy male *Mus norvegicus* (Wistar strain) weighing 150-200 gm were procured from Hoffkin Institute, Parel, Mumbai and housed in specially made plastic cages. They were acclimatized under the ambient laboratory conditions (temperature 28±2°C; photoperiod 14L:10 D) for 10 days, fed *ad libitum* on rat feed (Lipton, Hindustan Lever Ltd., Bangalore) and clean water was provided for drinking. 60 male rats were randomly selected and divided into two equal groups - experimental and control. Heroin (85 per cent pure) was dissolved initially in small quantity of alcohol and diluted with physiological saline to prepare the test dose of 0.50 LD_{50} (13.5 mg/kg/day). The drug was administered through subcutaneous (s.c.) route to the experimental rats while the control rats received equal volume (0.2 ml/kg body weight) of the physiological saline. The animals were killed on day 30 and thyroid glands from both the groups were removed and washed in normal saline and fixed immediately in Bouin's fluid for light microscopy. After 24 hours, the tissues were dehydrated in ascending series of alcohol, cleared in xylene and embedded in paraffin wax at 60°C. The sections were cut at 5 µm and stained in hematoxylin-eosin (H&E).

For electron microscopy, the tissues were removed immediately after the sacrifice and sliced into 1 mm pieces to allow better penetration of fixative chemical. Tissues were fixed in 3 per cent ice-cold glutaraldehyde for 12 hours followed by 4 hours in 0.1.M cacodylate buffer. They were rinsed in buffer and post-osmicated in 1 per cent osmium tetraoxide (OsO_4) for 1-2 hours. Thereafter, the tissues were dehydrated in ascending alcohol grades followed by propylene oxide and embedded in resin polymerized at 60°C. The blocks were prepared in araldite. 1 µm thin sections were cut with glass knife on an LKB-2000 ultra-microtome. The sections were mounted on glass slide and stained with buffered toluidine blue for light microscopic studies. Ultra-thin sections of the selected blocks were cut with glass knife, picked up on copper grids and stained with uranyl acetate and lead citrate for final observation under ZEIM-EM-109 electron microscope.

Results and Discussion

Thyroid gland of control *Mus norvegicus* consisted of spherical or round follicles with low cuboidal and columnar epithelium. The follicles were mostly filled with eosinophilic colloid (Figures 3.1, 3.2). There were two distinct types of cells, follicular and parafollicular, in thyroid gland of the rat. The follicular cells are also known as principal cells that synthesize thyroid hormones. Ultrastructurally, the thyroid follicular cells showed the presence of round nucleus in the cytoplasm. Basal lamina of the follicular cells was often in close association with the basal lamina of the endothelium of capillaries. The mitochondria were polymorphic-rod, filamentous or oval in shape. The Golgi complex was located on the apical side of the nucleus which was dilated during cell activity. The cytoplasm contained lipid droplets of different sizes and smooth as well as rough endoplasmic reticulum (RER). Lysosomes were present at the apical region of nucleus (Figure 3.5).

Sublethal heroin administration for 30 days elicited degenerative changes in thyroid gland of the rat. Light microscopic study of the gland showed degenerative changes in the follicular epithelial cells as evident by the vacuolization of cytoplasm, pycnotic nuclei and reduced colloid material. Staining intensity of the colloid material was also markedly reduced (Figures 3.3, 3.4). Ultrastructurally, the thyroid follicular cells showed indented nucleus with heavy deposition of chromatin material on inner membrane of nucleus and dilated rough RER (Figure 3.6). Groups of follicular cells also exhibited enlarged nuclei with slight crenation and chromatin margination in the inner membrane of nucleus. Along with this, clumps of chromatin material were also seen inside the nucleus. Rough endoplasmic reticulum was prominent with large number of ribosomes scattered in the cytoplasm. Many small electron dense round secretory granules distributed uniformly in the cytoplasm were also seen. Infiltrations of RBCs were also encountered in some follicles (Figures 3.6, 3.7).

The cytoplasm of follicular cell of heroin treated *Mus norvegicus* exhibited vacuolization. The rough endoplasmic reticula were prominent in one area whereas in other area they were fragmented (Figures 3.8, 3.9). This may probably be due to injury caused to the tissue by the drug. Slight alteration in nucleo-cytoplasmic ratio was seen in a group of cells along with indented nuclear membrane and

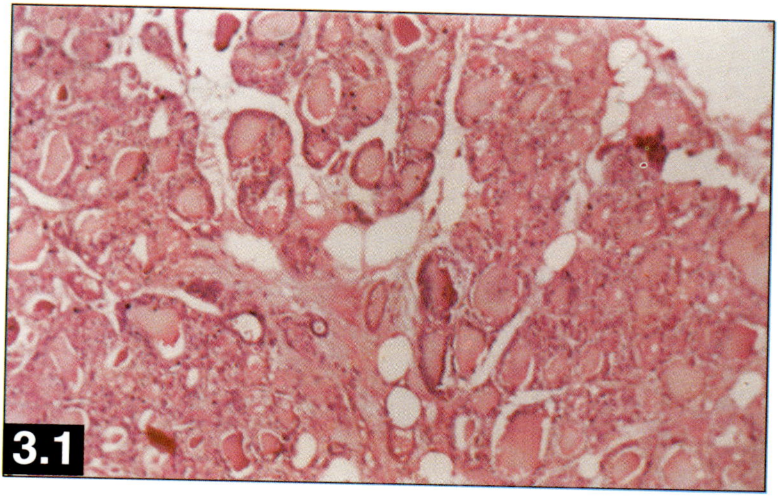

Figure 3.1: Thyroid Gland of Control *Mus norvegicus* Showing Follicles Lined with Squamous to Columnar Epithelial Cells and Lumina Filled with Eosinophilc Colloid Material. H&E. x 100.

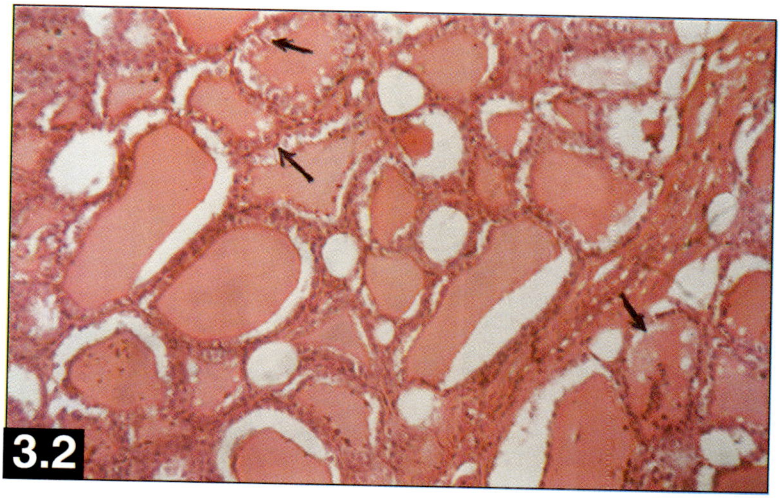

Figure 3.2: Magnified View of the Thyroid Gland of Control Rat Exhibiting Follicles with Colloid Material. Mark some follicles with vacuolated colloid (arrow). H&E. x 250.

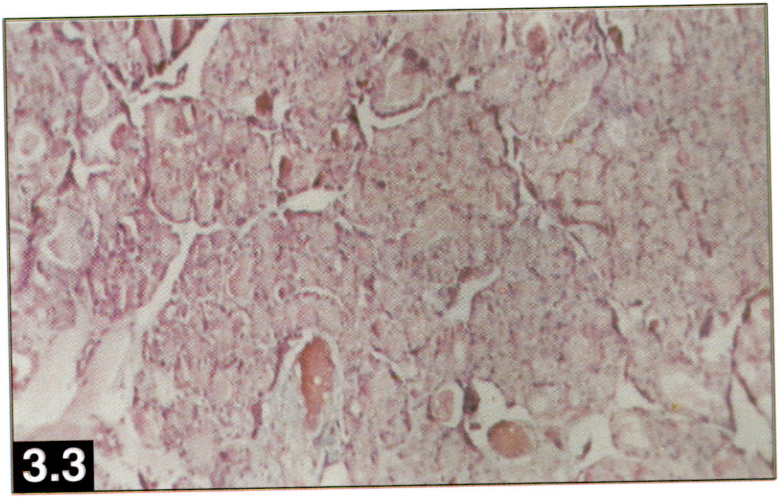

Figure 3.3: Thyroid Gland of the Rat on Day 30 of Heroin Administration Depicting Degenerative Changes in the Follicles. Mark the decline in staining affinity of the colloid. H&E. x 100.

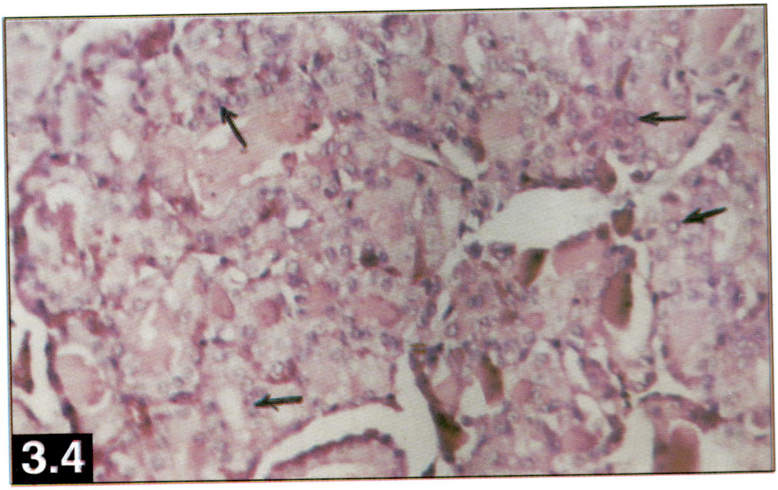

Figure 3.4: Magnified View of Thyroid Gland of Heroin-Treated *Mus norvegicus* on Day 30 Showing Epithelial Lining Cells with Pycnotic Nuclei (arrow) and Reduction in the Colloid Content. H&E. x 250.

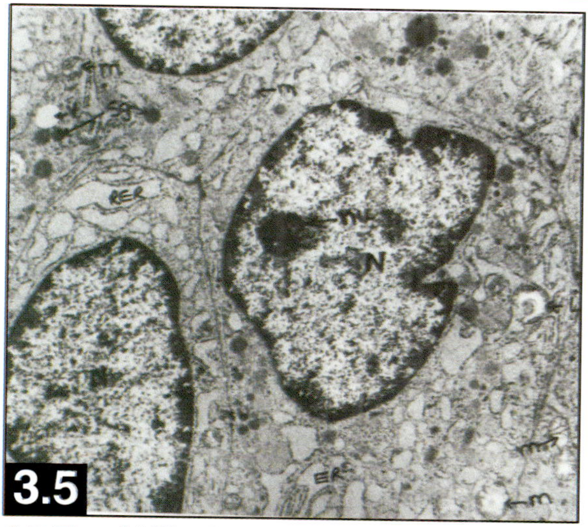

Figure 3.5: Thyroid Gland of Control Rat Exhibiting Nucleus (N), Nucleolus (nu), Lysosome, Rough Endoplasmic Reticulum (RER), Secretory Granules (sg) and Mitochondria (m). x 7,200.

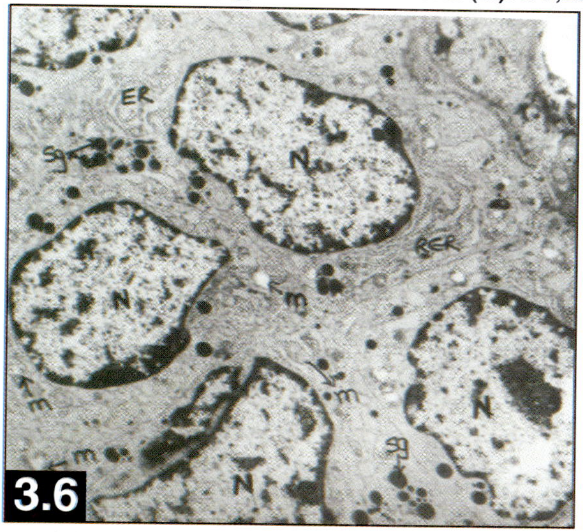

Figure 3.6: Thyroid Gland of the Rat on Day 30 of Heroin Administration Depicting Follicular Cells with Rough Endoplasmic Reticulum (RER), Secretory Granules (sg) and Mitochondria (m). Mark the indentation of chromatin material on inner surface of the nucleus (N). x 7,200.

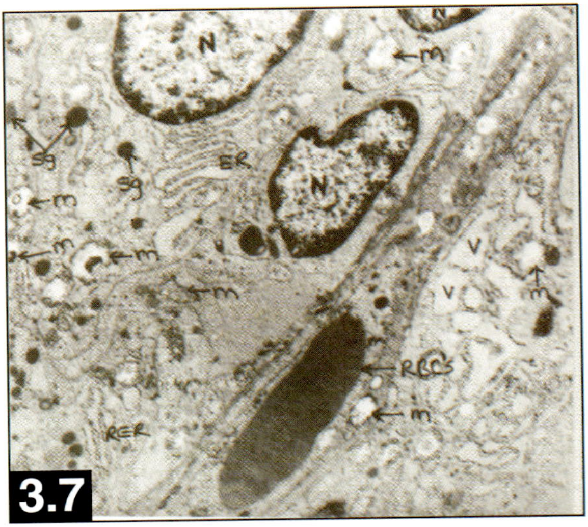

Figure 3.7: Thyroid Gland of Heroin Treated *Mus norvegicus* on Day 30 Showing Damaged Nucleus (N), Mitochondria (m), Secretory Granules (sg), Rough Endoplasmic Reticulum (RER), Vacuoles in the Cytoplasm (v) and RBC Infiltration (RBC). x 10,700.

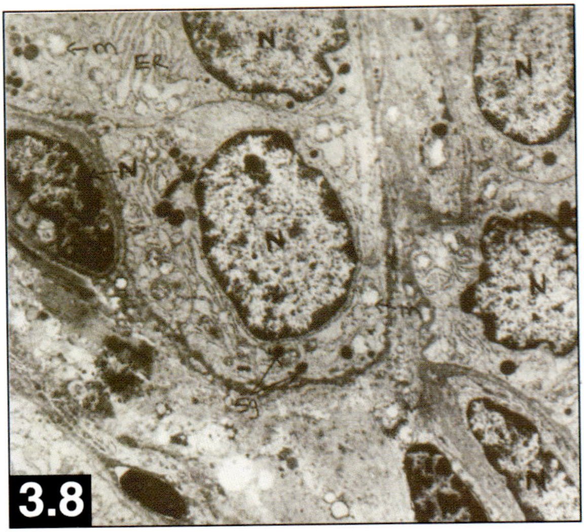

Figure 3.8: Group of Thyroid Cells of Heroin Treated Rat on Day 30 Exhibiting Degenerative Changes in Nucleus (N), Mitochondria (m), Secretory Granules (sg) and Endoplasmic Reticulum (ER). x 6,600.

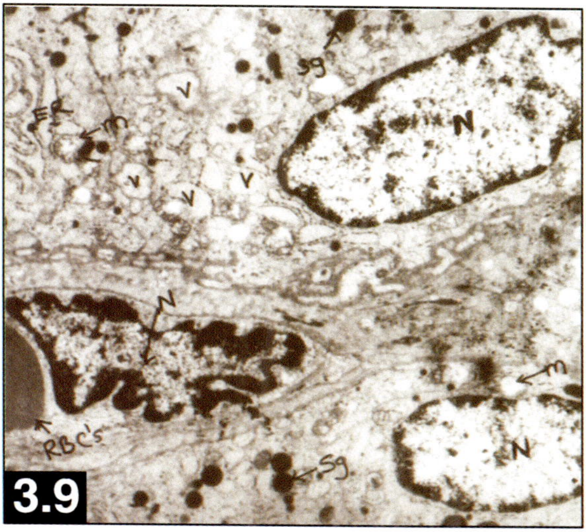

Figure 3.9: Thyroid Gland of Heroin Treated Rat on Day 30 Depicting Marked Degenerative Changes in Nucleus (N) and Cytoplasm. Mark the secretory granules (sg), large number of vacuoles (v) and damaged mitochondria (m) as well as endoplasmic reticulum (ER). RBC infiltration (RBC) is also seen. x 11,400.

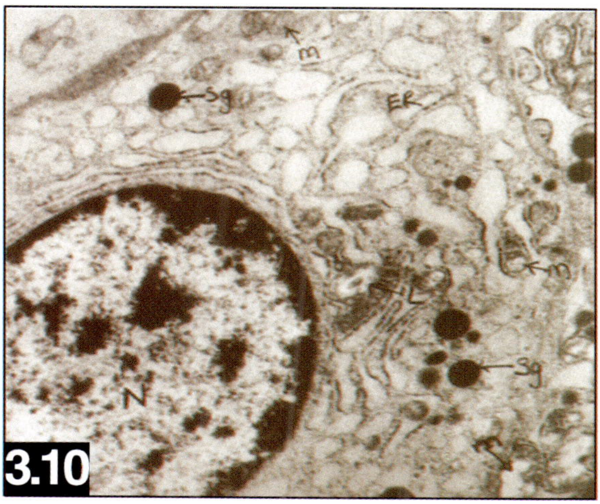

Figure 3.10: Thyroid Cells of Heroin Treated Rat on Day 30 Showing Nucleus (N), Dilated Endoplasmic Reticulum (ER), Mitochondria (m), Secretory Granules (sg) and Lysosome (L). x 18,000.

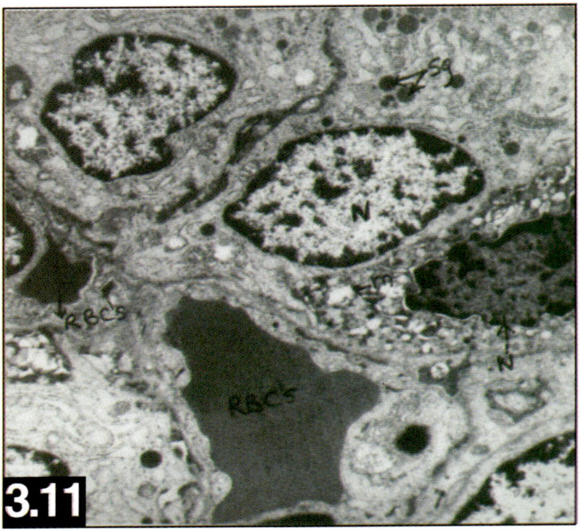

Figure 3.11: Group of Thyroid Cells of Heroin Treated Rat on Day 30 Exhibiting Degenerative Changes in Cytoplasm and Nucleus (N). Mark the damaged mitochondria (m), RBC infiltration (RBCs), secretory granules (sg) and endoplasmic reticula. x 10,700.

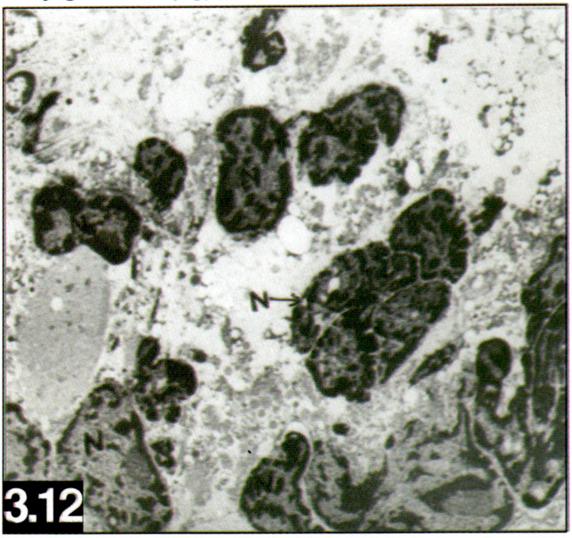

Figure 3.12: Thyroid Gland of Heroin Treated Rat on Day 30 Depicting Necrotic Nucleus (N) and Extensive Vacuolization in the Cytoplasm. x 6,500.

prominent chromatin margination. Small mitochondria were encountered at some while RER is seen in other places. Interestingly, both smooth and rough endoplasmic reticula were seen at the same time.

Some follicular cells showed more indentation of nuclear membrane with heavy deposition of chromatin material. Effects of heroin were so severe that caused injury to the cell body and to cope up with this adverse effect, more blood was diverted to the affected cellular region which entered the intercellular space created due to edema caused by the drug administration (Figure 3.10). Along with blood infiltration, mitochondria were seen vesiculated due to loss of cristae. Some cells showed enlarged nucleus with large amount of chromatin deposition. Diffused electron dense material was seen at periphery of cell body. RBCs infiltration was prominently seen in the major areas of the section (Figure 3.11). Heroin treatment caused cellular necrosis and fragmentation of cytoplasmic organelles were also seen (Figure 3.12).

Secretion of TSH from the anterior pituitary gland is regulated by the feedback control of thyroid hormone and neural control by the hypothalamus (Reichlin, 1978). Morphine decreases serum TSH levels of male rats while this inhibition was antagonized by naloxane (George and Kokka, 1976; Bruni *et al.*, 1977). Morphine failed to reduce TSH level in urethane-anesthetized rats that may probably be related to the effects of anesthesia. In man, morphine did not alter the serum TSH level. It is suggested that inhibitory effect of morphine on both the basal and stimulated TSH levels may be mediated by the activation of opiate receptors and the opioid peptide neurons may play a modulating role in the regulation of serum TSH (Tolis *et al.*, 1975). It is believed that morphine acts on hypothalamus to exert its inhibitory effect on TSH secretion (George and Kokka, 1976). However, it has not actually been demonstrated that morphine reduced the release of hypothalamic TRH. The failure of morphine to inhibit TSH release induced by exogenous TRH suggested that morphine lacks, unlike thyroxine, the ability to block the effect of TRH on the anterior pituitary gland to release TSH (Martin and Reichlin, 1972).

Kalant (1997) suggested decreased secretion of thyroid gland in opium-addicted men. Rasheed and Tareen (1995) observed that heroin does not affect the serum levels of T_4 and TSH but slightly increase T_3 level. Dammann *et al.* (1994) found no clinically relevant changes in levels of TSH, T_4 and T_3 with any other endocrine parameters due to pentoprazol administration. Opiate derivatives and synthetic drugs elicited varied effects on hormonal levels in man. D-Ala^2-mepha^4-met enkephalin (DAMME) and methodone stimulated the release of growth hormone (GH) but not TSH. Morphine stimulated TSH but not GH while pentazocine had no effect on either GH or TSH. Brambilla *et al.* (1980) studied the effects of chronic heroin addiction on pituitary-thyroid function in man. Data were obtained from 10 male addicts, aged 18-24 years with history of addiction to heroin lasting from 8 months to 4 years and 9 controls (matched for sex and age). The plasma levels of TSH, T_4 and T_3 revealed no difference between addicts and controls in basal levels of these hormones.

Sublethal heroin administration in *Mus norvegicus* for 30 days inflicted severe degenerative changes in thyroid gland as observed under light as well as electron microscopic studies. The follicular epithelial cells of the heroin treated rat showed vacuolization with pycnotic nuclei, increased number of lysosomes, polymorphic mitochondria without cristae and reduced number of secretory granules. The heroin administration in rat ultimately led to necrosis of the thyroid follicular epithelial cells as confirmed by electron microscopic studies resulting in hypothyroid status. It appears that the observed degenerative changes in thyroid gland are mediated by the decreased secretion of TSH in heroin treated rat resulting in the low output of T_4 and T_3 (Fujita, 1986; Ekholm, 1990).

Acknowledgements

One of us (KKB) is thankful to N.M. Wadia Trust, Mumbai for partial financial assistance to carry out this work. We are grateful to Hon'ble Justice Mrs. K.K. Baam, the then High Court Judge, Bombay for permitting us to work on heroin and to Mr. Rahul Rai Sur, the then Deputy Commissioner of Police, Narcotics Cell, Greater Mumbai for the procurement of the drug. Help extended by the staff of Tata Cancer Research Centre, Mumbai in electron microscopy is acknowledged.

References

Al-Gommer, O., George, S. and Haque, S., 2007. Sexual dysfunctions in male opiate users: a comparative study of heroin, methadone and buprenorphine. *Addict. Disord. Their Treat.*, 6: 137-143.

Becu, D. and Libertum, C., 1982. Comparative maturation of the regulation of prolactin and thyrotropin by seratonin and thyrotropin releasing hormone in male and female rats. *Endocrinology*, 110: 1879-1884.

Blesener, N., Albrecht, S., Schwager, A., Wecbecker, K., Litchermann, D. and Lingmuller, D., 2005. Plasma testosterone and sexual function in men receiving buprenorphine maintenance for opioid dependence. *J. Clin. Endocrinol. Metab.*, 90: 203-206.

Brambilla, F., Nobile, P., Zanoboni, A., Muciaccia, W. and Meroni, P.L., 1980. Effects of chronic heroin addiction on pituitary-thyroid function in man. *J. Endocrinol. Invest.*, 3: 251-255 (1980).

Brambilla, F., Bellodi, L., Aracio, C., Nobile, P., Perna, C.R. and Murphy, D.L., 1984. Effects of clonidine on secretion of anterior pituitary hormones in heroin addicts and normal volunteers. *Psychiatry Res.*, 13, 295-304.

Brown, T.T., Wisniewski, A.B. and Dobs, A.S., 2006. Gonadal and adrenal abnormalities in drug users: cause of consequence of drug use behaviour and poor health. *Am. J. Infect. Dis.*, 2: 130-135.

Bruni, J.F., von Vugt, D., Marshall, S. and Meites, J., 1977. Effects of naloxone, morphine and methionine enkephlin on serum prolactin, luteinizing hormone, follicle stimulating hormone, thyroid stimulating hormone and growth hormone. *Life Sci.*, 21: 461-466.

Budziszewska, B., Jaworska-Feil, L. and Lason, W., 1996. The effects of repeated amphetamine and cocaine administration on adrenal, gonadal and thyroid hormone levels in rat plasma. *Exp. Clin. Encorinol. Diabet.*, 104, 334-338.

Cami, J. and Farre, M., 2003. Drug addiction. *N. Eng. J. Med.*, 349: 975-986.

Cushman, P. Jr., Bordier, B. and Hilton, J.G., 1970. Hypothalamic-pituitary-adrenal axis in methadone treated heroin addicts. *J. Clin. Endocrinol. Metab.*, 30: 24-29.

Dammann, H.G., Bethke, T.H., Burkhardt, F., Wolf, N., Khalil, H. and Luchmann, R., 1994. Effects of pentoprazol on endocrine function in healthy male volunteers. *Aliment Pharmacol. Therp.*, 8: 549-554.

Ekholm, R., 1990. Biosynthesis of thyroid hormones. *Int. Rev. Cytol.*, 120: 243-288.

Fujita, H., 1986. Functional morphology of the thyroid. *Int. Rev. Cytol.*, 113: 145-185.

George, R. and Kokka, N., 1976. The effects of narcotics on growth hormone, ACTH and TSH secretion. In: Tissue Response to Addictive Drugs (Eds.) D.H. Ford and D.H. Clonet. pp. 527-540. Spectrum Publication. New York.

George, S., Murali, V. and Pullickal, R., 2005. Review of neuroendocrine correlates of chronic opiate misuse: dysfunctions and pathophysiological mechanisms. *Addict. Disord. Their Treat.*, 4: 99-109.

Goldberger, B.A., Cone, E.J., Grant, T.M., Caplan, Y.H., Levine, B.S. and Smialek, J.E., 1994. Disposition of heroin and its metabolites in heroin-related deaths. *J. Anal. Toxicol.*, 18: 22-28.

Grossman, A. and Jones, A.E., 1983. Opiate receptors: endorphins and enkephalins. *Clin. Endocrinol. Metab.*, 12: 31-56.

Humbert, T., 1994. Neuroendocrine effects of benzodiazepines. *Ann. Med. Psychcol.*, 153: 261-171.

Jenkines, A.J., Keenan, R.M., Henningfield, J.E. and Cone, E.J., 1994. Pharmacokinetics and pharmacodynamics of smoked heroin. *J. Anal. Toxicol.*, 18: 317-330.

Kalant, H., 1997. Opium revisited: a brief review of its nature, composition, non-medical use relative risk. *Addiction*, 92: 276-277.

Koob, G.F. and Le Moal, M., 1997. Drug abuse: hedonic homeostatic dysregulation. *Science*, 278: 52-56.

Kotani, M., Onaya, T. and Yamada, T., 1973. Acute increase of thyroid hormone secretion in response to cold and its inhibition by drug which acts on the autonomic and central nervous system. *Endocrinology*, 93: 288-294.

Kringsholm, B. and Christoffersen, P., 1989. Morphological findings in fatal drug addiction. An investigation of injection marks, endocrine organs and kidneys. *Forensic Sci. Int.*, 40: 15-24.

Kuhn, C.M. and Bartolome, M.A., 1985. Effect of chronic methadone administration on neuroendocrine function in young rats. *J. Pharmacol. Exp. Ther.*, 234: 204-210.

Martin, J.B. and Reichlin, S., 1972. Plasma thyrotropin (TSH) response to hypothalamic electrical stimulation and injection of synthetic thyrotropin releasing hormone (TRH). *Endocrinology*, 90: 1079-1085.

Martin, W.R., 1984. Pharmacology of opioids. *Pharmacol. Rev.*, 35: 283-323.

Pechnick, R.N., 1993. Effects of opioids on the hypothalamo-pituitary-adrenal axis. *Ann. Rev. Pharmacol. Toxicol.*, 33: 353-382.

Rasheed, A. and Tareen, I.A.K., 1995. Effects of heroin on thyroid function, cortisol and testosterone levels in addicts. *Polish J. Pharmacol.*, 47: 441-447.

Reichlin, S., 1978. Neuroendocrine control. In: *The Thyroid* (Eds.) S.C. Werner and S.C. Ingbar. pp. 151-173. Harper & Row, New York.

Sawynok, J., 1986. The therapeutic use of heroin: a review of the pharmacological literature. *Can. J. Physiol. Pharmacol.*, 64: 1-6.

Shakir, K.M.M., Kroll, S., April, B.S., Drake, A.J. III and Eisold, J.F., 1995. Nicotine acid decreases serum thyroid hormone levels while maintaining a euthyroid state. *Mayo Clin. Proc.*, 70: 556-558

Sporer, K.A, 1999. Acute heroin overdose. *Ann. Intern. Med.*, 130: 584-590.

Tolis, G., Hickey, J. and Guyda, H., 1975. Effects of morphine on serum growth hormone, cortisol, prolactin and thyroid stimulating hormone in man. *J. Clin. Endocrinol. Metab.*, 41: 797-800.

Chapter 4

Seasonal Fluctuation of Nutrients and Phytoplankton in Igoumenitsa Bay (NW Greece)

Beza Paraskeyi[1,2] *Vasilis Moussis*[2], *Cosmas Nathanailides*[1]*, *Alkiviadis Bougiouklis*[2], *Maria Mpoti*[1] *and Giorgos Ioannou*[1]

[1]*Department of Aquaculture and Fisheries, TEI of EPIRUS, Igoumenitsa, Greece*

[1]*Department of Chemistry, University of Ioannina, Ioannina, Greece*

ABSTRACT

Water samples were collected from different sampling locations over a period of twelve months in Igoumenitsa bay. Temperature followed the expected seasonal pattern with the lowest values in March (14.88°C to 15.55°C) and the highest in August (26.66°C). Salinity values ranged from 37 psu (April) to 38.95 psu (August). The pH varied from 7.59 to 8.60. Oxygen ranged between 7.1 mg/l (May) to 12.5 mg/l on March. Annual min and max values of total ammonical nitrogen (TAN) = <0.01–0.62 mg-at $l^{-1,}$ PO4: <0.01–0.14 mg-at l^{-1}. Chl-a content (mg m^{-3}) ranged from 0.0189 at surface waters, to 26.63 (April) and 7.23 (June) at 5m depth. The results indicate a medium

* Corresponding author: E-mail: cosmasfax@yaoo.com

status of Eutrophication in the bay, but during summer, at the location of fish farms, PO_4 levels peaked with values, which correspond to high eutrophic status. Further monitoring and data are required for management decision regarding the ecosystem of the bay.

Keywords: *Seasonal fluctuation, Nutrients, Phytoplankton, Igoumenitsa bay.*

Introduction

Eutrophication is a significant problem in many estuaries and coastal zones. High levels of chlorophyll- *a*, hypoxi-anoxic condition and toxicity of the waters caused by algal blooms, have economic and social costs (Turner *et al.*, 1998). A combination of natural and anthropogenic parameters influences algal growth (Cloern, 2001; Andersen *et al.*, 2006) with serious ecological and economic consequences in coastal marine ecosystems (Segerson and Walker, 2002). Nutrient concentrations such as total nitrogen and total phosphorus and algal growth are essential parameters in efforts to manage and monitor coastal zone eutrophication (Elliot and Jange, 2002; Smith, 2007). Igoumenitsa is a coastal city in NW Greece. The bay has an important ferry terminal, which handles approx. 100 vessels annually. During the summer the frequency of ferry connections to Corfu and Italy exceeds 30 per day. The bay is exposed to point sources of pollution such as a nearby sewage treatment plant (estimated annual inflow of domestic household wastewater exceeds one million m^3) intense marine transportation activities and fish farms. The purpose of this work was to monitor the variability of eutrophication indices in Igoumenitsa bay.

Materials and Methods

The study was carried out between April 2007 and April 2008. Seawater samples were collected on the 1st week of every month, at noon, from 5 locations (Figure 4.1) at 0m(surface) and 5m, at the opening of Igoumenitsa bay (B), the fish farms (FF), the centre of the bay (K), the north side of the bay (BP) and the mouth of a lagoon in the bay (L). Oxygen, temperature, pH and salinity was measured *in situ* with portable multi-parameter YSI equipment. Water samples collected in glass containers and transferred to the laboratory for determination of total ammonia nitrogen (TAN) and phosphate

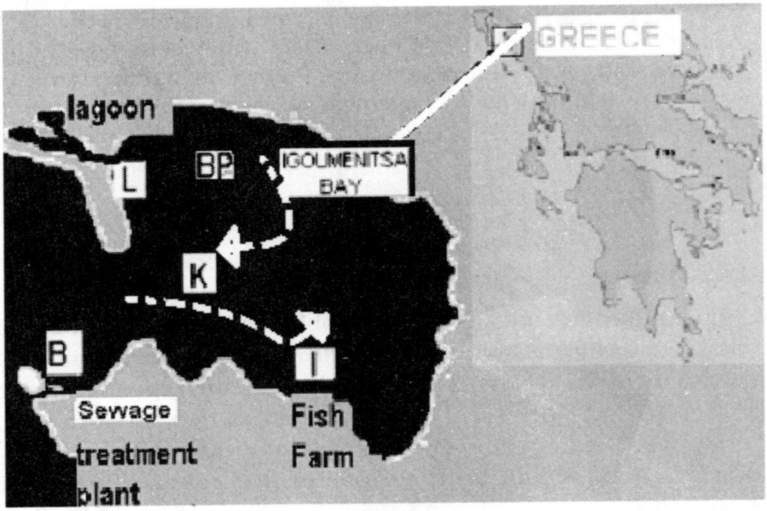

Figure 4.1: Sampling Locations in Igoumenitsa Bay

L: Lagoon channel; BP: Gas station; K: Highest depth; B: Sewage treatment, I: Fish farms. Dotted arrows indicate the direction of the prevailing surface water currents.

(PO_4^{-3}) according to APHA (1985) standard procedures using a Jasco 630-UV model spectrophotometer. Chl-*a* was measured according to Strickland and Parsons (1968). The significance of differences was estimated using ANOVA ($P<0.05$).

Results and Discussion

Temperature followed the expected seasonal pattern with the lowest values in March (14.88°C to 15.55°C) and the highest in August (26.66ºC). Salinity values ranged from 37 psu (April) to 38.95 psu (August). The pH varied from 7.59 to 8.60. Oxygen ranged between 7.1 mg/l (May) to 12.5 mg/l on March. Annual min and max values of total ammonical nitrogen (TAN) = <0.01–0.62 mg-at l^{-1}, PO4: <0.01–0.14 mg-at l^{-1}.

Chl-a ranged from 0.0189 at surface waters, to 26.63 (April) and 7.23(June) at 5m depth. The results indicate a medium status of Eutrophication in the bay, with a significant seasonal variability (Table 4.1) in all monitored parameters. For example, during summer,

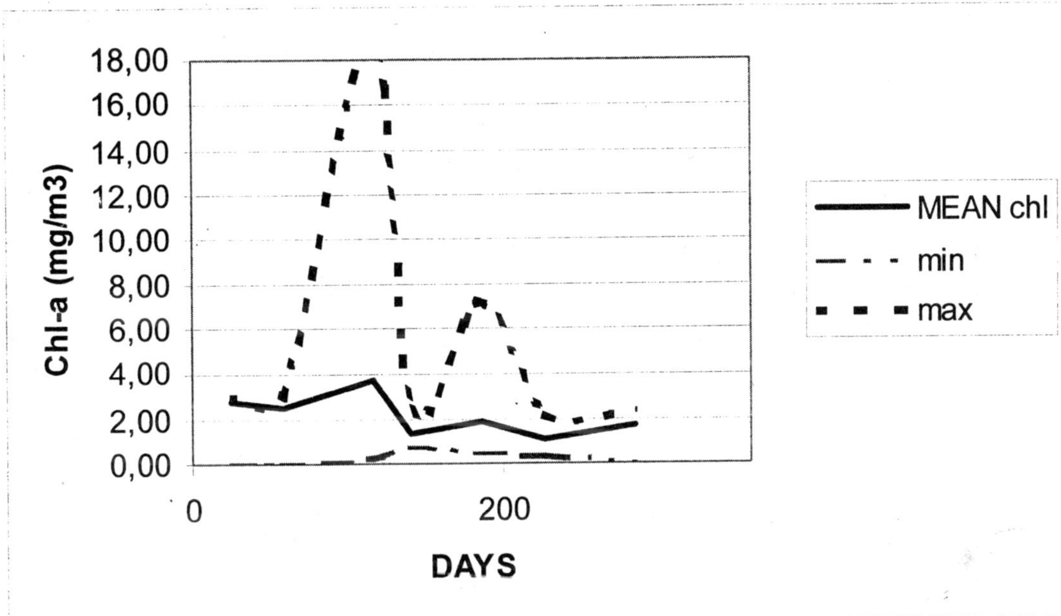

Figure 4.2: Average (Solid line) and Min-Max (Dotted lines) of Chlorophyll-*a* Content (mg/m³) in Igoumenitsa Bay Over the Period of the Study

at the location of fish farms, PO_4 levels peaked with values, which correspond to high eutrophic status. Coastal waters eutrophication, results from a combination of natural and anthropogenic influences (Cloern, 2001). Anthropogenic enrichment of water with nutrients can result in increased growth of algae and higher forms of plant life that can result in deviation of structure, function and stability of the ecosystem and to the quality of water (Andersen *et al.*, 2006).

Table 4.1: Variability in DO_2; TAN; PO_4; Chl-*a* and Salinity in Igoumenitsa Bay

Parameter	*Seasonal Variability*	
Depth (m)	*0*	*5*
PO_4	P<0,001	NS
TAN	P<0.001	P<0,001
DO_2	P<0.001	P<0.001
Chl-a	P=0,041	NS
Salinity	P<0,001	P<0,001

Apart from the ecological issue, eutrophication is a serious economic problem in coastal marine ecosystems world-wide (Segerson and Walker, 2002). A key element of nutrient inflow-outflow in a bay involves the natural flow of nutrients from the land and the outflow of nutrients to the open sea. In addition to a natural flow of nutrients, anthropogenic sources result in an increased nutrient content of bay's aquatic ecosystems. Agricultural runoff of nitrate, olive processing plants, domestic sewage plants, can result in increased nutrient content and Eutrophication. Furthermore, resuspension of the sediment by water currents and winds and decomposition of algae can further increase the available nutrients for primary production. Measurements of nutrient concentrations such as total nitrogen, total phosphorus and algal growth are essential parameters in efforts to manage and monitor coastal zone eutrophication (Bricker *et al.*, 2003; Smith, 2007). The results indicate a seasonal element of high primary productivity of the Bay, with a strong spatial element of variability attributed to the sampling locations of the fish farming site. Fish farms can generate nutrient waste (uneaten food and metabolic waste). Changes in the Chl-*a* content can be an indicator of changes in plankton primary productivity. The levels of Chl-*a* observed here indicate a medium status of Eutrophication in the bay (Bricker *et al.*, 2003). PO_4 levels at the fish farms peaked during summer, this peak is usually an indication of increased nutrient loading of the nearby water bodies (Lupatsch and Kissil, 1998, Belias *et al.*, 2007). Chl-a is a reliable indicator of organic enrichment in a bay. Currents may wash and dilute a point source of organic enrichment but increased primary production remains over a significant period and is reflected in the Chl-*a* content of the samples. Changes in the Chl-*a* content can be an indicator of changes in plankton primary productivity.

Apart from the fish farm, two other sites exhibited high levels of some parameters. Fish farms samples from depth 0 and 5m exhibited the higher average chl-a content, but samples form the lagoon opening also exhibited high annual average increased primary productivity with values reaching levels above 20 mg chl-a/l. In the same manner the samples form the centre of the Bay exhbited some high peaks in TAN, phospohorus and chl-a content. Prevailing surface water currents in the Bay exhibit a round path, entering the bay, passing the location of the fish farms and turning North to complete a circular motion after passing from the North side, meeting a second (outshone) current and exiting the bay (unpublished data based on seasonal observations). The location of the fish farms and the path of water currents may result in the diffusion of nutrient from point sources such as the fish farms to the rest of the Bay. The chemical parameters monitored in this study indicate minimal environmental effect of aquaculture activity in the Bay, nevertheless there is risk of environmental degradation, especially if the fish farming activity is intesified. In the future, the focus of monitoring should include the benthic biota below and near the fish cages. A combination of benthic ecosystem and water body physico-chemical indexes can offer a complete ecological monitoring tool for studying the aquatic ecosystem in coastal zones (Mavraganis *et al.*, 2010). This information can be useful for the management of development in Igoumenitsa Bay.

References

Andersen, J.H., Schlüter, L. and Aertebjerg, G., 2006. Coastal eutrophication: Recent developments in definitions and implications for monitoring strategies. *J. Plankton Res.*, 28: 621–628.

APHA, 1985. *Standard Methods of Examination of Water and Wastewater*, 16th Edition. American Public Health Association, Washington D.C., 1193 pp.

Belias, C., Dassenakis, M. and Scoullos, M., 2007. Study of the N, P and Si fluxes between fish farm sediment and seawater. Results of simulation experiments employing a benthic chamber under various redox conditions. *Mar. Chem.*, 103: 266–275.

Bricker, S.B., Ferreira, J.G. and Simas, T., 2003. An integrated methodology for assessment of estuarine trophic status. *Ecol. Modell.*, 169: 39–60.

Cloern, J., 2001. Our evolving conceptual model of the coastal eutrophication problem. *Mar. Ecol. Progr. Ser.*, 210: 223–253.

Lupatsch, I. and Kissil, G.Wm., 1998. Predicting aquaculture waste from gildhead sea bream culture using a nutritional approach. *Aquat. Liv. Res.*, 11: 265–268.

Mavraganis, T., Telfer, T. and Nathanailides, C. 2010. A combination of selected indexes for assessing the environmental impact of marine fish farms using long term metadata analysis. *Int. Aquat. Res.*, 2: 167–171.

Segerson, K. and Walker, D., 2002. Nutrient pollution: An economic perspective. *Estuaries*, 25: 797–808.

Strickland, J.D.H. and Parsons, T.A., 1972. *A Practical Handbook of Sea Water Analysis* , 2nd edn. Bull. Fish. Res. Board of Canada Bulletin 168, 310 pp.

Turner, R.K., Georgiou, S., Gren, I., Wulff, F., Barrett, S., Söderqvist, T., Bateman, I.J., *et al.*, 1999. Managing nutrient fluxes and pollution in the Baltic: an interdisciplinary simulation study. *Ecol. Econ.*, 30: 333–352.

Val Smith, H., 2007. Using primary productivity as an index of coastal eutrophication: the units of measurement matter. *J. Plankton Res.*, 29: 1–6.

Chapter 5

Phagocytosis of Charcoal Particulate as Immunological Marker of Azadirachtin Exposure

*Suman Mukherjee, Mitali Ray and Sajal Ray**

Aquatic Toxicology Laboratory, Department of Zoology, University of Calcutta. 35, Ballygunge Circular Road, Kolkata – 700 019, India

ABSTRACT

Lamellidens marginalis (Mollusca: Bivalvia) is distributed in the freshwater habitat of different states of India including West Bengal. This species is an important member of freshwater ecosystem. Through filter feeding, the species keeps the water body clean and is considered as a potential bioaccumulator of diverse xenobiotics. This edible species of economic importance demands thorough toxicological screening under the exposure of azadirachtin, a pesticide of new generation. Freshwater ecosystem bears the risk of contamination by pesticide residues during monsoon and other seasons. Haemocytes, the chief immunoeffector cells of blood of bivalve are capable of performing diverse physiological functions including phagocytosis, encapsulation, nonself surface adhesion and wound healing Phagocytosis of nonself particulates is

* Corresponding author: E-mail: raysnailmail@rediffmail.com

considered as a classical immunological response in bivalve. We have examined phagocytosis of charcoal particles by haemocytes of animal exposed to 0.006, 0.03, 0.06 and 0.09ppm of azadirachtin for 1, 2, 3, 4 and 7 days *in vivo*. For *in vitro* treatment, isolated haemocytes were exposed to 0.0004ppm, 0.0005ppm, 0.0006ppm, 0.0007ppm and 0.0008 ppm of azadirachtin for 1 hour for determination of phagocytosis of charcoal particles. Alteration in phagocytic response is suggestive of cellular stress which may lead to decline in population of *L. marginalis* in xenobiotic affected districts of West Bengal.

Keywords: *Azadirachtin, Lamellidens marginalis, Haemocytes, Phagocytosis.*

Introduction

L. marginalis is an economically important species and considered as an important dietary item of human. Human population of urban and rural sector of India consumes the species in various forms. Flesh of *L. marginalis* constitutes an important supplement of fish and poultry feed (Chakraborty *et al.*, 2009). This species faces an ecological threat due to shrinkage of its natural habitat. Rapid urbanization and contamination of water bodies by xenobiotics are regarded as major threats to bivalves and allied species of aquatic ecosystem. Azadirachtin based biopesticides are efficient in controlling pest population and considered as relatively less hazardous in relation to environmental stability and bioaccumulation (Schmutter, 1990). Multineem is an azadirachtin based biopesticide which is being frequently used by the farmers for the purpose of protection of crops from pests. Azadirachtin (steroid-like limonoid) is the active ingredients of neem based pesticide, capable of controlling insect pest at low concentration. Chemically azadirachtin is a tetraterpenoid which is present in the seed kernal of neem tree (*Azadirachta indica*). During monsoon, agricultural runoff containing azadirachtin often contaminates the freshwater ponds and lakes inhabited by *L. marginalis*. Bivalve molluscs represents interesting specimen to perform immunotoxicological studies since they exist in direct contact with contaminated aquatic sediments. In bivalve molluscs, haemocytes circulating in the haemolymph represent the main component of their immune system (Cheng and Sullivan, 1984, Adema *et al.*, 1991). A discrete population of haemocyte of bivalve is capable of recognition and adhesion to various nonself surfaces and considered as immune response (Armstrong, 1980; Martin *et al.*, 1999). Aquatic invertebrates distributed in polluted environment engulf invading microorganisms and parasites (Hose *et al.*, 1990; Martin *et al.*, 1999). Phagocytic responses of *in vivo* and *in vitro* toxin exposed haemocytes have already been screened under the sublethal exposure of azadirachtin in *L. marginalis* (Armstrong and Levin, 1979; Ehlers *et al.*, 1992).

Materials and Methods

Live adult healthy *L. marginalis* were collected from selected wetlands of district of South 24 Parganas of West Bengal with minimum anthropogenic interference. After manual collection, species were transported to the laboratory in plastic aquaria with proper care (Mukherjee *et al.*, 2006). Animals were acclimatized for seven days in laboratory conditions prior to experimentation. During acclimatization, animals were fed with chopped *Hydrilla* sp. and received uniform ration of illumination (Raut, 1991). Commercial formulation of azadirachtin was procured from the market under brand name of Multineem (Azadirachtin E.C. 0.03 per cent, Multiplex Private Limited, India). Separate sets of acclimatized animals were exposed to different concentrations of pesticide formulation to determine LC_{50} (Shibu Vardhanan and Radhakrishnan, 2002, Vijayavel and Balasubramanian, 2006). For *in vivo* treatment of haemocytes, *L. marginalis* of uniform size were exposed to pesticide at a concentration of 0.006, 0.03, 0.06 and 0.09 ppm of azadirachtin along with a batch of control. The control set received olive oil exposure of identical volume. Each batch comprised of ten individuals and span of exposure ranged as 1, 2, 3, 4 and 7 days. For *in vitro* treatment, isolated hemocytes were exposed to 0.0004ppm, 0.0005ppm, 0.0006ppm, 0.0007ppm and 0.0008 ppm of azadirachtin for 1hour. Haemolymph was collected from posterior adductor muscles using sterile syringe with 23 gauge needle (Brousseau *et al.*, 1999). Enumeration of haemocyte was done microscopically using Neubauer hemocytometer. For identification of haemocytes, cells were fixed with methanol and stained with Giemsa's stain (Cheng *et al.*, 1996). Cell viability was estimated with 2 per cent trypan blue following the principle of dye exclusion. Experiments were carried out with cell suspension with greater than 95 per cent viable cell.

For assay of phagocytosis, fixed numbers of viable haemocytes were challenged with charcoal particles (0.1mg/ml). For short term culture of cells, haemocytes were maintained in RPMI-1640 (Himedia, India) and were challenged with charcoal particles. Cells which received as *in vivo* exposure of azadirachtin were maintained in short term culture for 6 hours. Cells which received an *in vitro* exposure of azadirachtin were maintained in short term culture for 2 hours (Sauve *et al.*, 2002). Percentage of phagocytic haemocytes was determined microscopically (Ehlers *et al.*, 1992; Adamowicz and Wojtaszek, 2001) following the formula:

$$\text{Phagocytic haemocytes (\%)} = \frac{\text{Total number of phagocytosed cells}}{\text{Total number of cells}}$$

Post incubated cell samples were processed and were fixed and stained in Giemsa's stain (Himedia, India) or Haematoxylin-Eosin (Himedia, India) for microscopic analyses.Examination and photo documentation of haemocytes were carried out by light microscope (Axiostar plus, Zeiss, Germany) fitted with digital camera.

Results

Phagocytosis is considered as the primary line of defence in molluscs (Feng, 1988) and therefore, any factor that affects this activity would greatly influence the general immune status of the animal. Haemocytes of invertebrates phagocytose invading microorganisms and pathogen under proper elicitation. Haemocytes of invertebrates were able to phagocytose charcoal particles. They were characterised by having one or more vacuoles that displayed both the cytoplasm and nucleus to the periphery. Some phagocytic haemocyte can contain a lot of charcoal particles in their interior. In the controlled population, percentage of phagocytic haemocytes which received *in vitro* and *in vivo* exposure of azadirachtin were 60 per cent and 80 per cent after 3 hours and 6 hours of incubation with charcoal particles in short term culture respectively. *L. marginalis* receiving *in vivo* exposure of 0.09 ppm/7days of azadiarchtin expressed a lowest value of percent phagocytic haemocyte as 13 per cent in comparison to control value of 80 per cent. A sharp decrease of phagocytic haemocyte percentage was recorded against 0.06 ppm of azadirachtin for 1, 2, 3, 4 and 7 days of *in vivo* exposure as 60 per cent, 59 per cent, 53 per cent, 41 per cent and 24 per cent respectively.

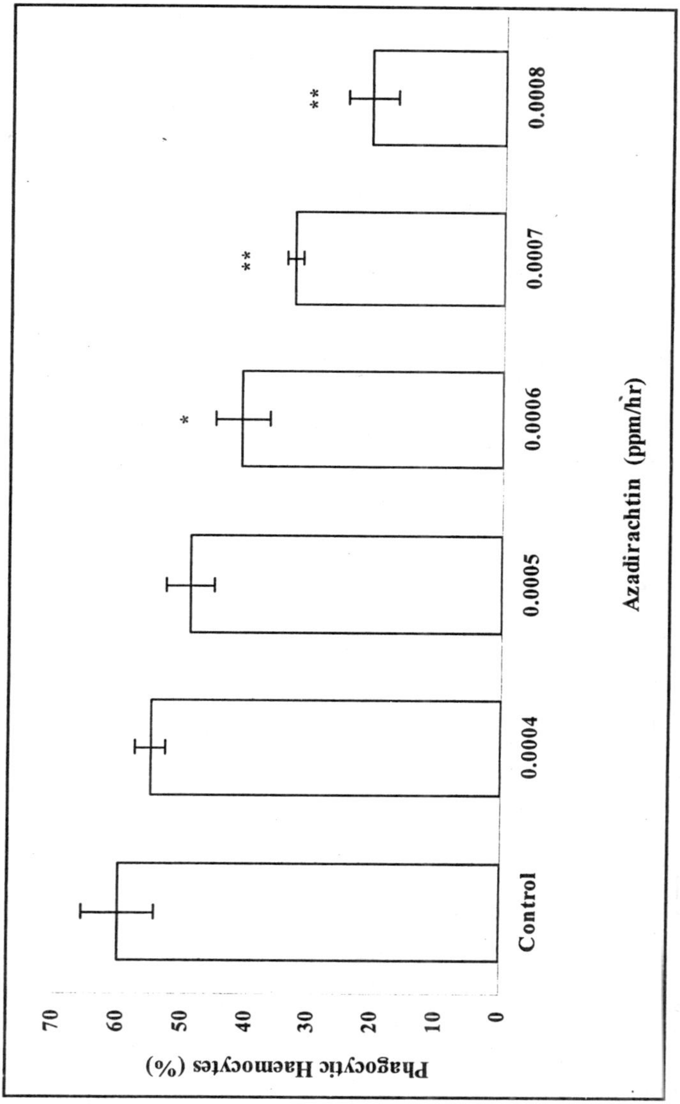

Figure 5.1: Percentage of Phagocytic Haemocytes Challenged with Charcoal Particles of *L. marginalis* Exposed to Azadirachtin *in vitro*. Data is represented as Mean ± S.D. Statistical significance is shown at $P<0.05^{*}$, $P<0.01^{**}$, $P<0.001^{***}$. (n=5).

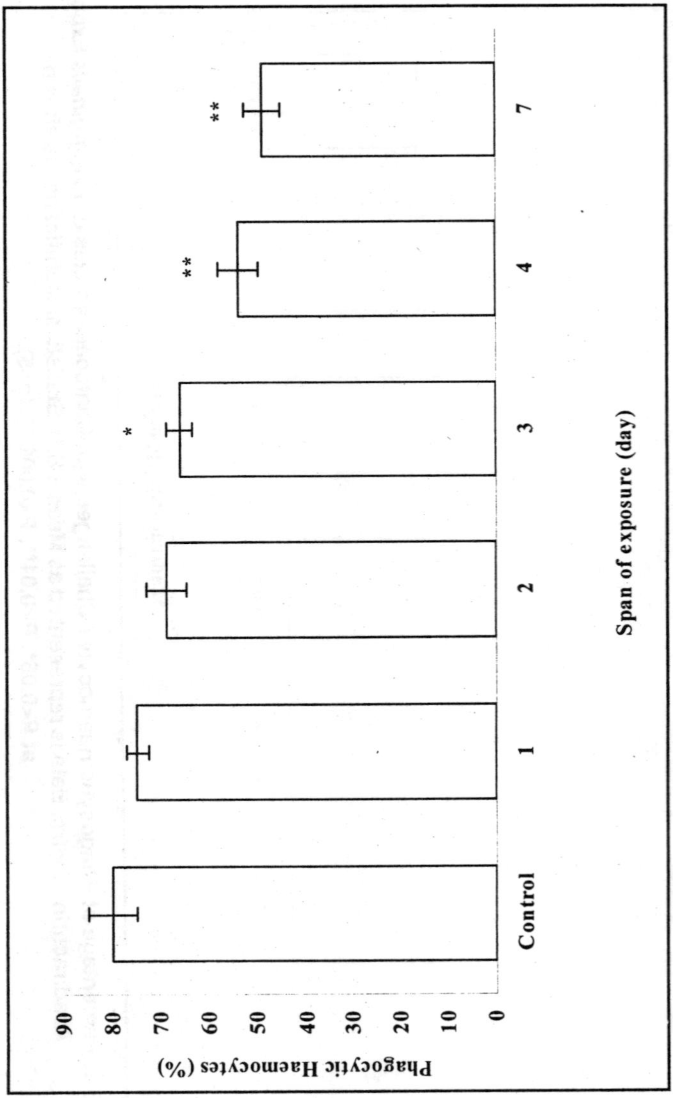

Figure 5.2: Percentage of Phagocytic Haemocytes Challenged with Charcoal Particles of *L. marginalis* Exposed to 0.006 ppm of Azadirachtin *in vivo*. Data is represented as Mean ± S.D. Statistical significance is shown at $P<0.05^*$, $P<0.01^{**}$, $P<0.001^{***}$. (n=5).

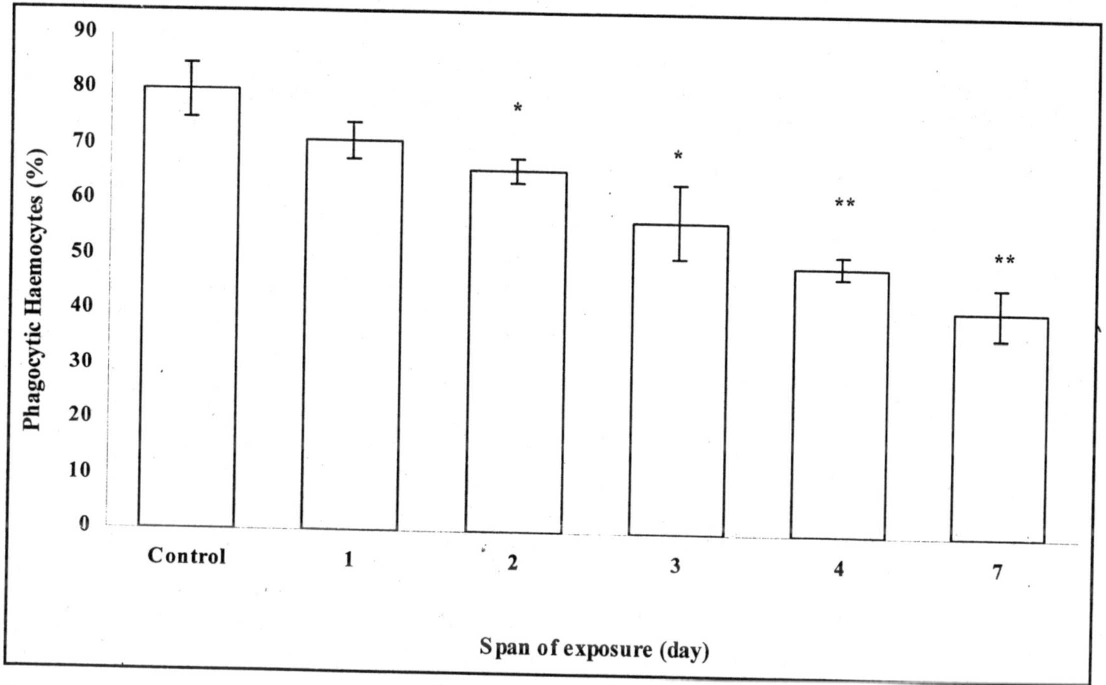

Figure 5.3: Percentage of Phagocytic Haemocytes Challenged with Charcoal Particles of *L. marginalis* Exposed to 0.03 ppm of Azadirachtin *in vivo*. Data is represented as Mean ± S.D. Statistical significance is shown at P<0.05*, P<0.01, P<0.001***. (n=5).**

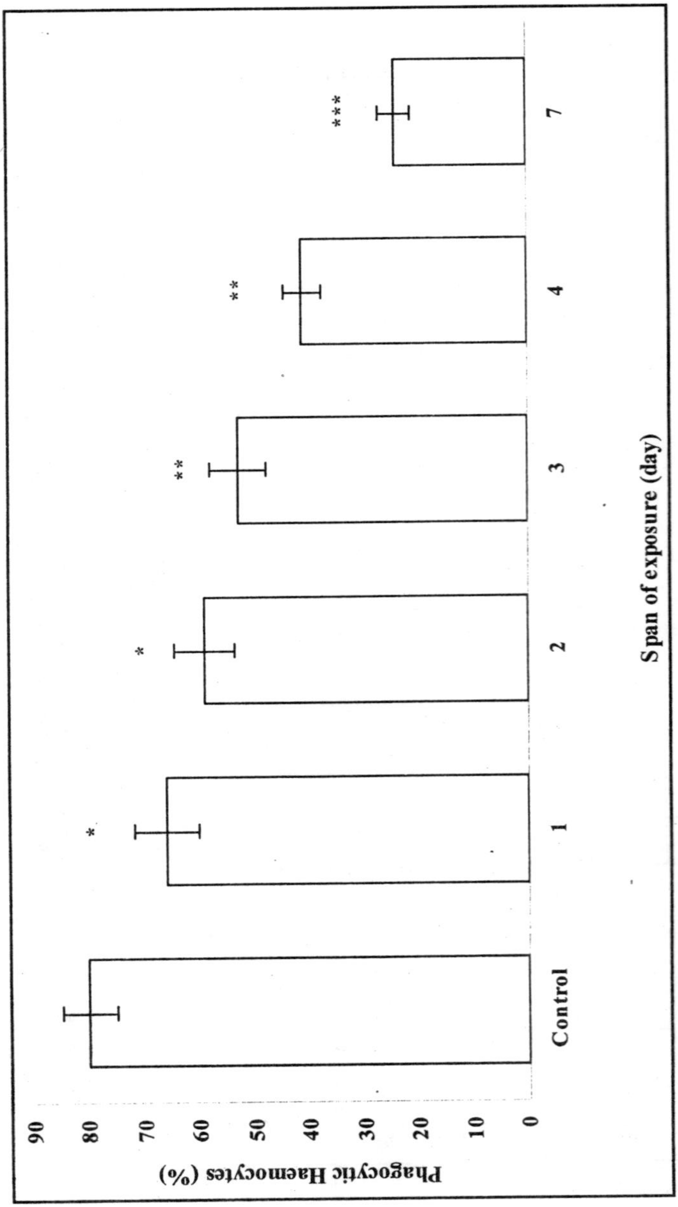

Figure 5.4: Percentage of Phagocytic Haemocytes Challenged with Charcoal Particles of *L. marginalis* Exposed to 0.06 ppm of azadirachtin *in vivo*. Data is represented as Mean ± S.D. Statistical significance is shown at P<0.05*, P<0.01, P<0.001***. (n=5).**

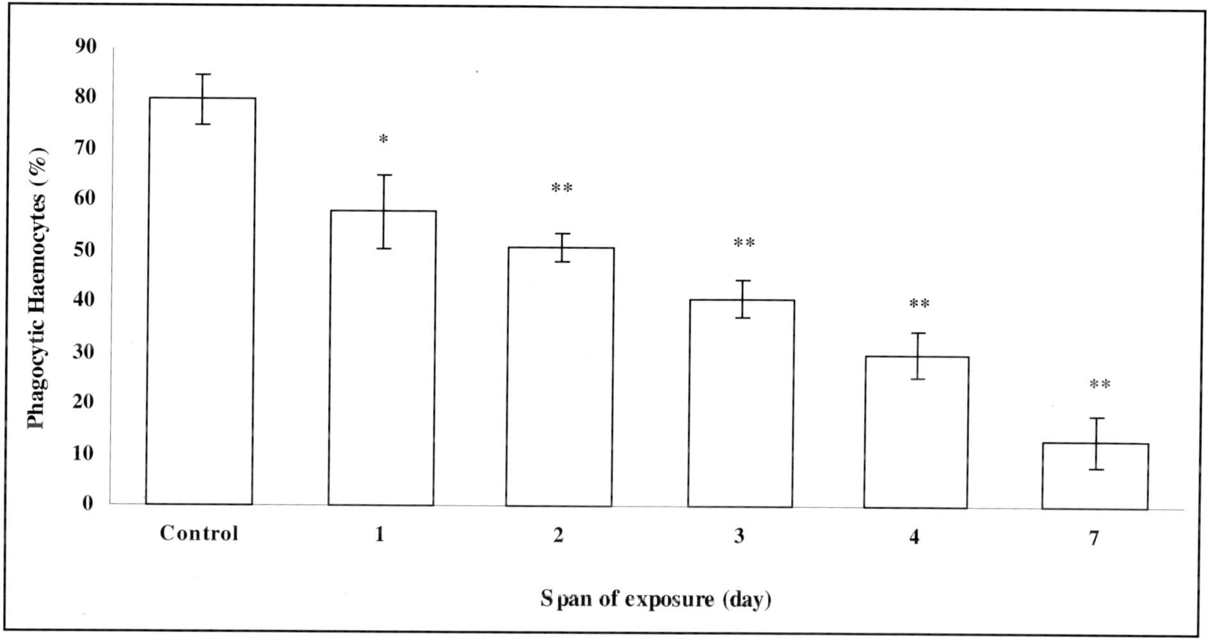

Figure 5.5: Percentage of Phagocytic Haemocytes Challenged with Charcoal Particles of *L. marginalis* Exposed to 0.09 ppm of azadirachtin *in vivo*. Data is represented as Mean ± S.D. Statistical significance is shown at $P<0.05^*$, $P<0.01^{**}$, $P<0.001^{***}$. (n=5).

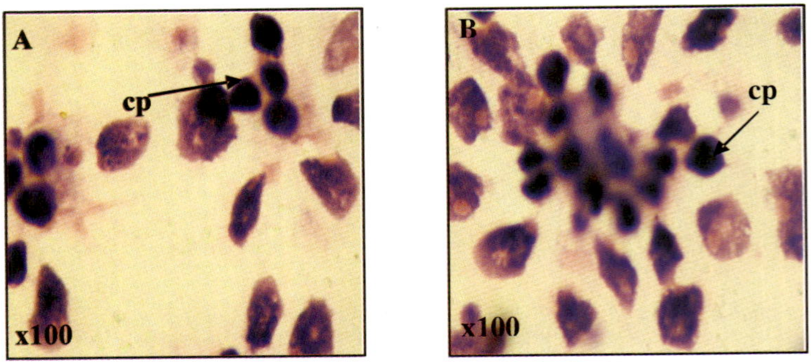

Figure 5.6: Phagocytosis of Charcoal Particles (CP) by Haemocytes (A and B) of *L. marginalis*

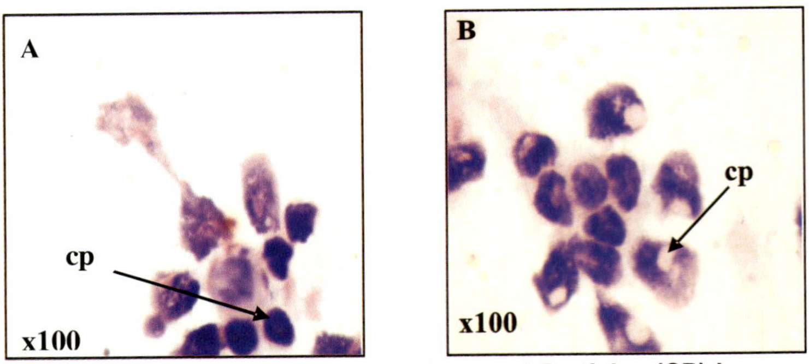

Figure 5.7: Phagocytosis of Charcoal Particles (CP) by Haemocytes of *L. marginalis* Exposed to Azadirachtin [(A). 0.0008 ppm/hr, (B).0.0004 ppm/hr] *in vitro*.

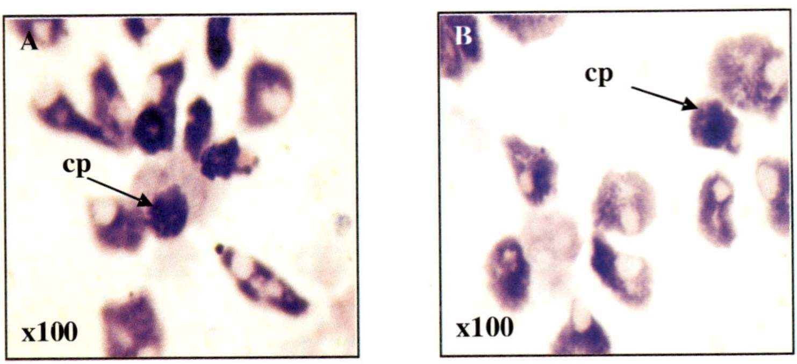

Figure 5.8: Phagocytosis of Charcoal Particles (CP) by Haemocytes of *L. marginalis* Exposed to Azadirachtin [(A). 0.006 ppm/7 day, (B). 0.09 ppm/7 day] *in vivo*.

Animals exposed to 0.03 ppm of azadirachtin *in vivo* for various span of exposure exhibited depletion in percentage of phagocytic haemocyte. The studies on the effects of azadirachtin exposure (*in vivo*) on the phagocytic activity of haemocytes of freshwater mussel exhibited an inhibition of this activity in animals compared with control values. The haemocytes collected from *L. marginalis* have been exposed to various azadirachtin concentrations *in vitro*. The doses were selected on the basis of NOEL value. The cells showed a classical pattern of *in vitro* toxicity of azadiarchtin with a drastic decrease in percentage of phagocytic haemocytes. Indeed the impairment of phagocytic activity was manifested at sublethal concentrations.

Discussion

Phagocytosis is a very important immunological parameter and widely acknowledged as an effective biomarker of aquatic pollution (Oliver *et al.*, 1999; Sauve *et al.*, 2002 and Chakraborty *et al.*, 2009). Indeed the central role of phagocytosis in immune defence and the sensitivity of this biological function to environmental xenobiotics may be considered as cellular stress in organisms. Phacocytic cells are highly conserved throughout evolution and by using the phagocytic activity of these cells, it has been possible to address the comparative sensitivity of these cells to environmental contaminants. The present study involves *in vivo* and short term *in vitro* exposure of haemocytes to azadiarchtin and demonstrates the potential detrimental effect of sublethal concentrations azadirachtin to the immune system of bivalve molluscs. Azadirachtin induced impairment of phagocytic response and also indicated a physiological state of immune suppression. These conditions, which represent reality for organisms living in contaminated sediments, could eventually render the bivalves much more vulnerable to opportunistic parasites and infection, thus drastically affecting their survival and the equilibrium of aquatic ecosystems.

References

Adamowicz, A. and Wojtaszek, J., 2001. Morphology and phagocytotic activity of coelomocytes in *Dendrobaena veneta* (Lumbricidae). *Zoologica Poloniae.*, 46(1 – 4):91 – 104.

Adema, C.M., Van der Knaap, W.P.W. and Sminia, T., 1991. Molluscan hemocyte mediated cytotoxicity: The role of reacting oxygen intermediates. *Rev. Aquat. Sci.*, 4: 201–223.

Armstrong, P.B. and Levin, J., 1979. *In vitro* phagocytosis by *Limulus* blood cells. *J. Invert. Pathol.*, 34: 145–151.

Armstrong, P.B., 1980. Adhesion and spreading of *Limulus* blood cells to artificial surface. *J. Cell Sci.*, 44: 243–262.

Brousseau, P., Payette, Y., Tryphonas, H., Blakley, B., Boernaus, H., Flipo, D. and Fournier, M., 1999. *Manual of Immunological Methods*. CRC Press, Boca Raton, FL.

Chakraborty, S., Ray, M. and Ray, S., 2009. Evaluation of phagocytic activity and nitric oxide generation by molluscan haemocytes as biomarkers of inorganic arsenic exposure. *Biomarkers*, 14(8): 539–546.

Cheng, T.C. and Sullivan, J.T. 1984. Effects of heavy metals on phagocytosis by molluscan hemocytes in response of marine organisms to pollutants (Ed.) J.J. Stegemen. Elsevier; Amstardam, pp. 305–315.

Cheng, J.H., Yang, H.Y.Y., Peng, S.W., Chen, Y.J. and Tsai, K.Y., 1996. Characterization of abalone (*Haliotis diversicolor*) hemocytes *in vitro*. *Biol. Bull. NTNU.*, 31(1): 31–38.

Ehlers, D., Zosel, B., Mohrig, W., Kauschke, E. and Ehlers, M., 1992. Comparison of *in vivo* and *in vitro* phagocytosis in *Galleria mellonella* L. *Parasitol. Res.*, 78: 354–359.

Feng, R. and Isman, M.B., 1998. Selection for resistance to azadirachtin in the green peach aphid *Myzus persicae*. *Experimentia*, 51: 831–833.

Hose, J.E., Martin, G.G. and Gerand, A.S., 1990. A decapod hemocyte classification scheme integrating morphology, cytochemistry and function. *Bio. Bull.*, 178: 33–45.

Martin, G.G., Lin, J.H. and Luc, C., 1999. Reexamination of hemocytes in brine shrimp (Crustacea, Branchiopoda). *J. Morphol.*, 242: 283–294.

Mukherjee, S., Ray, M. and Ray, S., 2006. Azadirachtin induced modulation of total count of haemocytes of an edible bivalve *Lamellidens marginalis*. *Proc. Zool. Soc. Calcutta*, 59(2): 37–39.

Oliver, L.M. and Fisher, W.S., 1999. Appraisal of prospective bivalve immunomarkers. *Biomarkers*, 4(10): 510–530.

Raut, S.K., 1991. Laboratory rearing of medically and economically important molluscs. In *Snails, Flukes and Man*, (Ed.) M.S. Jairajpuri. Zoological Survey of India Pub., pp. 79–83.

Sauve, S., Brousseau, P., Pellerin, J., Morin, Y., Senecal, L., Goudreau, P. and Fournier, M., 2002. Phagocytic activity of marine and freshwater bivalves: *in vitro* exposure of hemocytes to met (Ag, Cd, Hg and Zn). *Aquat Toxicol.*, 508(3–4): 189–200.

Schmutterer, H., 1990. Properties and potential of natural pesticides from the neem tree *Azadirachta indica*. *Annu. Rev. Entomol.*, 35: 271–297.

Shibu Vardhannan, Y. and Radhakrishnan, T., 2002. Acute toxicity of copper, arsenic and HCH to paddy field crab, *Paratelphusa hydrodromus* (Herb.). *J. Environ. Biol.*, 23(4): 387–392.

Vijayavel, K. and Balasubramanian, M.P., 2006. Fluctuations of biochemical constituent and marker enzymes as a consequence of naphthalene toxicity in the edible estuarine crab, *Scylla serrata*. *Ecotoxicol. Environ. Saf.*, 63(1): 141–147.

Chapter 6

Ecological Analysis of Digenetic Trematode Parasites of Two Populations/Anuran Hosts, *Rana tigrina* and *Rana cyanophlyctis* in Jammu Province, J&K State

G.D. Singh[1*], *Anil K. Verma*[2], *Ranvijay Singh*[3] *and V.K. Gupta*[1]

[1]*Indian Institute of Integrative Medicine (CSIR), Canal Road, Jammu, J&K, India*

[2]*Govt. Degree College, Rajouri, J&K, India*

[3]*Govt. Degree College, Kathua, J&K, India*

ABSTRACT

The present work was undertaken with the main purpose of examining the trematode fauna in the intestine of two sympatric frog species from two ponds in Jammu province and assess the distributional patterns of the infecting taxa among the two hosts besides evaluating their ecological status and any interspecific interactions between them at infra-community level.

Earlier studies on trematode parasites of the frogs and toads in Jammu province of J&K state, describe only taxonomic

* Corresponding author: E-mail: singh_gd@iiim.ac.in

descriptions of trematodes from vertebrate hosts. The present investigation deals with the interactions between the co-occurring species, the study uses the parameters of the host size, the intestinal length and the host gender in relation with prevalence intensity, and the ecological analysis of the infecting species in the two ranids, *Rana tigrina* and *Rana cyanophlyctis* in Jammu province. A total of 256 adult frogs, *Rana tigrina* and 548 *Rana cyanophlyctis* were collected from two populations in Jammu district of J&K state. Of the total sample sizes from the two populations of the host from the two sites (Station I and Station II), only 125 adults of *Rana tigrina* and 270 of *Rana cyanophlyctis* from the two populations were included for analysis of their enteric helminthes. As many as seven species of trematodes have been found infecting the frogs in the two populations, one at Bishnah (CB population) and the other at Ramgarh (CR population). A total of three thousand two hundred and sixty six (3,266) above mentioned trematodes were recovered from 283 frogs (both sexes of the two species) out of a total 395 host individuals collected and screened for infection during the months of April to October, (1990 – 1992) and analyzed for ecological studies. The overall mean Snout Vent Length (SVL) of *Rana cyanophlyctis* examined during the study was found to be significantly smaller than that of *Rana tigrina* This size difference was maintained even at the level of the gender of the hosts; the males and the females of *Rana tigrina* were found to be respectively larger than that of females of *Rana cyanophlyctis* from the two populations. One hundred and eighty nine (189) frogs of *Rana cyanophlyctis* and ninety four (94) frogs of *Rana tigrina* from the two populations were found infected with one or more species of the seven trematodes.

While *Rana cyanophlyctis* was found infected with all the seven trematode species, *Rana tigrina* on the other hand, was found infected by only four. The location of the various enteric trematodes taxa constituting the infra-community as found infecting each of the two frog species from the two populations were located along the length of the intestines (duodenum; small intestines and the large intestine) corresponding to segments (I), (II, III, IV) and (V) respectively, and this interactions at community levels have been discussed.

Keywords: *Rana tigrina, Rana cyanophlyctis, Ganeo tigrinus, Diplodiscus mehrai, Trematode, Ecology, Parasite-Intensity.*

Introduction

Innumerable studies on helminth parasites of vertebrates world-over have been published almost regularly during the last century or more. While most of these publications are purely taxonomic in nature (Yamaguti, 1971), only a few concern the life-history of the parasites (Yamaguti, 1975), and very little or no information is available about helminth diversity of the different vertebrate hosts at the community level of the infecting taxa. This paucity of information on ecological aspect of helminthes is wide-spread and apparently neither restricted in geography nor limited to some specific groups or hosts.

Although Rankin (1945) was the pioneer helminthologist to have studied the ecology of helminthes of some North American amphibians and reptiles, yet his study surprisingly failed to evoke sufficient interest of parasitologists and/or of ecologists in this interesting aspect of parasite communities. This is obvious from the fact that a long spell of quiet followed his publication, till Crofton (1971) published an interesting quantitative study of animal parasites. With this publication, interest in the ecology of animal parasites resurged and a steady increase was registered in the number of publications on different aspects of the ecology of parasites in general and helminthes in particular, from different parts of the world. Indian contribution in this fascinating field of parasitology has been and continues to remain pathetically meager.

The parasites of vertebrates and their community ecology have attracted interest of many workers in the last few decades, thus developing a conceptual frame work for the hierarchical structure of parasites community (Poulin, 2001). There are several reports on ecological studies on helminth infra-communities of marine mammals, birds, fish etc. (Goater *et al.*, 1987; Goater and Bush, 1988; Edwards and Bush, 1989; Kennedy and Williams, 1989; Balbuena and Raga, 1993). However, relatively limited number of investigations have been made on the helminth communities of amphibians and reptiles.

Although there are numerous reports and general ecological studies on the helminthes of amphibians and reptiles, a very less amount of work has been done on specific studies utilizing community measures (species richness, helminth intensity and diversity indices). Aho (1990) opined, helminth communities of amphibians and reptiles are highly variable, depauperate, and have traits characteristics of non interactive community structure.

Few reports exist on the helminth fauna of amphibians and reptiles from various parts of the world. Lees (1962) studied the incidence of helminth parasites in amphibians and reptiles. Frandsen (1974) investigated the parasite fauna of Danish amphibians, and Baker (1987) provided a synopsis on the helminth fauna of amphibians and reptiles.

A report of parasitic nematodes from amphibians and reptiles was given by Moravec *et al.* (1987). Goldberg and Bursey (1991) studied helminthes of toads, *Bufo alvarius* and *Bufo cognatus* (Bufonidae) from Southern Arizona. Goldberg *et al.*(1998) reported gastrointestinal helminth of four gekkonid lizards from Marina Island, Micronesia. Goldberg *et al.*(2000) studied helminth fauna of plain leopard frog, *Rana blairi*. Goldberg and Bursey (2002) reported helminthes of 10 species of anurans from Honshu Island, Japan. Recently Goldberg *et al.* (2004b) made an exhaustive survey on the helminthes of six species of snakes from Honshu Island, Japan.

Although surveys, reports and investigations on helminth fauna of amphibians from across the globe are numerous but unfortunately there is limited number of investigations on parasite fauna from India. A review of literature on parasites of Indian amphibians and reptilians hosts revealed a paucity of information. Surveys and general taxonomic studies on the helminthes of Indian amphibians, particularly the reninine forms, are too many to be cited here, but specific studies on helminth communities in frogs and/or toads are either not undertaken or not published. Narain's (1989) seems to be only study on parasites (protozoan and platyhelminth) of aquatic vertebrates of J&K State, which has dealt with some ecological parameters of the helminthes of the frog, *Dicroglossus* (=*Rana*) *cyanophlyctis* from Kashmir water-bodies. Whereas in water bodies of Jammu province, seasonal occurrence of *Indopleurogenes orientalis* in *Rana cyanophlyctis* was studied by Ranvijay *et al.* (1993). Recently Pooja *et al.* (2005) investigated helminth-infracommunities of amphibians and reptilians hosts from Rohilkhand region of India.

Earlier studies on trematode parasites of the frogs and toads in Jammu province of J&K State, describe only taxonomic descriptions of trematodes from vertebrate hosts (Sudan, 1979; Verma, 1988;

Singh, 1989; Pandoh, 1992). However, apart from the studies very little is known about the ecological analysis of trematode parasites in their definite hosts. The present investigation deals with the interactions between the co-occurring species and the study uses the parameters of the host size, the intestinal length and the host gender in relation with prevalence intensity, and the ecological analysis of the infecting species in the two ranids, *Rana tigrina* and *Rana cyanophlyctis* in Jammu province.

Materials and Methods

A total of 256 adult frogs, *Rana tigrina* and 548 *Rana cyanophlyctis* (Table 6.1) were collected from two populations from two well defined and discrete sites called hereafter as Station I, designated as CB (Bishnah), Jammu (Figure 6.1) located 30 kilometers in the south of Jammu city and Station II as CR (Ramgarh, Jammu (Figure 6.2) located 40 kilometers in the south of Jammu city and some 10 km further south of Station I respectively (Figure 6.3). The collections were made between late March, 1990 to October, 1992. Frogs were collected with the use of strong steel mesh hand net, transported to the laboratory alive, measured in their snout-vent-length (SVL), weighed, sexed and only adult specimens necropised within 24 hr of collection after anaesthetized. The intestinal tract (from pyloric end of stomach to vent) was removed and then thoroughly scanned for the trematodes infection. The trematodes recovered were processed for fixation in A.F.A. or hot 70 per cent alcohol, stained in Mayer's acetocarmine (Gray, 1973) and mounted in D.P.X. The identification of trematodes was done according to Yamaguti (1971).

Table 6.1: Showing Gender Wise Numerical Data of the Two Hosts from Two Stations from March, 1990 to November, 1992

Host Species	*Station–I*			*Station–II*		
	Sex	*Number*	*Total*	*Sex*	*Number*	*Total*
Rana tigrina	M	68	125	M	61	131
		F	57		F	70
Rana cyanophlyctis	M	138	306	M	108	242
		F	168		F	134

M: Male; F: Female.

Figure 6.1: A View of Pond at Bishnah (Station No. 1)

Figure 6.1: A View of Pond at Ramgarh (Station No. 2)

Of the total sample sizes from the two populations of the host from the two sites (Station I and Station II), only 125 adults of *Rana tigrina* and 270 of *Rana cyanophlyctis* from the two populations (Table 6.1) were included for analysis. This shortened sample size of the two host from the two populations represented the adult hosts that

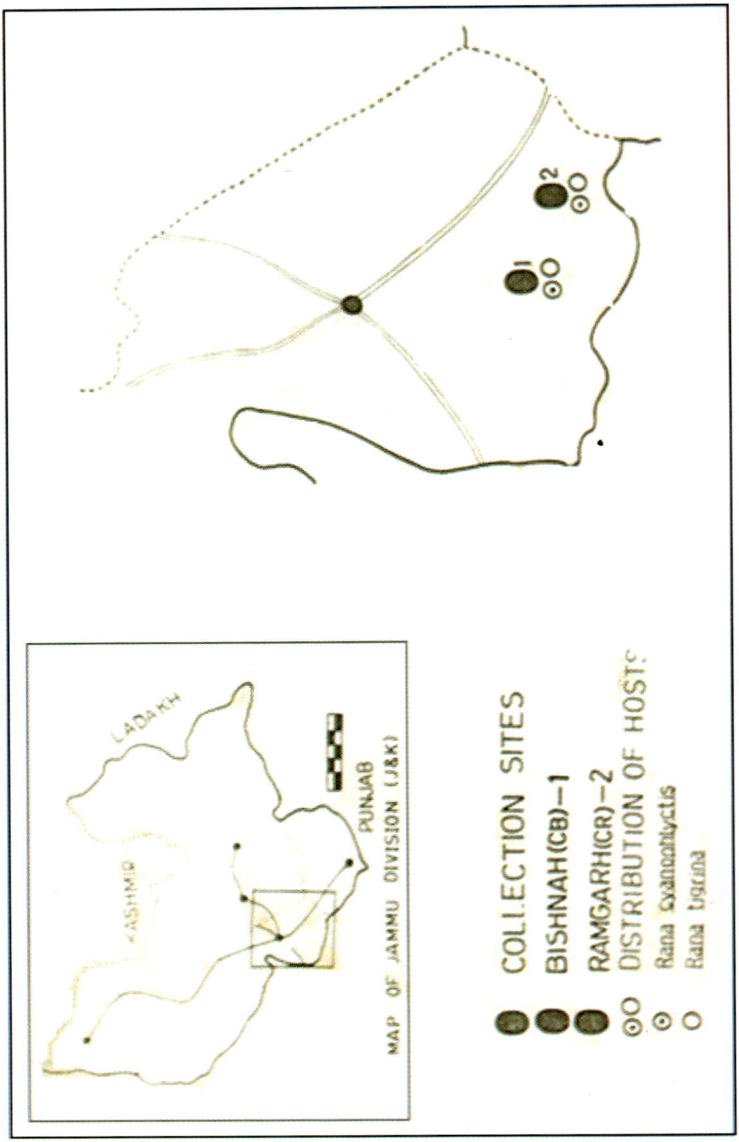

Figure 6.3: An Outline Map of Jammu Province Showing Location of Collection Sites

Table 6.2: Showing the Sex-wise Data Ob SVL of the Two Host Species Population-wise Screened for Trematode Infection and Used for Ecological Analysis

Host Species	*CB Population*		*CR Population*		*Collected SVL for Both Population Sex-wise*			
	Sex(n)	*Mean±SE (Range)*	*Sex (n)*	*Mean±SE (Range)*	*Sex (n)*	*Mean±SE (Range)*	*Species Total (n)*	*Mean±SE (Range)*
Rana tigrina	M (33)	103.2±1.24 (90 – 120)	M (30)	92.8±1.40 (70 – 114)	M (63)	98.25±0.96 (70 – 120)	125	102.56±0.87 (70 – 130)
	F (28)	101.03±1.23 (80 – 120)	F (34)	111.8±1.10 (90 – 130)	F (62)	106.9±0.98 (80 – 130)		
R. cyanophlyctis	M (68)	42.94±0.86 (34 – 51)	M (53)	38.47±0.64 (32 – 47)	M (121)	40.98±0.72 (32 – 51 0)	270	45.01±0.67 (32 – 57)
	F (83)	48.10±0.47 (42 – 57)	F (66)	48.53±0.6 (39 – 54)	F (149)	48.29±0.59 (39 – 57)		

M: Male; F: Female

CB = (Station – I, Bishnah); CR = (Station – II, Ramgarh); SVL = Snout-Vent Length.

were collected over as short a period as possible (*i.e.* from April to August, 1990–1992) in order to minimize the effects of seasonal variations on the helminth infra-community structure, available within the host. Ecological analysis of trematodes were done according to Margolis *et al.* (1982), and Chi-square (X^2) analysis for two samples were used to determine significant differences. Values were considered statistically significant at $P< 0.05$. The values following mean in the tables or text are standard errors, unless otherwise stated.

Results

As many as seven trematodes species (Figures 6.4–6.10) namely *Diplodiscus mehrai* (Pandey, 1937a); *Ganeo tigrinus* (Mehra and Negi, 1928); *Prosotocus himalayai* (Pandey, 1937b); *Indopleurogenes yamagutii* (Sudan, 1979); *Gorgoderina elliptica* (Dwivedi, 1968); *Loxogenes jammuensis* (n.sp.) and *Mehraorchis ranarum* (Srivastava, 1934) have been found infecting the two frogs in the two populations. A total of three thousand two hundred and sixty six (3,266) above mentioned species were recovered from 283 frogs (both sexes of the two species) out of a total 395 host individuals collected and screened for infection and analyzed for ecological studies. However, the overall collection of the enteric trematodes was made from 804 frogs of both species spread over a period of two years from 1990/1991 and 1991/1992 obtained between late march, 1990 to October, 1992 from the above said collection stations.

Population-wise data on snout-vent length (SVL) in mm of the adult (mature) frogs of either sex of the two species, *Rana tigrina* and *Rana cyanophlyctis* are presented in Table 6.2. The overall mean SVL of *Rana cyanophlyctis* (45.01±0.67 mm; n = 270) examined during the study was found to be significantly smaller than that of *Rana tigrina* (102.56±0.87 mm; n = 125). This size difference was maintained even at the level of the gender of the hosts; the males (98.25±0.96 mm; n = 63) and the females (106.94±0.98mm; n = 62) of *Rana tigrina* were found to be respectively larger than males (40.98±0.72 mm; n = 121) and females (48.29±0.59 mm; n = 149) of *Rana cyanophlyctis* from the two populations.

Infection Patterns

One hundred and eighty nine (189) frogs of *Rana cyanophlyctis* (70 per cent of the total number of frogs examined) and ninety four

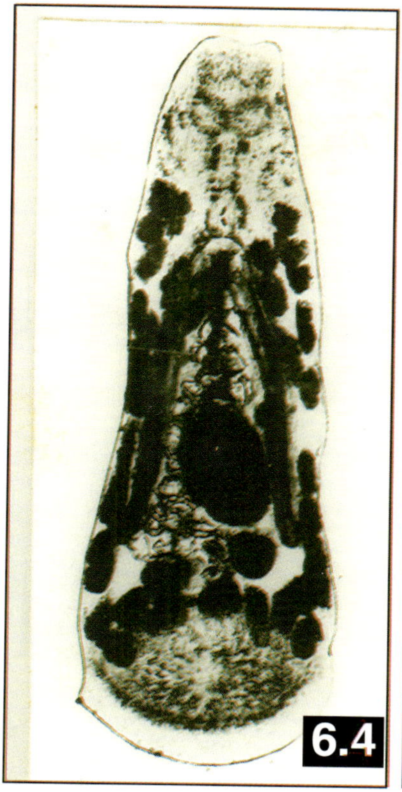

Figure 6.4: *Diplodiscus mehrai* (Pande, 1971) Showing General Body Organization

Figure 6.5. *Ganeo tigrinus* (Mehra and Negi, 1928) Showing General Body Organization

(94) frogs of *Rana tigrina* (75.2 per cent of the total number examined) from the two populations were found infected with one or more of the seven trematode species (Table 6.3).

While *Rana cyanophlyctis* was found infected with all the seven trematodes species, *Rana tigrina* on the other hand, was found infected by only four, which therefore, constituted the shared species. However, all the trematodes recovered were never found simultaneously infecting the same host individual, only about 1 to 5 different species were, if ever, found parasitizing the same individual. Three thousand two hundred and sixty six (3,266) trematodes (Table

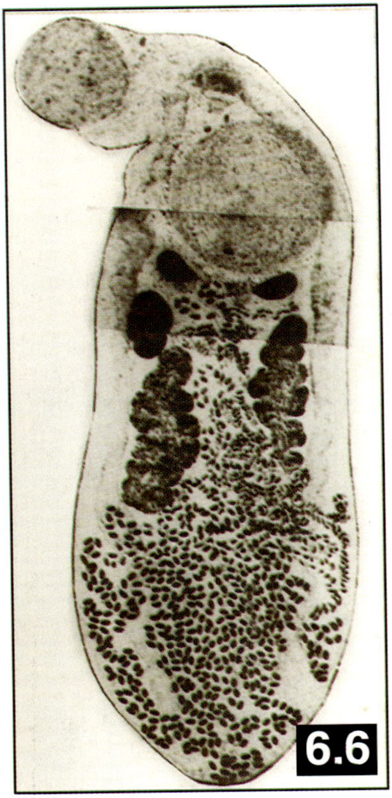

Figure 6.6: ***Gorgoderina elliptica*** **(Dwivedi, 1968) Showing General Body Organization**

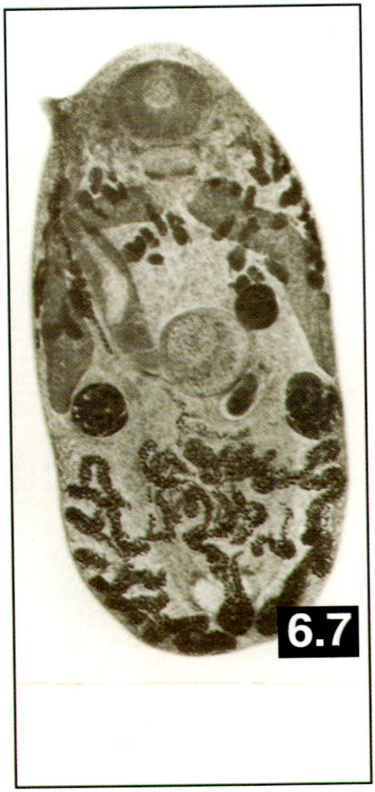

Figure 6.7: ***Indopleurogenes yamagutii*** **(Sudan,1979) Showing General Body Organization**

6.4) were recovered from a total of two hundred and eighty three (283) infected frogs representing the combined tally of two host species from the CB and CR populations.

Of the 3,266 intestinal trematodes obtained, 1,141 (about 35 per cent) were recovered in the two host species from CR population at an overall relative intensity (RI) of 8.5 and the remaining 2,125 (about 65 per cent) in the two host species from CB population (Table 6.4) at an overall relative intensity of 14.24.The trematodes recovered from the two ranid species in CR population were less evenly distributed (Table 6.5) in their host (44.3 per cent in *Rana cynophlyctis* and 55.66

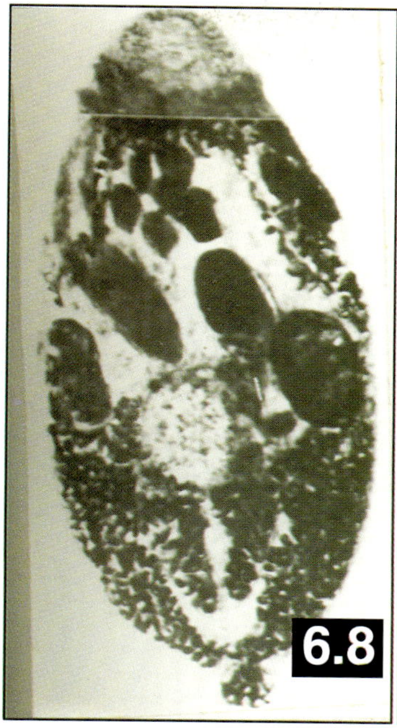

Figure 6.8: ***Loxogenes jammuensis*** **n.sp. Showing General Body Organization**

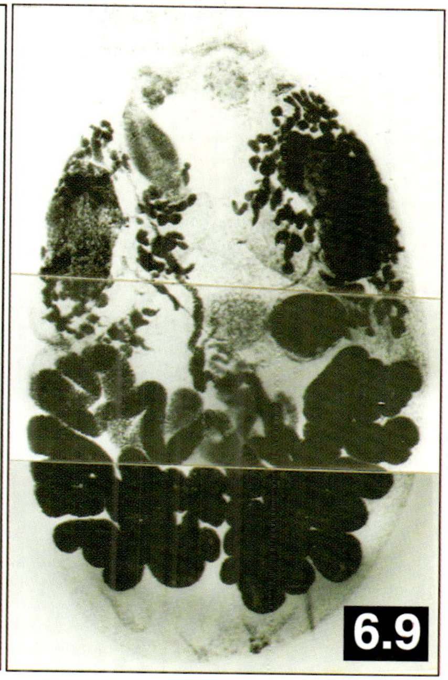

Figure 6.9: ***Mehraorchis ranarum*** **(Srivastava, 1934) Showing General Body Organization**

per cent in *Rana tigrina*) than these were in the two ranids of the other population CB (48.02 per cent in *Rana cyanophlyctis* and nearly 52 per cent in *Rana tigrina*), although in either population, the members of the host *Rana tigrina* were found to show a relatively higher intensity than the members of the other host species.

Spatial Distribution of the Infecting Trematodes

The recovery of the various enteric trematodes taxa constituting the infra-community as found infecting each of the two frog species from the two populations was made along the length of the intestines (duodenum; small intestines and the large intestine) corresponding to segments (I), (II, III, IV) and (V) respectively.

The distributional limit or the location of an infecting species of the trematode was measured by the position of the individual parasite of the infra-community from one end of its distribution to

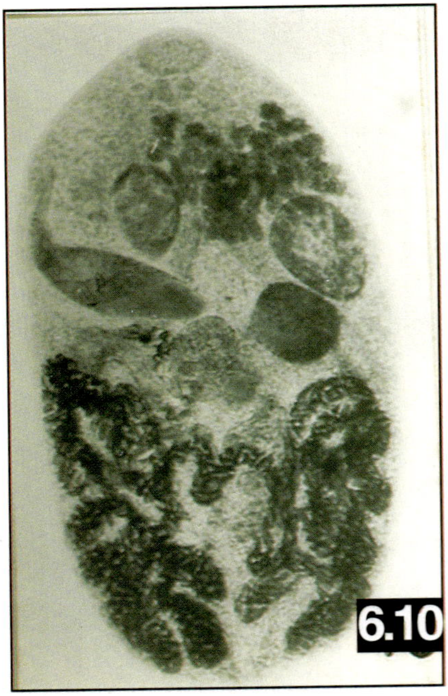

Figure 6.10: *Prosotocus himalayai* (Pande, 1937) Showing General Body Organization

the other along the length of the intestine. Even if only a single individual of a trematode taxon occupied any segment in continuity with other than that segment was included in the limit of its location.

In *Rana tigrina*, only 4 specimens of *Mehraorchis ranarum* were found in the mid-intestinal region, corresponding to the segments II, III and IV of the intestine, in a single male from CR population, the other three enteric trematodes *Ganeo tigrinus, Prosotocus himalayai* and *Diplodiscus mehrai* were found located in specifically different niche along the intestine of the host. *Ganeo tigrinus* was found consistently occupying the post-duodenal and pre-cloacal portion of the intestine corresponding to the segments II, III and IV of host in either population whether found alone or in association with *Diplodiscus mehrai, Prosotocus himalayai, Indopleurogenes yamagutii* or *Gorgoderina elliptica.*

Table 6.3: Showing the Presence (+) or Absence (–) of Different Intestinal Trematode Species Found in the Two Ranids, *Rana tigrina* and *Rana cyanophlyctis* in Two Populations

Trematode species	*CB Population*				*CR Population*			
	R. cyanophlyctis		*R. tigrina*		*R. cyanophlyctis*		*R. tigrina*	
	M	*F*	*M*	*F*	*M*	*F*	*M*	*F*
Diplodiscus mehrai	+	+	+	+	+	+	+	+
Ganeo tigrinus	+	+	+	+	+	+	+	+
Gorgoderina elliptica	+	+	–	–	+	+	–	–
Indopleurogenes *yamaguti*	+	+	–	–	+	+	–	–
Loxogenes jammuensis	–	+	–	–	+	–	–	–
Mehraorchis ranarum	+	+	–	–	–	–	+	–
Prosotocus himalayai	–	+	+	+	–	+	+	+

+: Presence; –: Absence; M: Male; F: Female.

Table 6.4: Sex-wise Data on the Number of Intestinal Trematodes Recovered from the Ranid Hosts Along with Relative Intensity of the Flukes (RI) in Each Host

Hosts	*CR Population (n = 134)*			*CR Population (n = 134)*			*Total No. of Trematodes*	*Overall Intensity*
	Sex (n)	*Trematode Recovered*	*RI*	*Sex (n)*	*Trematode Recovered*	*RI*		
Rana tigrina	M (23)	390	16.9	M (23)	832	36.1	1222 (n = 46)	26.3
	F (26)	245	9.4	F (22)	271	12.3	516 (n = 48)	10.75
Rana	M (39)	188	4.82	M (36)	183	5.0	371 (n = 75)	4.9
cyanophlyctis	F (46)	318	6.9	F (68)	839	12.3	1157 (n = 46)	10.1

M: Male; F: Female; RI: Relative Intensity.

Table 6.5: Overall Distribution Pattern of the Trematode Parasites in the Two Ranid Hosts from Two Populations

Hosts	CR Population (n= 134)		CB Population (n= 149)		Overall Trematode Load (per cent)	Overall (n)	Absolute Intensity
	Trematode Load (per cent)	(n)	Trematode Load (per cent)	(n)			
Rana cyanophlyctis	506 (44.34)	85	1020 (48.00)	104	1525 (46.80)	189	8.08
Rana tigrina	635 (55.66)	49	1103 (51.98)	45	1738 (53.30)	94	18.49

N: Total no. of hosts

Prosotocus himalayai when infecting the *Rana tigrina* alone was always recovered from duodenum (segment I) and rarely in the post-duodenal intestinal portion (corresponding to the segment II). On the other hand, whenever recovered in association with *Ganeo tigrinus* in double or in triple infection, it would remain confined to the duodenal region (*i.e.* segment I) only. However, when found as member of double infection with *Diplodiscus mehrai, Prosotocus himalayai* was invariably recovered from segments I and II (and once even from I, II and III segment) of the intestine.

Diplodiscus mehrai infecting the *Rana tigrina* from the two populations was recovered from the last third of the small intestine (segment IV) and the last segment (segment V) of the intestine. Even in the presence of any other trematode (*Prosotocus himalayai* or *Ganeo tigrinus*) in double or triple infection, the location of the trematode in the intestine *i.e.* segment III and IV, remained unaltered.

In *Rana cyanophlyctis* the enteric infection due to trematodes enumerated above was comparatively richer and more diverse. Of the seven infecting trematodes, *Mehraorchis ranarum* and *Loxogenes jammuensis* despite very low abundance and erratic distribution were always recovered from the mid-intestinal region corresponding to the segments III and IV of the host intestine. The location distribution of the two trematodes was invariable of the sex or the population of the host from which recovered.

Of the remaining five infecting trematodes, *Prosotocus himalayai* was recovered from only the females of the host species in the two populations (once from CB population and twice from CR population in double infection). When recovered alone (*i.e.* in CB population or in association with *Diplodiscus mehrai* in double infection (*i.e.* in CR population), it was found well distributed over the segment I (*i.e.* duodenum) and the II segment of the host intestine.

However, in association with *Ganeo tigrinus* in a double infection found in the host, *Prosotocus himalayai* was found confined to the duodenal segment *i.e.* I segment of the host intestine. Of the remaining infecting trematodes, *Prosotocus himalayai* recovered only from females of the host (once alone in CB population and twice in double infection in CR population) was found well distributed over the I. II and III segments of the host intestine; particularly so when alone, otherwise whenever in association with *Diplodiscus mehrai* or

Ganeo tigrinus; the trematode showed its distribution limited to the I segment (*i.e.*, duodenal region) alone.

Ganeo tigrinus was recovered from the segment II, III and IV of the intestine of the *Rana cynophlyctis*, irrespective of co-occurrence with other infecting trematodes (*Diplodiscus mehrai, Prosotocus himalayai, Indopleurogenes yamagutii* or *Gorgoderina elliptica*) in either sex of the host or either population of the host. However, in one instance, where *Ganeo tigrinus* and *Mehraorchis ranarum*, co-occurred only a single specimen of the former was found spatially separated from 3 specimens *Mehraorchis ranarum* occupying segment IV of the intestine.

Diplodiscus mehrai was recovered from the IV and V segments of the intestine of the host in individual or mixed infections, except when the mixed infection was comprised of *Gorgoderina elliptica* as the associated or co-occurring infection. In such a situation, *Diplodiscus mehrai* was found restricted to the V segment *i.e.* the terminal segment of the intestine, closest to the cloaca in both sexes and in both populations of the host.

Indopleurogenes yamagutii often occupied the I and II segments and occasionally the III segment of the host intestine, whether occurring alone or in association with *Diplodiscus mehrai* or *Gorgoderina elliptica* or both. Both in presence of *Ganeo tigrinus* as a co-occurring infecting taxon, the spread of the trematode, *Indopleurogenes yamagutii* was confined to the I segment (*i.e.* duodenal region) of the intestine of the host. The spread of this trematode over to segments II and III of the host intestine was found whenever the number of this trematode was 19 or more in a host. Whereas *Gorgoderina elliptica* was always found to occupy the anterior half of the V segment (*i.e.* the pre-cloacal segment) of the host intestine, whether it occurred as a source of solitary infection or as a member of mixed infection of more than one infecting trematodes. In one of the infected host of the species (a female in CB population) the worm was found to lie distributed over the entire length of the terminal (V) segment of the intestine, possibly because of its very high numbers (152) in the host.

Discussion

Present investigation has revealed that the seven trematodes found infecting the intestines in the two hosts populations, only 4, namely *Diplodiscus mehrai, Ganeo tigrinus, Indopleurogenes yamagutii* and *Mehraoechis ranarum* were found to infect both host species in both populations. However, of these only *Ganeo tigrinus* and *Diplodiscus mehrai* were found to infect both sexes of both species in both the populations. Therefore, for purposeful comparison, only these two species were analyzed for their infection intensity in relation to the size of the infected hosts.

The two host species examined for enteric infections of trematodes from the two populations were significantly different in their snout-vent lengths (SVL); the *Rana tigrina* being the larger host by a factor of 1.5 (Table 6.2). The overall infection level of the different enteric trematodes including *Ganeo tigrinus* and *Diplodiscus mehrai* in *Rana tigrina* was found to be consistently higher than that of the comparable infecting taxa obtained from *Rana cyanophlyctis*, the smaller of the two hosts. While this finding *per se* reflects a higher sustenance for infection in the larger host, but no significant change in the infection intensity was evident with the changing SVL of either host (Tables 6.6a and 6.6b). These results thus directly support the conclusions of Plasota (1969), Hollis (1972), Anderson (1978), Mishra (1978) and Holmes and Price (1986) who has maintained that the helminth abundance and prevalence was directly proportional to the size of the host. However, no correlation could be established between the species diversity and size of the host, contrary to that maintained by Goater, Esch and Bush (1987) and Kuc and Sulgostowska (1988).

Seven enteric trematode species infecting the frogs in the two stations, only 4 species; namely *Diplodiscus mehrai, Ganeo tigrinus, Prosotocus himalayai* and *Mehraorchis ranarum* were infecting the *Rana tigrina* of which *Mehraorchis ranarum* was obtained only once during the study period from a single male of *Rana tigrina,* whereas the remaining three trematode species infected both sexes of this frog in both populations.

Out of seven species of the enteric trematodes found infecting the two frogs in the two populations, only 4 trematode namely *Diplodiscus mehrai, Ganeo tigrinus, Indopleurogenes yamaguti* and *Gorgoderina elliptica* were found infecting *Rana cynophlyctis* in both sexes and populations, whereas *Mehrarochis ranarum* was found infecting both the sexes of the frog, *Rana cynophlyctis* only in CB population. This trematode was not ever recovered in this frog species

Table 6.6a: Monthly and Overall Mean of the Trematode Species Richness, Trematode Intensity in Two Ranid from Bishnah (CB) Population

Hosts	*Months*	*No. of Host Infected (Sex)*	*Trematode Species Richness*	*Trematode Intensity*	*Mean SVL of Hosts (mm)*
Rana cyanophlyctis	April	13 (2M, 11F)	1.51±0.43 (1–2)	2.46±1.26 (1–4)	41.30±10.44 (22–66)
	May	14 (6M, 8F)	2.69±1.08 (1–4)	6.85±77.69 (1–152)	45.57±9.78 (35–78)
	June	14 (6M, 8F)	2.69±1.08 (0–5)	6.85±7.69 (1–19)	45.57±9.78 (30–63)
	July	14 (8M, 6F)	1.77±0.65 (1–3)	2.75±1.42 (1–6)	47.00±12.26 (30–70)
	August	7 (4M, 3F)	1.36±0.40 (1–2)	2.33±1.50 (1–5)	44.16±13.34 (18–52)
	September	13 (4M, 9F)	2.60±0.51 (1–2)	6.38±5.00 (1–16)	50.61±9.57 (39–70)
	October	13 (4M, 9F)	1.90±0.50 (1–2)	2.69±1.31 (1–4)	39.76±8.95 (20–54)
	Overall	104 (36M,68F)	2.24±0.65 (0–5)	6.83±8.38 (1–152)	45.84±10.82 (18–78)
Rana tigrina	April	2 (1M, 1F)	2.25±0.50 (1–2)	3.00±2.00 (1–3)	113.5±2.12 (112–115)
	May	3 (1M, 2F)	1.77±0.47 (1–2)	4.66±4.49 (1–10)	98.33±7.63 (90–105)
	June	6 (3M, 3F)	1.77±0.51 (1–2)	8.33±4.03 (1–11)	97.50±7.89 (88–110)
	July	14 (8M, 6F)	2.02±0.75 (1–3)	8.50±8.60 (1–27)	115.35±18.13 (70–140)
	August	9 (5M, 4F)	2.41±0.52 (1–2)	18.66±14.38 (1–43)	101.33±19.77 (60–120)
	September	7 (2M, 5F)	2.02±0.53 (1–2)	48.14±101.99 (1–275)	85.85±17.26 (62–110)
	October	4 (3M, 1F)	3.06±0.82 (1–3)	102.50±128.2 (1–309)	108.75±18.15 (80–130)
	Overall	45 (23M, 22F)	2.18±0.58 (1–3)	27.68±37.67 (1–309)	102.94±12.99 (60–140)

M: Male; F: Female; SVL: Snout-Vent Length.

Table 6.6b: Monthly and Overall Mean of the Trematode Species Richness, Trematode Intensity in Two Ranid from Ramgarh (CR) Population

Hosts	*Months*	*No. of Host Infected (Sex)*	*Trematode Species Richness*	*Trematode Intensity*	*Mean SVL of Hosts (mm)*
Rana cyanophlyctis	April	14 (5M, 9F)	1.46±0.42 (1–2)	3.21±1.67 (1–6)	46.35±12.05 (30–72)
	May	11 (4M, 7F)	2.66±0.88 (1–4)	9.09±9.70 (1–24)	48.18±15.48 (28–80)
	June	16 (7M, 9F)	2.05±0.62 (1–3)	3.87±3.73 (1–9)	46.50±9.78 (30–62)
	July	12 (5M, 7F)	1.77±0.49 (1–2)	8.66±12.33 (1–44)	50.50±9.77 (38–64)
	August	10 (6M, 4F)	1.69±0.48 (1–2)	2.80±1.81 (1–6)	36.60±6.34 (28–45)
	September	14 (8M, 6F)	1.91±0.50 (1–2)	9.46±12.22 (1–27)	40.15±6.37 (28–50)
	October	8 (4M, 4F)	1.88±0.51 (1–2)	3.62±2.06 (1–4)	35.62±8.71 (24–50)
	Overall	85 (39M,46F)	1.91±0.54 (1–4)	5.81±6.21 (1–44)	43.41±9.78 (24–80)
Rana tigrina	April	2 (2F)	2.25±0.50 (1–2)	7.00±4.00 (1–10)	105.5±5.00 (100–110)
	May	5 (1M, 4F)	1.96±0.54 (1–2)	8.6±4.87 (1–10)	99.00±14.96 (70–110)
	June	10 (5M,5F)	3.61±0.73 (1–3)	20.6±17.39 (1–51)	91.6±13.90 (80–120)
	July	9 (5M, 4F)	2.76±0.70 (1–3)	12.5±15.40 (1–46)	113.00±39.07 (60–170)
	August	12 (7M, 5F)	3.66±10.66 (1–3)	12.08±11.61 (1–33)	109.16±11.04 (90–130)
	September	5 (2M, 3F)	3.24±0.44 (1–2)	10.40±8.87 (1–22)	128.00±32.34 (80–185)
	October	6 (3M. 3F)	2.25±0.83 (1–3)	10.52±6.47 (1–18)	88.00±19.11 (62–111)
	Overall	49 (23M, 26F)	2.81±0.62 (1–3)	11.66±9.80 (1–51)	108.82±19.06 (60–170)

M: Male; F: Female; SVL: Snout-Vent Length.

in CR population. The infection pattern of *Loxogenes jammuensis* was also ecologically bizarre in infecting only a male and a female of the frog from two populations.

While *Prosotocus himalayai* was found as a parasite equally compatible in the intestines of both sexes of *Rana tigrina* in the two populations, this trematode was recovered only in the female members of the *Rana cyanophlyctis* in the either population and never from its male members. On the other hand, *Indopleurogenes yamagutii* and *Gorgoderina elliptica* likewise were found parasitic only in *Rana cynophlyctis* but never in *Rana tigrina* in either population despite their sympatry.

Perusal of the Table 6.7 indicates the location of the common species in each host. It may be noted that in each host species, the trematode species are spread neither too evenly nor throughout the intestine. Analysis of the enteric trematodes of the two frog hosts (Table 6.7) clearly demonstrate a spatial stratification of niches in the intestine occupied by different trematode species. Examining the natural infection of the hosts, it is apparent that *Diplodiscus mehrai* was spatially displaced from its typical site in the posteriomost segment of intestines in the two hosts, by the presence of another core species *Gorgoderina elliptica* in the host *Rana cyanophlyctis*. This finding supports the observations of Stock and Holmes (1987, 1988) that have shown that in some grebes the presence of one of the core species was usually associated with the spatial displacement of the other core species.

The spatial displacement of *Diplodiscus mehrai* more posteriowards under the influence of *Gorgoderina elliptica* not only qualifies *Diplodiscus mehrai* for being designated as a generalist species (Colwell and Fuentas, 1975) and *Gorgoderina elliptica* as species specialist, but also that given an opportunity to distribute itself, *Diplodiscus mehrai* has a greater habitat range or niche breadth than the other trematode, *Gorgoderina elliptica* whose habitat range is much smaller in comparison. Clearly, therefore, *Gorgoderina elliptica* is suitably adapted to the topography of its preferred site closer to the vent and a feeder upon that is about to be rejected by the host.

With regards to the other dominant species of the enteric trematode recovered in the two host species, the longitudinal movements of the trematodes along the alimentary tract is more or less confined to the anterior or the middle segment of the intestine.

Table 6.7: Showing the Niche of the Various Enteric Trematatode Species Inhabiting the Two Ranid Hosts, *Rana tigrina* and *Rana cyanophlyctis* from the Two Populations (CB and CR)

Hosts	*Trematode Species*	*Segments of Intestine (Niche)*				
		I	*II*	*III*	*IV*	*V*
Rana	*Ganeo tigrinus*	–	+	+	+	–
tigrina	*Prosotocus himalayai*	+	+	+	–	–
	Diplodiscus mehrai	–	–	–	+	+
	Mehraorchis ranarum	–	+	+	+	–
Rana	*Ganeo tigrinus*	–	–	+	+	–
cyanophlyctis	*Prosotocus himalayai*	+	+	+	–	–
	Diplodiscus mehrai	–	–	–	+	+
	Mehraorchis ranarum	–	–	+	+	–
	Loxogenes jammuensis	–	–	+	+	–
	Indopleurogenes yamagutii	+	+	+	–	–
	Gorgoderina elliptica	–	–	–	–	+

I: Duodenal part of the intestine; II: Post duodenal part of the intestine; III: Mid-intestinal region of the intestine; IV: Last segment of the intestine.

While *Prosotocus himalayai* and *Indopleurogenes yamagutii* are usually confined to the duodenum or anterior segment of the intestine immediately behind the duodenum, *Ganeo tigrinus, Mehraorchis ranarum* and *Loxogenes jammuensis* remain invariably confined to the middle or pre-cloacal region of the intestine. Obviously therefore, there is an ecological stratification down the intestine length, duodenum backwards which offers distinct and diverse micro-environments for different trematodes living inside. These micro-environments would reasonably reflect only mucosal differences because the other histological elements of the intestine in different vertebrates remain not only the same and similarly disposed but also do not come in contact with the parasites found inside the lumen of the intestine. It could, therefore, be envisaged that greater the species diversity among the trematodes inhabiting the intestine, the greater would be mucosal differences within the intestine of the host. Logically the greater the extension and ramification of the mucosal lining of small intestine, greater the food available for the species of the trematodes which browse upon the mucosa (Hayunga,

1991) and therefore, potentially more micro-environment present for enteric trematodes (Esch *et al.*, 1990). In the small intestine of the amphibian (Hayunga, pp. 866 *op. cit.*, and Crompton, 1973), the mucosal diversity is the least and therefore, the little species diversity among the parasites found in the two frogs is expected (Stock and Holmes, 1988).

Perusal of the data presented above indicates that although the enteric trematodes in the two frogs remained more or less restricted to their specific sites inside the intestines of the host, no section of the intestine were always unoccupied. These vacant sites increased towards the hinder segments of the intestines (IV and V). Moreover, both frogs were characterized by a reduction in the number of infecting taxa towards the posterior end of the intestines. The maximum species richness (at least in *Rana cyanophlyctis*, the host showing comparatively a richer diversity of enteric trematodes in comparison to *Rana tigrina*) occurred in the mid-intestinal region (segments II, III and IV) or in the anterior segment (I and II). Such a regressing species richness along the anterio-posterior gradient of intestines was also shown in other hosts (Kennedy *et al.*, 1986; Goater *et al.*, 1987; Stock and Holmes, 1988).

A careful scrutiny of the patterns of distribution along the intestinal length in the two frogs reveals that the trematodes resided largely in predictable locations (barring minor overlaps found when the number of a co-occurring species was large) along the length of the small intestine. Individual trematodes occupied a high proportion of the intestines and only mildly varied in position in the two host species (Table 6.7). This finding suggests some degree of tolerance for conditions along the intestines. In individual frog, the trematodes were much more restricted in distribution, overlapped considerably less than their overall ranges would suggest and there was evidence of interference by at least one core species (the species that are common and more abundant) all suggesting that interactions are important in these communities as also maintained for trematode communities in grebes by Stock and Holmes (1988). Nevertheless the core species were not distributed more evenly than would be expected through random placement. This also suggests that important resource may not be distributed evenly along the intestinal gradient. What is, however, evident is that the mid-intestinal region(s) of the host is far the richest in resource potential and therefore, harbours much richer variety of trematodes than any other segment of the intestine.

Acknowledgement

The first author is grateful to DOE (Department of Environment), Govt. of India, New Delhi for financial assistance.

References

Aho, J. M., 1990. Helminth communities of amphibians and reptiles: Comparative approaches to understanding patterns and processes. In: *Parasite Communities: Patterns and Processes*, (Eds.) G.W. Esch, A.O. Bush, and J.M. Aho. Chapman and Hall, London, U.K., p. 157–196.

Anderson, R.M., 1978. The regulation of host population growth in parasitic species. *Parasitology*, 76: 119–157.

Baker, M.R., 1987. Synopsis of the Nematoda parasitic in amphibians and reptiles. Memorial University of Newfoundland. *Occasional Papers in Biology*, 11: 1–325.

Chandra, Pooja, Gupta, Neelam and Bhaskar, Manju, 2005. The incidence of helminth parasites in amphibian and reptilian population from Rohilkhand Region, India. *In: Disaster, Ecology and Environment*, (Ed.) Arvind Kumar. Daya Publishing House, New Delhi, India, pp. 1–367

Colwell, P.K. and Fuentas, E.R., 1975. Experimental studies of the niche. *Review of Ecology and Systematics*, 6: 281–310.

Crofton, H.D., 1971. A quantitative approach to parasitism. *Parasitology*, pp. 179–193.

Crompton, D.W.T., 1973. The sites occupied by some parasitic helminth in the alimentary tract of vertebrates. *Biological Reviews*, 48: 27–83.

Dwivedi, M.P.,1968. Three new species of *Gorgoderina* Looss, 1902. *Ibid*, 19(2): 132–140.

Edwards, D.D. and Bush, A.O., 1989. Helminth communities in avocets: Importance of compound community. *J. Parsitol.*, 98: 439–445.

Esch, G.W., Bush, A.O. and Aho, J.M. (Eds.), 1990. *Parasite Communities: Patterns and Processes*. Chapman and Hall, New York, 335p.

Goater, C.P., and Bush, A.O., 1988. Intestinal helminth communities in long-billed curlews. The importance of Congeneric host specialists. *Holarctic. Ecology,* 11: 140–145.

Goater, T.M., Esch, G.W., and Bush, A.O., 1987. Helminth parasites of sympatric salamanders: Ecological concepts at infra-community, component and compound community level. *Amrican Midland Naturalist,* 118: 289–300.

Goldberg, S.R. and Bursey, C.R., 2002. Helminths of 10 species of Anurans from Honshu Island, Japan. *Comparative Parasitology,* 69(2): 162–176.

Goldberg, S.R. and Bursey, C.R., 1991. *Parapharyngodon kartana* in two skinks, *Emoia nigra* and *Emoia samoense* (Sauria: Scincidae), from Samoa. *J. Helminthol. Soc. Wash.,* 58: 265–266.

Goldberg, S.R., Bursey, C.R. and Pltz, J.E., 2000. Helminths of the Plains Leopard Frog, *Rana blairi* (Ranidae). *The Southwestern Naturalist,* 45(3): 362–366

Goldberg, S. R., Bursey, C.R. and Cheam, H., 1998. Gastrointestinal helminths of four gekkonid lizards, *Gehyra mutilata, Gehyra oceanica, Hemidactylus frenatus* and *Lepidodactylus lugubris* from the Mariana Islands, Micronesia. *J. Parasitol.,* 84: 1295–1298.

Goldberg, S.R. and Bursey, C.R., 2002a. Helminth parasites of seven anuran species from Northwestern Mexico. *Western North American Naturalist,* 62: 160–169.

Goldberg, S.R., Bursey, C.R. and S. Telford, J.R., 2004b. Helminths of six species of snakes from Honshu Island, Japan. *Comparative Parasitology,* 71: 49–60.

Gray, P., 1973. *Encyclopedia of Microscopy and Microtechniques,* New York, p. 638.

Hayunga, E.C., 1991. Morphological adaptation of intestinal helminthes. *J. Parsitol.,* 77(6): 865–873.

Hollis, P.D., 1972. A survey of parasites of the bull-frog *Rana catesbeiana* Shaw, in Central-East Taxas. *South-Western Naturalist,* 17: 198–200.

Holmes, J.C. and Price, P.W., 1986. Communities of parasites. In: *Community Ecology: Patterns and Processes,* (Eds.) D.J. Anderson and J. Kikkawa. Blackwell Scientific Publications, Oxford, pp. 187–213.

Balbuena, J.A. and Raga, J.A., 1993. Intestinal helminth communities of the long-finned pilot whale (*Globicephala melas*) off the Faroe Islands. *Parasitology*, 106: 327–333. Cambridge University Press.

Kennedy, C.R. and Williams, H.H.,1989. Helminth parasite Community Diversity in a marine Fish, *Raja batis* L. *J. Fish Biol.*, 34: 971–972

Kennedy, C.R., Bush, A.O. and Aho, J.M., 1986. Patterns in helminth communities: Why are birds and fishes different. *Parasitology*, 93: 205–215.

Kuc, I. and Sulgostowska, T., 1988. Helminth fauna of *Rana ridibuna* Pallas, 1771 from Goclawski Canal in Warszawa (Poland). *Acta Parasitol. Polonica*, 33: 101–105.

Lees, E., 1962. The incidence of helminth parasites in a particular frog population. *Parasitology*, 52: 95–102.

Margolis, L., Esch, G.W., Holmes, J.C. Kuris, A.M. and Schad, G.A., 1982. The use of ecological terms in parasitology. (Report of an adhoc committee of the American Society of Parasitologists). *Parasitol.*, 68: 131–133.

Mehra, H.R. and Negi, P.S., 1928. Trematode parasites of the Pleurogenetinae from *Rana tigrina* with a revision and synopsis of the subfamily. *Allahabad University Stud.*, 4: 63–118.

Mishra, T.N., 1978. The occurrence of *Acanthocephalus lucii* in the fishes of the Shropshire Union Canal, Cheshire, England, U.K. *Annales of Zoology (Agra)*, 14: 181–188.

Moravec, F. and Køie, M., 1987. *Daniconema anguillae* gen. et sp. n., a new nematode of a new family Daniconematidae fam. n. parasitic in European eels. *Folia Parasitologica*, 34: 335–340.

Pande, B.P., 1937a. Two new fish trematodes from Allahabad. *Proc. Nat. Acad. Sci., India*, 7(2): 111–115.

Pande, B.P., 1937b. *Prosotocus himalayai* (n.sp.) from a frog. *Ibid*, 7(2): 202–204.

Pandoh, B.R., 1992. Ecological studies on trematodes of some aquatic vertebrates of Jammu province. Unpublished *Ph.D. Thesis*, Jammu University, Jammu.

Plasota, K., 1969. The effect of some ecological factors on the parasitofauna of frogs. *Acta Parasitologica Polonica*, 16: 47–60.

Poulin, R., 2001. Interactions between species and the structure of helminth communities. *Parasitology,* 122: S3–S11.

Rankin, J.S. Jr., 1945. An ecological study of the helminth parasites of amphibians and reptiles of Western Massachusetts and vicinity. *Journal of Parasitology,* 31: 142–150.

Singh, Gurdarshan, 1989. Studies on Lecithodendriid infection in frogs in Jammu. Unpublished *M.Phil Dissertation,* University of Jammu, Jammu, pp. 1–99.

Srivastava, H.D., 1934. On a new genus *Meharaorchis* and two new species *Pleurogenes* (Pleurogenetimae) with a systematic discussion and revision of the family Lecithodendendriidae. *Bull, Acad. Sci., U.P., Agr. and Oudh,* 3(4): 239–256

Stock, T.M. and Holmes, J.C., 1987. Host specificity and exchange of intestinal helminths among four species of grebes (Podicipedidae). *Canadian Journal of Zoology,* 65: 669–676.

Stock, T.M. and Holmes, J.C., 1988. Functional and relationships and microhabitat distributions of enteric helminthes of grebes (Podicipedidae): The evidence for interactive communities. *J. Parasit.,* 74(2): 214–227.

Sudan, O.S., 1979. Studies on trematode fauna of Jammu province. Unpublished *Ph.D. Thesis,* Jammu University, Jammu.

Verma, A.K., 1988. Polymorphism in *Polydistomides gregarinum. M.Phil. Dissertation,* University of Jammu.

Yamaguti, S., 1971. *Synopsis of Diagenetic Trematodes of Vertebrates.* Keigaku Publishing Co., Tokyo, Japan, Vols. 1 and 2: 1–1074.

Yamaguti, S., 1975. *A synoptical Review of Life Histories of Diagenetic Trematodes of Vertebrates.* Keigaku Publishing Co., Tokyo, Japan, Vols. 1 and 2: 1–1074.

Chapter 7

Phenotypic Plasticity of *Drosophila melanogaster* at Different Temperature

B.R. *Guru Prasad* and S.N. *Hegde*[*]

Department of Studies in Zoology, Manasagangotri, University of Mysore, Mysore – 570 006, India

ABSTRACT

Influence of temperature on sexual behavior of both male and female of *D. melanogaster* collected at top of Chamundi hill, Mysore and maintained at 12°C, 17°C, 22°C, 27°C, and 32°C was studied. The general sexual behavioral acts such as courtship latency, mating latency, copulation duration and the quantitative behavioral acts of males such as tapping, circling, scissoring, vibration and licking and female quantitative sexual behaviors such as extruding, decamping and ignoring were recorded. The results showed that 22°C is the optimum temperature for mating in *D. melanogaster*. The males showed higher vigor and females showed higher receptivity at this temperature. Although mating was fast at 22°C, it did occur at other temperatures with higher courtship activities. The fast mating observed in the present studies at 22°C with optimum

* Corresponding author: E-mail: gurup2006@yahoo.co.in; malerhegde@yahoo.com

performance of courtship activities in comparison to mating in other temperatures suggests the sexual behavioral plasticity of *D. melanogaster* at different temperature.

Keywords: *Courtship, Drosophila, Mating, Temperature, Plasticity.*

Introduction

Phenotypic plasticity is the ability of organisms to alter its physiology, morphology, behavior in response to changes in its environment. The capacity of a given genotype to produce different phenotypes in different environments is of growing interest among evolutionary biologists (David *et al.*, 2006; Gibert *et al.*, 2004; Moreteau *et al.*, 2003; Karan *et al.*, 1999; Scheiner, 1993; Scheiner and Lyman, 1989; Schlicting and Pigliucci, 1998; Steigenga *et al.*, 2005; West and Packer, 2002; Via, 1993). Lenski and Bennet (1993) and Partridge *et al.*(1994a, 1995) demonstrated that in laboratory, natural selection of populations exposed to different temperatures can suggest a fitness trade off among environments because populations perform better in the environment. Females of *D. ananassae* reared to adult hood at 18°C showed a significant increase in body weight compared to those females reared at 25°C (Sisodia and Singh, 2002).

With reference to temperature several workers (David *et al.*, 1983; Hoffmann and Parsons, 1991; Hollingsworth and Bowler, 1966; Smith, 1956) have demonstrated that *Drosophila* is a widely used and well suited model system for studying evolutionary response to extreme temperatures. Adult characteristics may also be responsive to thermal evolution with changes demonstrated in life history, morphological and physiological measures (Azevedo *et al.*, 1996; Gilcrist *et al.*, 1996; Partridge *et al.*, 1994b, 1995). In *D. serrata* exposure to non lethal cold shock can affect viability, development time and productivity in progeny and effects have the potential to persist across multiple generations (Hoffmann *et al.*, 2003). Alpatov and Pearl (1929) studied that *D. melanogaster* reared at 18°C outlived flies reared at 28°C at three temperatures during adult stage. Burcombe and Hollingsworth (1970); Lints and Lints (1971) obtained similar results. Flies reared at 15°C had higher cold resistance than those reared at 23°C, but rearing temperature did not affect heat resistance. Kilias *et al.* (1980) have demonstrated sexual isolation as a by product of adaptation to different temperature and humidity regimes in *D. melanogaster*.

In *Drosophila melanogaster*, Hoffmann *et al*. (2003) showed that repeated mild stress had a detrimental effect on fertility and fecundity when flies were exposed to the stress, but not thereafter and also found that the repeated exposures led to increased longevity and heat resistance. Crossley (1974) observed that even changes in photoperiod causes changes in mating behavior in two mutant strains of *D. melanogaster*. Laudien *et al*. (1980) demonstrated temperature dependence of courtship in male guppies *Poecila reticulate*. West and Packer (2002) demonstrated temperature dependent phenotypic plasticity in the Lion's mane and its significance in sexual selection. David *et al*. (2006) showed phenotypic plasticity of the body size in a natural population of *Drosophila melanogaster*. All these illustrations stress upon parallelism between the effects of plasticity and genetic variation and are an argument for an adaptive interpretation of the different genotypes. This emphasizes the fact that phenotypic plasticity is also a target of natural selection.

Different aspects of sexual behavior such as courtship latency, mating latency, copulation duration and the activities performed by the males and females are good estimates of reproductive fitness of both the sexes. These behavioral traits are also genetically determined. Survey of literature on phenotypic plasticity indicates that the expression of these genetic traits is influenced by the environment particularly temperature. Although both phenotypic and behavioral traits are genetically controlled, the latter have lower heritability than the first. Hence the obvious question that arises is, like phenotypic traits, whether these behavioral traits are also influenced by temperature or not. Authors have tried to address this question using *Drosophila melanogaster* flies collected from Chamundi Hill, Mysore. The behavioral plasticity has been studied by maintaining these flies at 12°C, 17°C, 22°C, 27°C and 32°C.

Materials and Methods

To study the effect of different temperature on sexual behavior, natural population of *D. melanogaster* collected at the top of Chamundi hill, Mysore were used. The flies were collected by keeping the quarter pint milk bottles containing meshed banana in kitchen and stores of a few houses at the top of the Chamundi hill, Mysore. After two days the bottles were plugged with cotton so as to close the mouth and then brought to the laboratory. Then the flies were sexed, and the females were individually placed in vials containing food so as to develop isofemale lines. When progeny appeared, equal number of them from each isofemale line were separately distributed to different culture bottles and reared under different temperature regimes 12°C, 17°C, 22°C, 27°C and 32°C. When adults emerged, virgin and bachelor flies were separated with in six hours of eclosion, maintained separately until they reach five days and then used to study the sexual behavioral plasticity.

For observation of sexual behavior a virgin female and bachelor male were introduced into an Elens-Wattiaux mating chamber (5 cm x 5 cm circular glass chamber with a lid to facilitate easy observation). Because maximum mating occurs during morning hours, observation was made between 7 and 11 a.m. Sexual behavioral acts such as courtship latency (time between introduction of male and female together into mating chamber and orientation of male towards female), mating latency (time between introduction of males and females into mating chamber and initiation of copulation of each pair), copulation duration (time between initiation and termination of copulation of each pair) were recorded. The quantitative behavioral acts of males such as tapping, circling, vibration and licking and female quantitative sexual behavior such as extruding, decamping and ignoring were recorded. The terminologies are used as per the description of (Hegde and Krishnamurthy, 1979). A minimum of 25 pairs involving each isofemale line were observed. To identify the difference in sexual behavior at different temperatures, one way analysis of variance (ANOVA) followed by Duncan's multiple range test (DMRT) was applied using SPSS 10.5 software.

Observations

The data on qualitative sexual activities such as courtship latency, mating latency and copulation duration of *D. melanogaster* at five different temperatures has been reported in Table 7.1. The courtship latency was highest at 12°C (119.2 ± 10.8). The courtship latency increased up to 22°C and then decreased with increasing temperature. Mean courtship latency at different temperatures such as 12°C, 17°C, 22°C, 27°C and 32°C was statistically significant by one way ANOVA (F value = 28.63; df = 4, 120; P<0.001; Table 7.4). Application of Duncan's multiple range test (DMRT) confirmed that courtship latency is dependent on temperature.

Table 7.1: Qualitative Sexual Behavior of *D. melanogaster* at Different Temperatures (values are Means ± SE)

Temperatures	*Parameters*		
	Courtship Latency (in seconds)	*Mating Latency*	*Copulation Duration*
12 ± 1°C	119.20 ± 10.80^{c}	16.64 ± 0.57^{c}	34.44 ± 1.05^{d}
17 ± 1°C	70.76 ± 5.75^{b}	10.32 ± 0.72^{b}	28.64 ± 1.15^{c}
22 ± 1°C	31.56 ± 2.80^{a}	7.20 ± 0.85^{a}	19.76 ± 0.62^{b}
27 ± 1°C	40.12 ± 3.01^{a}	7.56 ± 0.89^{a}	18.72 ± 1.51^{ab}
32 ± 1°C	68.16 ± 6.25^{b}	7.84 ± 0.99^{a}	15.80 ± 1.84^{a}

Same alphabet as superscript in each column (for temperatures) is non-significant by DMRT.

Table 7.1 also shows that the mating latency was shortest at 22°C (7.20 ± 0.85) and longest at 12°C (16.64 ± 0.57). The mating latency at 22°C was significantly different when compared to lower temperatures while it was non significant when compared with higher temperatures. In contrast to this, the copulation duration was lowest at 32°C (15.80 ± 1.84) which increased with the decreasing temperature. Thus the duration of copulation was longest at 12°C (34.44 ± 1.05) which was significantly (by ANOVA and DMRT) different from all other temperatures (Table 7.4).

Table 7.2 shows the quantitative courtship activities performed by males reared at different temperatures. Tapping was lowest at 27°C (14.44 ± 0.72) while it was highest at 22°C (20.04 ± 0.83). The application of one way ANOVA showed that tapping activity differed significantly among males reared at different temperatures (F value = 7.57; df = 4, 120; P<0.001; Table 7.5). However tapping at 12, 17 and 22°C were non-significant and similarly tapping by males at 27 and 32°C were also non-significant by DMRT.

The mean circling activity performed by males reared at different temperatures also varied significantly (F value = 25.32; df = 4, 120; P<0.001; Table 7.5). The mean circling was lowest at 32°C (2.32 ± 0.21) which increased until 22°C (5.60 ± 0.17). It decreased again

Table 7.2: Male Quantitative Sexual Behavior of *D. melanogaster* at Different Temperature (values are Means ± SE)

Temperature	*Parameters*				
	Tapping	*Circling*	*Vibration*	*Scissoring*	*Licking*
12 ± 1°C	19.08 ± 0.53^{c}	2.52 ± 0.22^{a}	14.76 ± 0.44^{bc}	8.24 ± 0.42^{a}	10.08 ± 0.28^{a}
17 ± 1°C	17.96 ± 0.36^{bc}	4.08 ± 0.27^{b}	16.12 ± 0.51^{c}	9.12 ± 0.32^{a}	12.08 ± 0.56^{b}
22 ± 1°C	20.04 ± 0.83^{c}	5.60 ± 0.17^{d}	19.12 ± 0.76^{d}	12.12 ± 0.29^{b}	15.60 ± 0.70^{c}
27 ± 1°C	14.44 ± 0.72^{a}	4.80 ± 0.31^{c}	13.72 ± 0.67^{b}	8.72 ± 0.43^{a}	13.64 ± 0.53^{b}
32 ± 1°C	16.12 ± 1.31^{ab}	2.32 ± 0.21^{cd}	11.16 ± 0.45^{a}	8.04 ± 0.05^{a}	10.80 ± 0.62^{a}

Same alphabet as superscript in each column (for temperatures) is non-significant by DMRT.

with the decrease in temperature. In other words the circling was highest at the optimum temperature of 22°C and lowest at the two extreme temperatures.

The mean wing vibration bouts per minute, scissoring of wings and licking performed by males reared at different temperatures also showed the same trend as that of circling. All these parameters were highest at 22°C and decreased at both low and high temperatures. The differences in these acts at different temperatures were also statistically significant from one another. These results thus show that the males are very active at 22°C compared to either low or high temperatures.

Data on the rejection responses of female towards the courting males reared at different temperatures is provided in Table 7.3. These are the behaviors exhibited by non-receptive females. The non receptive females maintained at 22°C showed least decamping than others. Females maintained at 12 and 32°C showed higher decamping activities than others. The differences in the mean values of decamping by these females were significant by ANOVA (F value = 5.36; df = 4, 120; P<0.001; Table 7.6). However, the DMRT showed that decamping performed by the females at 12 to 27°C were non significant with each other while between these and 32°C were significant.

Table 7.3: Female Quantitative Sexual Behavior of *D. melanogaster* at Different Temperature (values are Means ± SE)

Temperatures	*Parameters*		
	Decamping	*Extruding*	*Ignoring*
12 ± 1°C	3.20 ± 0.16^{a}	9.40 ± 0.45^{c}	19.28 ± 0.67^{b}
17 ± 1°C	2.48 ± 0.12^{a}	6.92 ± 0.60^{b}	14.08 ± 0.69^{a}
22 ± 1°C	2.36 ± 0.33^{a}	4.28 ± 0.51^{a}	13.80 ± 1.01^{a}
27 ± 1°C	3.28 ± 0.30^{a}	7.64 ± 0.52^{b}	13.84 ± 0.77^{a}
32 ± 1°C	4.24 ± 0.50^{b}	10.4 ± 0.66^{c}	25.20 ±1.54^{c}

Same alphabet as superscript in each column (for temperatures) is non-significant by DMRT.

Like decamping, significant differences were noticed with respect to extruding by females maintained at different temperatures. Females of 22°C group extruded least compared to others while

Table 7.4: One Way ANOVA for Qualitative Sexual Behavior of *D. melanogaster* at Different Temperature

Parameters	*Sources*	*Sum of Squares*	*df*	*Mean Square*	*F Value*	*P Value*
Courtship latency	Between groups	117836	4	29459	28.63	0.000*
	Within groups	123459	120	1028.82		
	Total	241295	124			
Mating latency	Between groups	1634.67	4	408.668	24.04	0.000*
	Within groups	2039.92	120	16.999		
	Total	3674.59	124			
Copulation duration	Between groups	6055.63	4	1513.91	35.45	0.000*
	Within groups	5123.52	120	42.696		
	Total	11179.2	124			

*$P<0.001$

Table 7.5: One Way ANOVA for Quantitative Male Sexual Behavior of *D. melanogaster* in Different Temperature

Parameters	*Sources*	*Sum of Squares*	*df*	*Mean Square*	*F Value*	*P Value*
Tapping	Between groups	510.59	4	127.64	7.57	0 .000**
	Within groups	2022.56	120	16.85		
	Total	2533.15	124			
Circling	Between groups	151.56	4	37.89	25.32	0.000**
	Within groups	179.52	120	1.49		
	Total	331.08	124			
Vibration	Between groups	866.68	4	216.67	24.85	0 .000**
	Within groups	1046.24	120	8.71		
	Total	1912.92	124			
Scissoring	Between groups	275.47	4	68.86	16.73	0 .000*
	Within groups	493.84	120	4.11		
	Total	769.31	124			
Licking	Between groups	493.23	4	123.3	15.75	0.000**
	Within groups	938.96	120	7.82		
	Total	1432.19	124			

*P<0.01; **P<0.001

Table 7.6: One Way ANOVA for Quantitative Female Sexual Behavior of *D. melanogaster* in Different Temperature

Parameters	*Sources*	*Sum of Squares*	*df*	*Mean Square*	*F Value*	*P Value*
Decamping	Between groups	56.832	4	14.208	5.36	0.000*
	Within groups	317.6	120	2.647		
	Total	374.432	124			
Extruding	Between groups	562.112	4	140.52	18.08	0.000*
	Within groups	932.64	120	7.77		
	Total	1494.75	124			
Ignoring	Between groups	2443.76	4	610.94	24.63	0.000*
	Within groups	2976.24	120	24.802		
	Total	5420	124			

$^*P<0.001$

females of 12 and 32°C extruded highest. The values of mean extruding at 22°C was significantly different (by DMRT) compared to others while extruding by females of 17 and 27°C were non significant. These mean values at 12 and 32°C were also not significantly different with each other.

The females maintained at 22°C ignored the males least compared to others. Ignoring was highest at 12 and 32°C. Application of ANOVA followed by DMRT showed that the ignoring performed at the two extreme temperatures were significantly different while the differences at 17, 22 and 27°C were non significant. Thus these results show that the females maintained at 22°C have greater receptivity than other temperatures. The females reared at low or high temperatures are non receptive than those at the optimum temperature.

Discussion

Sexually reproducing animals are endowed with special features, first to produce fertile offspring and second to adapt to a particular environment. The reproduction is preceded by a series of courtship acts wherein the males and females show unique rituals to attract each other, mate and produce the offspring. The courtship and mating although are genetic, but are also influenced by various factors. In the present study an effort has been made to study the effect of different temperatures on courtship and mating behaviors of *D. melanogaster*. In the present study, the courtship latency was shorter at 22°C, than either at high or low temperatures (Table 7.1). The differences in courtship latency at different temperatures was also statistically significant (By ANOVA and DMRT; Table 7.4). Courtship latency is measured as the time elapsed between the introduction of male and female flies into the observation chamber and initiation of courtship (Hegde and Krishnamurthy, 1979; Markow, 1985, 1988). It is the period during which the pairs acclimatize to the mating chamber and then start the courtship activities. It actually indicates the vigor of male Eastwood and Burnet (1977). A male with high vigor reacts quickly in the presence of female while a male with less vigor reacts slowly (Markow, 1985). Obviously shorter courtship latency indicates higher vigor of male. The shorter courtship latency noticed at 22°C thus suggests that the males at this temperature have higher vigor and, therefore, are quickly attracted by the females.

The mating latency was also shorter at 22°C compared high or low temperatures (Tables 7.1 and 7.4). Mating latency indicates both vigor of males and receptivity of females. It is the time required for males and females to initiate copulation. Higher the vigor of males and receptivity of females, shorter is the mating latency. During this period, courtship acts are performed mostly by males, to increase the receptivity of females and to make her sexually excited (Spieth, 1968a). A male with high vigor has to perform the same courtship act more number of times to a non-receptive female than to a receptive female. If she is receptive, only a few courtship acts are performed leading to quick pairing. The short mating latency at 22°C in the present studies, therefore, indicates that the males maintain high vigor and females maintain high receptivity at this temperature while at high or low temperatures, they cannot maintain the vigor and receptivity.

Courtship activity of the male or female culminates in copulation (Spiess, 1970). During copulation sperms from the male is transferred to the female reproductive tract and therefore the duration of copulation has a lot of significance in an animals' life. In the present study, the copulation duration was longest at 12°C than at other temperatures, and as temperature increased the copulation duration decreased. Thus the copulation duration was shortest at 32°C. At higher temperature perhaps the sperm transfer occurs quickly than at lower temperatures. According to Hegde and Krishna (1997) longer duration of copulation permits the transfer of more number of sperms by male to the female. Therefore, extension of copulation duration enhances the fitness of the male. It can also enhance the fitness of the females because the sperms received by a female can fertilize more number of eggs. This means at higher temperature (32°C) maximum fitness should occur. However, the courtship latency and mating latency are longer at this temperature. Therefore, it is unlikely that the longer copulation duration could enhance the fitness at 32°C. As the females have high receptivity and males have high vigor at 22°C, perhaps that is the most optimal temperature for *Drosophila* to copulate and produce offspring.

Different courtship activities of males at different temperatures exhibited similar trend, with maximum activity at 22°C and it decreased at high and low temperatures (Table 7.2). The differences in courtship acts at different temperatures were statistically significant (By ANOVA and DMRT; Table 7.5). This may imply that these courtship activities are directed by the same set of genes and that these traits are related genetically. This agrees with the observations of Ferveur *et al.* (1996); Parsons (1973); Spiess (1970); Spieth and Ringo (1983) that shows the genetic determination of certain components of sexual behavior in *Drosophila.* Though genetically determined there is every possibility for a change in sexual behavior because these activities are also influenced by changes in physical environment (Crossley, 1974; Gutzke and Crews, 1998; Luadien *et al.*, 1980; Raymond, 1983; Sisodia and Singh, 2002; West and Parker, 2002). The activities like circling, scissoring, vibration, licking etc by males were less at 12°C and 32°C compared to that at 22°C. According to Parsons (1973) temperature is a major stress for the survival and success of *Drosophila.* Perhaps this decrease in these activities may be due to stress caused by the low or high temperatures.

Among the male courtship acts, vibration is an important parameter because it serves two functions, species specific recognition and sexual excitation. Through this activity, the female will be able to recognize the courting male of its own from others and get sexually excited. The vibration bouts performed by males of different geographic origin have been found to vary Hegde and Krishnamurthy (1979). During wing vibration, male *Drosophila* emits a sound pulse which is recognized as 'courtship song' (Bastock and Manning, 1955; Bennet-Clark and Ewing, 1967; Chatterjee and Singh, 1987; Ewing, 1983; Limatainen *et al.*, 1992; Shorey, 1962; Spieth, 1952). Courtship song thus serves as an auditory stimulus conveyed by the male to stimulate the female. Any difference in the song pattern would indicate sexual isolation (Ewing, 1964, 1969a, 1969b; Koref-Santibanez, 1972; Spieth, 1968a, 1968b). In the present study, males at 22°C performed least number of vibrations than others. This indicates that the females can identify the males quickly at this temperature than others.

The authors in the present investigation have analyzed the behavior of non-receptive females of *D. melanogaster* at five different temperatures (Tables 7.3 and 7.6). Even with reference to the non-receptive behavior of females, 22°C seems to be congenial because at this temperature the females displayed least rejection responses. The females which are immature or those which have already mated exhibit these non-receptive behaviors (Singh, 1996; Singh and Chatterjee, 1987; Spieth, 1952). There may be other reasons for these non-responsiveness also. A given female can exhibit any one of these non-receptive behaviors so as to prevent male mating with her. Environmental conditions like temperature, humidity, day light, time of the day, and the physiological condition of the female may also influence the receptivity. Thus 22°C is the optimum temperature for mating of the drosophila flies. At this temperature the males have higher vigor, females have higher receptivity and, therefore, the pairs are able to mate with courtship acts being performed least. Present observations confirm the earlier studies on mating activities showing intra-specific differences in different species of *Drosophila* (Singh and Chatterjee, 1988a, 1988b). In tropical regions the temperature will be between 22°C to 29°C in most part of the year. Since mating speed is one of the fitness components and fast mating indicates the better adaptability of species, the flies seem to be adjusted to live in that temperature range. This is in conformity with the experiments involving abrupt and slow temperature changes of Laudien *et al.* (1980) which proved temperature compensations and stress effects in the courting behavior of the male guppies, *Poecila reticulata.* In Australian death adders, genus *Acanthophis* (Serpentes: Elapidae) though matings were observed at 12°C fast mating can be achieved at 22°C or above (Raymond, 1983). Although mating was fast at 22°C, it did occur at other temperatures which were facilitated by increased performance of other courtship activities. The fast mating activity observed in the present case at 22°C and mating in other temperatures with higher activity also agrees with the variation of morpho-phenotypic traits of *D. melanogaster* observed by David *et al.* (2006) demonstrating the phenotypic plasticity. The present observations of the authors thus demonstrates the sexual behavioral plasticity of *D. melanogaster* at different temperatures.

References

Alpatov, W.W. and Pearl, R., 1929. Experimental studies on the duration of life xii. Influence of temperature during the larval period and adult life on the duration of the life of the imago of *Drosophila melanogaster. Am. Nat.,* 63: 37–67.

Azevedo, R.B.R., French, V. and Partridge, L., 1996. Thermal evolution of egg size in *Drosophila melanogaster. Evolution,* 50: 2338–2345.

Bastock, M. and Manning, A., 1955. The courtship of *Drosophila melanogaster*. *Behavior*, 8: 85–111.

Bennet–Clark, H.C. and Ewing, A.W., 1967. Stimule provided by courtship of male *Drosophila melanogaster*. *Nature*, 215: 669–671.

Chatterjee, S. and Singh, B.N., 1987. Variation in mating behavior of beadox mutant and wild type *D. ananassae*. *Indian Journal of Experimental Biology*, 25: 278–280.

Crossley, S.A., 1974. Changes in mating behavior produced by selection for the ecological between and vestigial mutants of *Drosophila melanogaster*. *Evolution*, 28: 631–647.

David, J.R., Allemand, R., Van Herrewege, J., and Conet, V., 1983. *Ecophysiology; Abiotic factors in Asburner M*, Carson, H.L. Thompson jr., J.N (Eds). The genetics and biology of *Drosophila*. Academic press. London. 3d. pp105–170.

David, J.R., Legout, H., and Morteteau, B., 2006. Phenotypic plasticity of the body size in temperate population of *Drosophila melanogaster* when the temperature size rule does not apply. *Journal of genet*ica, 85(1): 9–23.

Eastwood, L. and Burnet, B., 1977. Courtship latency in male *D. malenogaster*. *Behaviour Genetic*, 7: 359–372.

Ewing, A.W., 1964. The influence of using area on the courtship behavior. *Animal Behavior*, 12 (2–3): 316–320.

Ewing, A.W., 1969. The genetic basis of sound production in *Drosophila pseudoobscura* and *Drosophila persimiles*. *Animal behavior*, 17: 555–560.

Ferveur, J.F.,Cobb, M., Boukella, H. and Jallen, J.M., 1996. World–wide variation in *Drosophila melanogaster* sex pheromone: behavioral effects, genetics bases and potential evolutionary consequences. *Genetika*, 120(1–3): 165–79.

Gibert, P., Capy, P., Imasheva, S., Moreteau, B., Morin, J.P.,Petavy, G. and David, J.R., 2004. Compartitive analysis of morphological traits among *D. melanogaster* and *D. Simulans*; genetic variability, clines and phenoplasticity. *Genetika*, 120: 165–179.

Gilchrist, G.W., Hury, R.B. and Partridge, L., 1996. Thermal sensitivity of *Drosophilia melanogaster* evolutionary responses of adults and eggs to laboratory natural selection at different temperatures. *Physiology Zoology*, 70: 403–414.

Gutzke, W.H. and Crews, D., 1998. Embryonic temperature determines adult sexuality in a reptile. *Nature*, 332 (1667): 832–834.

Hegde, S.N. and Krishna, M.S., 1997. Size–assortative mating in *Drosophila malerkotliana. Animal Behavior*, 54: 419–426.

Hegde, S.N. and Krishnamurthy, N.B., 1979. Studies on mating behavior in the *Drosophila bipectinata* complex. *Australian Journal of Zoology*, 27: 421–431.

Hoffmann, A.A. and Parsons, P.A., 1991. Evolutionary genetics and environment stress. Oxford university press, New York.

Hoffmann, A.A., Sorenson, J.G. and Loescheke, V., 2003. Adaptation of *Drosophila* to temperature extremes: bringing together quantitative and molecular approaches. *Journal of Thermal Biology*, 28: 175–206.

Hollinghworth, M.J. and Bowler, K., 1966. The decline in ability to withstand high temperature with increase in age in *Drosophila subobscura. Experiment of Gerantology*, 1: 251–257.

Karan, D., Manjal, A.K., Fibert, P., Mareteau, B., Parkash, R. and David, J.R., 1999. Growth temperature and reaction norms of morphological traits in a tropical *Drosophilid; Zaprianus inianus. Heredity*, 83: 398–407.

Kilias, G.S., Alahiotis, N. and Delecanos, M., 1980. A multifunctional investigation of speciation theory using *Drosophila melanogaster. Evolution*, 34: 730–737.

Koref–Santibenez, S., 1972. Courtship interaction in the semispecies of *Drosophila paulistorum. Evolution*, 26: 320–333.

Laudien, H., Fecher, W. and Schumann, W., 1980. Temperature dependence of Courtship in male guppies, *Poecilia reticulate.* Peters. Z. *Tieropaychol*, 52(1): 1–18.

Lenski, R.E. and Bennett, A.F., 1993. Evolutionary response of *Escherichia coli* to thermal stress. *American Nature*, S47–S64.

Limatainen, J. Hoikkala, A, Aspi, J and Welbergen, P.H., 1992. Courtship in *D. montana*: The effect of male auditory signals on the behavior of the flies. *Animal behavior*, 43: 35–48.

Lints, F.A. and Lints, C.V., 1971. Influence of preimaginal environment on fecundity in *Drosophila melanogaster* hybrids II Preimaginal temperature. *Experiment on Gerontolgy*, 6: 417–426.

Markow, T.A., 1985. A comparative investigation of the mating system of *Drosophila hydei*. *Animal Behavior*, 33: 775–781.

Markow, T.A., 1988. Reproductive behavior of *Drosophila melanogaster* and *Drosophila nigraspiracula* in the field and in the laboratory. *J. Comp. Phycology*, 102: 169–173.

Morteteau, B. Gibert, P. Dalpuech, J.M. Petavy, G. and David, J.R., 2003. Phenotypic plasticity of strenoplueral bristles number in temperate and tropical population of *Drosophila melanogaster*. *Genetica Research*. 81(1): 25–32.

Parsons, P.A., 1973. Behavioral and Ecological genetics: A study in *Drosophila*. Oxford: Clarenden Press: 1–55.

Partridge, l. Barrie, B. Barton, N.H. Fowler K and French, V. , 1994b. Evolution and development of body size and cell size in *Drosophila melanogaster* in response to temperature. *Evolution*, 48: 1269–1276.

Partridge, L. Barrie, B. Barton, N.H. Fowler, K and French, V., 1995. Rapid laboratory evolution of adult life-history traits in *Drosophila melanogaster* in response to temperature. *Evolution*, 49: 538–544.

Partridge, L. Barrie, B. Fowler, K and French, V., 1994a. Thermal evolution of pre adult life-history traits in *Drosophila melanogaster*. *Journal of Evolutionary biology*, 7: 645–663.

Raymond, T.H., 1983. Mating behavior in Australian death adders, genus; Acanthophis (Serentes: Elapidae). *Herpetology*, 8(1): 25–34.

Scheinar, S.M., 1993. Genetics and evolution of phenoplasticity. Ann. *Rev. Ecol. Syst.*, 24: 35–38.

Scheiner, S.M and Lyman, R.F., 1989. The Genetics of Phenoplasticity.I. *Heritabilty. Journal of Evoutioanary Biology*, 2: 95–105.

Schlicting, C.D and Pigliucci, M., 1998. Phenotypic evolution. A reaction norm perspective. Sineuar associates, Sunderland.

Shorey, H.H., 1962. Nature of the sound produced by *Drosophila melanogaster* during courtship. *Science*, 1376: 677–678.

Singh, B.N and Chatterjee, S., 1988a. Parallelism between male mating propensity and chromosomes arrangement frequency in natural populations of *D. ananassae*. *Heredity*, 60: 269–272.

Singh, B.N and Chatterjee, S., 1987. Variation in mating propensity and fertility in isofemale strains of *Drosophila ananassae*. *Genetica*, 73: 237.

Singh, B.N and Chatterjee, S., 1988b. Selection for high and low mating propensity in *Drosophila ananassae*. *Behaviour Genetics*, 18: 357–369.

Singh, B.N., 1996. Population and behavior genetics of *D. ananassae*. *Genetica*, 97: 371–329.

Sisodia, S and Singh, B.N., 2002. Effect of the temperature on longevity and trade productivity in *Drosophila ananassae*; evidence for adaptive plasticity and trade off between longevity and productivity. *Genetika*, 114(1): 95 –102.

Smith, J.M., 1956. Acclimatization to high temperature inbreed and outbreed *Drosophila subobscura*. *Journal of Genetics*, 54: 407–505.

Spiess, E.B., 1970. Mating propensity and its genetic basis in *Drosophila*. In essay in evolution and genetics in honour of the Dobzhansky (Eds. Hecht, M.T and Streere W.C) 315–379. Appleton–Century– Crafts, New York.

Spieth, H.T (1968b. Evolutionary implication of the mating behavior of the species of *Antopocerus* (Drosophilidae) in Hawaii: Studies in Genetics. University of Texas Publication, 4: 85–111

Spieth, H.T and Ringo, J.M., 1983. Mating behavior and sexual isolation in *Drosophila*. In the genetics and biology of *Drosophila* (Eds Ashburner, M. Carson, H.L. and Thompson, Jr. J.N) Vol. 3c., Pp 223–224. Academic Press, London.

Spieth, H.T., 1952. Mating behavior within Genus *Drosophila* (diptera). *Bull. Am. Mus. Nat. History*, 99: 395–474.

Spieth, H.T., 1968a. Evolutionary implication of several behavior in *Drosophila*. In evolutioanary biology (eds) Dobzhansky, Th. Hect, M.K and Steere, W.C. Appleton– Century– crofts, New York, pp157–193.

Steigenga, M.J. Zwaan, B.J. Brakeflied, P.M and Fischer. K., 2005. The Evolutionary genetics of egg size plasticity in a butterfly. *Journal Evolutionary Biology,* 28: 1–9.

Via, S., 1993. Adaptive phenotypic plasticity: Target or byproduct of selection in a variable environoment? *American Nature,* 142: 352–365.

West, P.M. and Packer, C., 2002. Sexual selection, temperature and the lion's mane. *Science*. 297: 1339– 1343.

Chapter 8

Effects of Sublethal Cypermethrin, Carbaryl and Monocrotophos Administration on Certain Biochemical Parameters in Muscles and Nerve Cord of *Periplanata americana*

B.G. *Kulkarni*[1], F.A. *Mohammed*[1], S.S. *Kupekar*[2] *and* A.K. *Pandey*[3*]

[1]*Department of Zoology, Institute of Science, 15 Madam Cama Road, Mumbai – 400 032, India*

[2]*Department of Zoology, Mahatma Phule A.S.C. College, Panvel, Raigad – 410 206, India*

[3]*National Bureau of Fish Genetic Resources, Canal Ring Road, Dilkusha, Lucknow – 226 002, India*

ABSTRACT

Periplanata americana were administered sublethal dose (less than the LD_{50} value for 96 hours) of cypermethrin (2.0×10^{-3} mg), carbaryl (5.0×10^{-2} mg) and monocrotophos (1.8×10^{-1} mg) for 24 and 96 hours. The treatments invariably induced depletion of glycogen content in thoracic and coxal muscles

* Corresponding author: E-mail: akpandey_cifa@yahoo.co.in

suggesting increased glycogenolysis. Increased level of lactic acid as compared to pyruvic acid in the treated cockroaches suggests severe respiratory stress under the given insecticidal treatments. Increase in both pyruvic and lactic acid contents in muscles indicates that the treated insects are resorting to both aerobic as well as anaerobic respiration in stressful conditions. There was significant depletion in protein content in muscle and nerve cord, probably due to increased proteolysis.

Keywords: *Cypermethrin, Carbaryl, Monocrotophos, Biochemical changes, Periplanata americana.*

Introduction

Most of the insecticides act on nervous system of insects thereby disturbing its proper functioning and resulting to their death. Moreover, the neuroendocrine system in insects modulates synthesis and breakdown of carbohydrates, proteins and fats which, in turn, regulate the concentration of various metabolites in tissues. This has been proved by assessing the effects of extracts of brain, corpora allata and corpora cardiaca on carbohydrate metabolism of insects (Ralph and McCarthy, 1964; Downer, 1979; Mandal *et al.*, 1984). There exist reports that insecticide treatments alter the biochemical metabolites in haemolymph of insects (Kulkarni and Mehrotra, 1973; Rakshpal and Singh, 1976; Rastogi and Miglani, 1976; Pandey and Mathur, 1990). Many investigators found that the proximate composition in fish, crustaceans and molluscs gets altered due to insecticide treatments (Biswas, 1986; Krishna *et al.*, 1987; Ghosh, 1989; Ferrando and Andreu-Moliner, 1991; Singh, 2004). However, there is dearth of literature on the effects of insecticides on metabolites as well as energy reserves in insect tissues. The energy metabolism plays a key role as the animal is forced to expand more energy to mitigate the toxic effects of the insecticides. Though the toxicity of cypermethrin, carbaryl and monocrotophos on cockroach has been evaluated recently (Kulkarni *et al.*, 2005), the present study was undertaken to assess the effect of these pesticides on energy reserves of the cockroach, *Periplanata americana.*

Materials and Methods

Adult male cockroach, *Periplanata americana* (Linnaeus) (total length 3.92±0.08 cm, average weight 0.924±0.04 gm) were selected and acclimatized to the laboratory conditions for 4 days prior to treatment. They were kept in wired/aluminum cages of 45 × 24 × 22 cm size and fed on roughage throughout the study period. Cockroaches were administered sublethal dose (less than the LD_{50} value for 96 hours) of cypermethrin (2.0×10^{-3} mg), carbaryl (5.0×10^{-2} mg) and monocrotophos (1.8×10^{-1} mg) for 24 and 98 hours by topical application on the thorax using an aglamicrometer syringe (experimental). A set of control was also maintained for the entire period of the treatment. The insects of both the groups were dissected at 24 and 96 hours in normal saline and muscles of the thorax and coxae were carefully excised together with the ventral nerve cord (for protein estimation).

Glycogen

The tissue glycogen levels were determined according to the method of Seifter (1950). Freshly excised muscles were digested in 3 ml 30 per cent KOH. Each KOH digest was diluted to 100 ml with distilled water. To 5 ml aliquot of this solution, 10 ml of freshly prepared anthrone reagent (0.2 per cent in concentrated H_2SO_4) was added gradually. Similarly, blank and standard tubes were also prepared using distilled water and glucose solution, respectively. The mixture was then boiled for 10 minutes in boiling water bath. After cooling, optical density of the mixture was read at 625 nm. Standard calibration curve was also prepared for calculating glucose values. The glucose values thus obtained were converted into glycogen using the Morrisons factor 1.11.

Pyruvic Acid

Pyruvic acid content of the tissues from the control as well as experimental insects were determined by the method of Friedmann and Haugen (1943, modified by Chaykin, 1966). Tissue of known weight was homogenized in 3 ml 30 per cent trichloroacetic acid. The mixture was centrifuged at 5,000 rpm for 5 minutes and clear protein-free supernatant was used for pyruvic acid assay. An appropriate aliquot of the deproteinized supernatant was diluted to 2 ml with distilled water and then treated with 0.5 ml 2,4-dinitrophenyl hydrazine reagent (0.1 per cent in 2 N HCl) followed by 3 ml 2.5 N NaOH. The reaction mixture was allowed to stand for 10 minutes and optical density was read at 540 nm. The amount of pyruvic acid was calculated using standard graph.

Lactic Acid

Lactic acid levels were determined according to the methods of Barker and Summerson (1941) following a preliminary procedure of van Slyke (1917). 2 ml aliquot of the deproteinized supernatant was treated with 1 ml 20 per cent $CuSO_4$ solution and then diluted to 10 ml with distilled water. 1 gm $Ca(OH)_2$ powder was added and the mixture was shaken vigorously and allowed to stand for 30 minutes. The mixture was then centrifuged at 5,000 rpm for 5 minutes. To 1 ml supernatant, 0.05 ml 4 per cent $CuSO_4$ solution was added followed by 6 ml concentrated H_2SO_4. The solution was heated in a boiling water bath for 5 minutes and cooled at room temperature. After cooling, 0.1 ml p-hydroxydiphenyl reagent (1.5 gm p-hydroxydiphenyl dissolved in 10 ml 5 per cent NaOH and diluted to 100 ml with distilled water) was added and the solution was incubated at 30°C for 30 minutes. The test tubes were then placed in a vigorously boiling water bath for 90 seconds and then allowed to cool down to room temperature. The optical density was read at 560 nm and the amount of lactic acid was calculated using a standard graph.

Soluble Proteins

Soluble proteins were assayed in muscles and ventral nerve cord by the method of Lowry *et al.* (1951). Freshly removed tissues were weighed and homogenized with chilled glass distilled water. The homogenates were centrifuged at 10,000 rpm for 10-15 minutes and the clear supernatants were used for estimation of soluble proteins. 0.2 ml supernatant was diluted to 1.0 ml by glass distilled water. 5 ml alkaline copper solution (prepared freshly by mixing 50 ml 2 per cent $NaHCO_3$ in 0.1N NaOH and 1 ml 0.5 per cent $CuSO_4$ in 1 per cent Na-K tartrate) was added. After incubation for 10 minutes at room temperature, 0.5 ml folin-phenol reagent was added rapidly and the solution was thoroughly mixed. After 30 minutes, optical density was read at 500 nm. A standard graph was also prepared using bovine serum albumen and protein content from different tissues was calculated from this graph.

Results and Discussion

The values for glycogen content of thoracic and coxal muscles of control as well as treated *P. americana* have been summarized in Table 8.1. Glycogen content of thoracic muscle of the control insects ranged between 5.48±0.32 to 5.64±0.24 mg/g whereas those of coxal muscle varied between 6.12±0.38 to 6.28±0.39 mg/g. Administration of all the three insecticides in the cockroach induced a progressive depletion of glycogen content in both the muscles. Though cypermethrin and monocrotophos treatments resulted in a more pronounced decline (P < 0.01) in the muscle glycogen content as compared to that of carbaryl (P < 0.05) at 24 hours, by 96 hours the glycogen content in muscle tissues registered further decline (P < 0.001) in all the treatments.

Table 8.1: Glycogen Content (mg/g) in Different Muscles of *P. americana* Treated with Different Pesticides

	Control	*Experimental*	
		24 Hours	*96 Hours*
	Cypermethrin		
Thoracic muscle	5.58 ±0.38	2.29± 0.19 (-58)[c]	2.00± 0.08 (-64)[c]
Coxal muscle	6.24 ± 0.43	3.12 ± 0.55 (-50)[b]	3.21±0.37 (-48)[c]
	Carbaryl		
Thoracic muscle	5.54 ±0.32	3.41± 0.17 (-38)[c]	2.54± 0.17 (-54)[c]
Coxal muscle	6.12 ± 0.38	5.20 ± 0.40 (-15)[a]	2.96±0.12 (-51)[c]
	Monocrotophos		
Thoracic muscle	5.64 ±0.24	3.12± 0.09 (-44)[c]	2.47± 0.23 (-74)[c]
Coxal muscle	6.28 ± 0.39	4.88 ± 0.21 (-22)[b]	3.62±0.59 (-42)[c]

Values are mean±SE of 5 determinations.

Significant responses: [a] P < 0.05, [b] P < 0.01, [c] P < 0.001.

Numbers in parentheses indicate per cent increase over control.

Changes in pyruvic acid content in control as well as treated cockroaches have been given in Table 8.2. Pyruvic acid content in thoracic muscle of the control insects varied between 23.42±1.32 to 23.63±1.48 mg/g whereas those of coxal muscle ranged between 9.08±0.32 to 9.24±0.28 mg/g. Though there was a significant (P < 0.05) decline in pyruvic acid content in both the muscles of cypermethrin as well as monocrotophos treated *P. americana* after 24 hours of administration which declined further (P < 0.001) in both the treated groups at 96 hours. However, in case of carbaryl treatment, there was a marked time–dependent elevation (P < 0.001) in pyruvic acid content in both the muscle tissues after 24 and 96 of the pesticide administration.

Table 8.2: Pyruvic Acid Content (mg/g) in Different Muscles of *P. americana* Treated with Different Pesticides

	Control	*Experimental*	
		24 Hours	*96 Hours*
	Cypermethrin		
Thoracic muscle	23.51±1.52	17.43± 1.68 (-25)[a]	3.32± 0.37 (-85)[c]
Coxal muscle	9.12 ± 0.28	7.70 ± 0.45 (-15)[a]	6.59±0.23 (-27)[c]
	Carbaryl		
Thoracic muscle	23.42±1.32	29.38± 0.31 (+25)b	38.31± 0.76 (+63)[c]
Coxal muscle	9.24 ± 0.28	22.84 ± 0.40 (+147)[c]	24.57±0.35 (+165)[c]
	Monocrotophos		
Thoracic muscle	23.63±1.48	14.82± 0.31 (-37)[c]	17.13± 1.12 (-27)[b]
Coxal muscle	9.08± 0.32	7.18 ± 0.19 (-20)[b]	6.72±0.59 (-26)[b]

Values are mean±SE of 5 determinations.

Significant responses: [a] $P < 0.05$, [b] $P < 0.01$, [c] $P < 0.001$.

Numbers in parentheses indicate per cent decrease (-) or increase (+) over control.

Variations in lactic acid content in muscles of control as well as treated *P. americana* have been shown in Table 8.3. Lactic acid content of thoracic muscle of the control cockroaches ranged between 7.74±0.65 mg/g while those of coxal muscle varied between 3.09±0.31 to 3.18±0.28 mg/g. There was a marked elevation ($P < 0.001$) in lactic acid content in thoracic muscle of cypermethrin-treated insects at 24 hours while the response was less marked ($P < 0.05$) in monocrotophos treated cockroaches. The coxal muscle showed marked increase in lactic acid content in all the three treatments at 24 hours. However, there was a decline in lactic acid content of thoracic muscle of the insects at 96 hours. Interestingly, lactic acid content of coxal muscle of the pesticides treated cockroaches showed a progressive increase.

Variations in soluble proteins in the muscles of control as well as treated cockroaches have been summarized in Table 8.4. Protein content of thoracic muscle of control insects vary between 98.08±4.34

to 98.34±4.45 mg/g, coxal muscle between 86.07±4.46 to 86.37±4.22 mg/g and nerve cord between 270.13±3.72 to 278.00±3.26 mg/g. Pesticide treatments in all the cases recorded in a progressive decline the tissues proteins at 24 and 96 hours of the administration. The depletion was more marked after carbaryl treatment at both the periods (24 and 96 hours) as compared to cypermethrin and monocrotophos administrations.

Table 8.3: Lactic Acid Content (mg/g) in Different Muscles of *P. americana* Treated with Different Pesticides

	Control	*Experimental*	
		24 Hours	*96 Hours*
		Cypermethrin	
Thoracic muscle	7.96±0.85	30.16± 0.84 (+278)[c]	10.15± 0.76 (+27)
Coxal muscle	3.09± 0.31	7.00 ±0.61 (+126)[c]	8.77±0.58 (+183)[c]
		Carbaryl	
Thoracic muscle	7.76±0.65	7.59± 0.21 (-2)	13.09± 0.55 (+68)[c]
Coxal muscle	3.12± 0.28	10.05 ± 0.36 (+222)[c]	11.62±0.28(+272)[c]
		Monocrotophos	
Thoracic muscle	7.84±0.90	10.32± 0.32 (+31)[a]	11.42± 0.76 (+45)[b]
Coxal muscle	3.18± 0.28	8.77 ± 0.27 (+175)[c]	8.90±0.17 (+179)[c]

Values are mean±SE of 5 determinations.

Significant responses: [a]$P < 0.05$, [b]$P < 0.02$, $P < 0.001$.

Numbers in parentheses indicate per cent increase over control.

A decrease in the glycogen content in tissues of experimental cockroaches seems to be the result of insecticide-induced stress. These results indicate impairment in synthesis of glycogen in addition to utilization of reserved glycogen for energy needs. These results are in agreement with many others who have reported depletion in glycogen content of tissues in fish, crustaceans and molluscs treated with pesticides (Biswas, 1986; Krishna *et al.*, 1987; Ferrando and Andreu-Moliner, 1991).

It has been shown that insects utilize glycogen under stress conditions such as flight (Downer and Parker, 1979; Rowan and Newsholme, 1979). Moreover, insects possess all the enzymes of

Embden-Meyerhof glycolytic cycle, hexomonophosphate shunt mechanism and Kreb's tricarboxylic acid cycle (Levenbook, 1961; Gilmour, 1961). The degradative metabolic pathways in insects such as glycolysis, fatty acid β-oxidation, tricarboxylic acid cycle (TCA) and amino acid oxidation serve to synthesize ATP which is used for the synthetic reaction, to provide energy for mechanical movements and to maintain ion and substrate gradients across membranes. The balance between the rates of these different pathways of energy depends upon a number of factors such as substrate availability, hormonal influences and physiological circumstances. Such factors are affected drastically by insecticides (Bose, 1991). Therefore, the depletion in glycogen during the present investigation may be attributed to impairment in the nervous functions due to pesticidal stress as well as in increased energy demand of the intoxicated insects.

Table 8.4: Protein Content (mg/g) in Tissues of *P. americana* Treated with Different Pesticides

	Control	*Experimental*	
		24 Hours	*96 Hours*
	Cypermethrin		
Thoracic muscle	98.34±4.45	60.51±0.68 (-38)[a]	50.79±1.68 (-48)[a]
Coxal muscle	86.07±4.46	54.75±3.99 (-36)[a]	36.49±2.28 (-57)[a]
Nerve cord	270.13±3.72	108.60±3.63 (59)[a]	105.79±1.89 (-60)[a]
	Carbaryl		
Thoracic muscle	98.08±4.34	20.62±2.43 (-79)[a]	17.70±0.37 (-81)[a]
Coxal muscle	86.27±4.24	18.93±1.40 (-78)[a]	19.36±0.72 (-77)[a]
Nerve cord	272.28±3.86	59.92±1.44 (-77)[a]	45.83±2.50 (-83)[a]
	Monocrotophos		
Thoracic muscle	98.14±4.21	60.40±1.26 (-38)[a]	57.85±0.94(-41)[a]
Coxal muscle	86.37±4.22	45.88±1.55 (-46)[a]	57.02±1.16 (-33)[a]
Nerve cord	278.00±3.26	147.94±1.86 (-46)[a]	128.79±2.19 (-53)[a]

Values are mean±SE of 5 determinations.

Significant response: [a] $P < 0.001$.

Numbers in parentheses indicate per cent decrease (-) or increase (+) over control.

The glycogen reserves undergo rapid breakdown to release glucose units in the haemolymph of insects. As the insects resort to glycolytic anaerobic conditions during stressful conditions, the observed depletion of glycogen and marked elevation of lactic acid in the muscle tissues of experimental *P. americana* shows a change over to anaerobic respiration to fulfill the increased energy demand. The elevated pyruvic acid levels in carbaryl treated cockroaches may be caused by the insects resorting to both aerobic and anaerobic respiration. According to Huckabee (1958) and Dange and Masurekar (1982), an upward trend in lactic acid relative to pyruvic acid as observed in the pesticide treated cockroaches indicate that oxygen supply in tissue is not adequate.

Insects possess high levels of free amino acids (aminoacidemia) in the haemolymph (Chen, 1985). This indicates that the free amino acids (FAAs) fulfill additional metabolic functions such as osmoregulation, neurotransmission, detoxification, synthesis of phospholipids and morphogenesis. Since these metabolic functions are drastically affected by insecticides (Kulkarni and Mehrotra, 1973; Bose, 19991), hence the cockroaches metabolize their tissue proteins into haemolymph FAAs to meet the energy requirements (Kulkarni and Mehrotra, 1973; Rao *et al.*, 1980; Shankara *et al.*, 1989).

The amino acid degradation takes place through specific pathways to give acetyl-Co A, pyruvate or an intermediate of the TCA cycle. In general, removal of the amino group is the first or an early step in the degradation and this process occurs through the action of transaminases. Elevated levels of transaminases were noted in the cockroaches treated with these insecticides (Kulkarni *et al.*, 2007) suggesting that the animal are resorting to degradation of amino acids for energy production by gluconeogenesis under stress conditions. Degradation of amino acids for energy production under insecticidal stress has been noted in *Diacrisia obligua* and *Hieroglyphus banian* (Pandey and Mathur, 1990) and *Periplanata americana* (Rastogi and Miglani, 1976). In both the cases, proline levels in the haemolymph were depleted suggesting possible utilization of this particular amino acid for energy purpose.

Acknowledgements

The senior author (FAM) is grateful to the Muslim Education and Welfare Association, Mombasa, Kenya for sponsoring her study in India. We are thankful to Prof. S.A. Suryawanshi, former Head, Department of Zoology and Director, Institute of Science, Mumbai for providing necessary facilities to carry out the work.

References

Barker, S.B. and Summerson, W.H., 1941. Colorimetric determination of lactic acid in biological material. *J. Biol. Chem.*, 138: 535-554.

Biswas, R., 1986. Toxic Effects of an Organochlorine Insecticide, Thiodan, on the Marine Crab, *Scylla serrata* (Forskal). *Ph.D. Thesis*. University of Bombay, Bombay.

Bose, C., 1991. Malathion induced biogenic amine levels and acetylcholinesterase activity in the cockroach, *Periplanata americana*. *Curr. Sci.*, 60: 707-709.

Chaykin, S., 1966. *Biochemistry Laboratory Techniques*. Wiley-Eastern Pvt. Ltd., New Delhi. 196 p.

Chen, P.S., 1985. Amino acid and protein metabolism. In: *Comparative Insect Physiology: Biochemistry and Pharmacology. Vol. 10.* (Eds.) G.A. Kerkut and L.I. Gilbert. pp. 177-217. Pergaman Press, Oxford, New York, Toronto and Sydney.

Dange, A.D. and Masurekar, V.B., 1982. Naphthalene induced changes in carbohydrate metabolism in *Sarotherodon mossambicus* Peters (Pisces: Cichlidae). *Hydrobiologia*, 94: 163-172.

Downer, R.G.H., 1979.Trehalose production in isolated fat body of the American cockroach, *Periplanata americana*. *Comp. Biochem. Physiol.*, 62C: 31-34.

Downer, R.G.H. and Parker, G.H., 1979. Glycogen utilization during flight in the American cockroach, *Periplanata americana*. *Comp. Biochem. Physiol.*, 64A: 29-32.

Ferrando, M.D. and Andreu-Moliner, E., 1991. Changes in selected biochemical parameters in the brain of fish, *Anguilla anguilla* (L.) exposed to lindane. *Bull. Environ. Contam. Toxiciol.*, 47: 459-464.

Friedmann, T.K. and Haugen, G.E., 1943. Pyruvic acid. II. The determination of keto-acids in blood and urine. *J. Biol. Chem.*, 147: 415-442.

Ghosh, T.K., 1989. Effect of cypermethrin on fish, *Labeo rohita*. *Ph.D. Thesis*. University of Bombay, Bombay.

Gilmour, D., 1961. *The Biochemistry of Insects*. Academic Press, New York and London.

Huckabee, W.E., 1958. Relationship of pyruvate and lactate during anaerobic metabolism. II. Exercise and function of oxygen debt. *J. Clin. Invest.*, 37: 255-263.

Krishna, P., Venkataramani, L., Ravi, C. and Indira, K., 1987. Metabolic consequences of methyl parathion exposure in the bivalve, *Lamellidens marginalis*. *Bull. Environ. Contam. Toxicol.* 38: 509-514.

Kulkarni, A.P. and Mehrotra, K.N., 1973. Effct of dieldrin and sumithion on the amino acid nitrogen and proteins in haemolymph of desert locust, *Schistcerca gregaria* Forsk. *Pest. Biochem. Physiol.*, 3: 420-434.

Kulkarni, B.G., Mohammad, F.A., Kupekar, S.S. and Pandey, A.K., 2005. Acute toxicity studies of cypermethrin, carbaryl and monocrotophos to the cockroach, *Periplanata americana* (Linnaeus). *J. Natcon.*, 17: 299-306.

Kulkarni, B.G., Mohammad, F.A., Kupekar, S.S. and Pandey, A.K., 2007. Alterations in enzymes activities in muscles and nerve cord of *Periplanata americana* induced by sublethal administration of cypermethrin, carbaryl and monocrotophos. *J. Ecophysiol. Occup. Hlth.*, 7: 31-37.

Levenbook, L., 1961. Organic acids in insects. II. Tricarboxylic acid cycle and related enzymes in the larvae of the southern armyworm, *Prodentia eridania*. *Arch. Biochem. Biophys.*, 92: 114-121.

Lowry, O.H., Rosenbrough, N.J., Farr, A.L. and Randall, R.J., 1951. Protein measurement with folin-phenol reagent. *J. Biol. Chem.*, 193: 265-275.

Mandal, S., Saha, L. and Choudhuri, D.K., 1984. Role of corpora allata and brain on carbohydrate level of haemolymph in *Lohita grandia* Gray (Pyrrocoridae: Heteroptera: Hemiptera). *Curr. Sci.*, 53: 277-279.

Pandey, R. and Mathur, Y.K., 1990. Amino acid alterations in insecticidal treated insects. *Indian J. Entomol.*, 52: 197-202.

Rakshpal, R. and Singh, A., 1976. Free amino acids in some tissues of *Periplanata americana* Linn. *Indian J. Entomol.*, 38: 171-175.

Ralph, C.L. and McCarthy, R., 1964. Effects of brain and corpus cardiacum extracts on haemolymph trehalose of the cockroach, *Perplanata americana*. *Nature*, 203: 1195.

Rao, S.P., Prasad, S. and Rao, V.R., 1980. Sublethal effect of methyl parathion on tissues proteolysis of the freshwater mussel, *Lamellidens marginalis* (Lamarck). *Proc. Indian Natl. Sci. Acad.*, 46B: 164-167.

Rastogi, S.C. and Miglani, R.K., 1976. Effects of insecticides on the free amino acids and nerve conduction in *Perplanata americana* L. *Indian J. Entomol.*, 38: 146-154.

Rowan, A.N. and Newsholme, E.A., 1979. Changes in the contents of adenosine nucleotides and intermediates of glycolysis and the citric acid cycle in flight muscles of locust upon flight and their relationship to the control of the cycle. *Biochem. J.*, 178: 209-216.

Seifter, S., Dayton, S., Novie, B. and Muntwyler, I., 1950. Estimation of glycogen with anthrone reagent. *Arch. Biochem.*, 25: 191-200.

Shankara, C.R., N.S.A. Babu, Indira K. and Rajendra, W., 1989. Metabolic resistance to methyl parathion toxicity in a bivalve, *Lamellidens marginalis*. *Curr. Sci.*, 58: 767.

Singh, S.K., 2004. Studies on toxicological effects of synthetic pesticides on freshwater fish, *Colisa fasciatus*. *Ph.D. Thesis*. Purvanchal University, Jaunpur (India).

van Slyke, D.D., 1917. Studies of acidosis. VII. Determination β-hydroxybutyric acid, acetoacetic acid and acetone in urine. *J. Biol. Chem.*, 32: 455-493.

Chapter 9

Improvement of Larval Duration and Survival in Commercially Important Swimming Crabs, *Portunus pelagicus* (Linnaeus) and *P. sanguinolentus* (Herbst) Through Enriched Live Feeds

P. Soundarapandian and S. Sowmiya*
Centre of Advanced Study in Marine Biology, Annamalai University, Parangipettai Tamil Nadu, India

ABSTRACT

The enrichment of live feeds to increase the nutritional efficiency of the live-feed has significantly contributed to the larval survival and quality. Therefore, the present experiment was designed to study the efficacy of enriched diet in the larval development of edible portunid crabs *Portunus pelagicus* and *P. sanguinolentus*. Enrichment of freshly hatched *Artemia nauplii* and laboratory cultured rotifers were carried out at a density of 200 and 600 individuals/ml respectively in a prepared emulsion of 0.025ml of lecithin and 0.05ml of cuttle fish liver in

* Corresponding author: E-mail: soundsuma@yahoo.com

one liter of seawater. Enriched live feeds were fed to zoeal and megalopal stages throughout the experimental period until the larval development were complete. The results of present study prove that the enrichment of live feed improves both survival and shortening of larval development duration in *P. pelagicus* and *P. sanguinolentus*.

Keywords: *Artemia nauplii, Portunus sanguinolentus, P. pelagicus, Brachionus plicactilis, Nutrient enrichment, Cattle fish liver.*

Introduction

In recent years, enrichment of live feed through the enriched diets (bioencapsulates) to increase the nutritional efficiency of the live-feed has significantly contributed to the larval survival and quality (Sorgeloos and Leger, 1992). Further more numerous experiments have been carried out on this aspect in marine and freshwater fishes and prawn species (Sorgeloos and Leger, 1992; Watanabe, 1993). The rotifers were enriched with fish and seal oils for 4-24 hours (Kannupandi *et al.*, 2003). The enriched rotifer of *B. plicatilis* was found to be required for better survival/development of marine fish larvae. Further it has also been felt that the enrichment of HUFA in rotifers is an essential process in the field of marine larval culture. However, studies on this aspect with regard to crab larval development is very much limited especially in case of blue swimming crabs (Levin and Sulkin, 1984; Kannupandi *et al.*, 2003). Therefore, the present experiment was designed to study the efficacy of enriched diet in the larval development of edible portunid crabs, *P. pelagicus* and *P. sanguinolentus*.

Materials and Methods

Nutrient Enrichment

Enrichment of freshly hatched *Artemia nauplii* and laboratory cultured rotifers were carried out at a density of 200 and 600 individuals/ml respectively in a prepared emulsion of 0.025ml of lecithin and 0.05ml of cuttle fish liver in one liter of seawater following the method of Leger *et al.* (1986) and Kannupandi *et al.* (2003). The lipid fraction from the cuttle fish liver oil was extracted by following the method of Folch *et al.* (1956). The enriched *Artemia nauplii* (OSI Brine shrimp eggs, USA) and rotifer *B. plicatilis* were harvested after 6 hours of incubation and were thoroughly rinsed with seawater before feeding. Enriched live feeds were fed to zoeal and megalopal stages throughout the experimental period until the larval development were complete. During feeding trials, the zoeal and megalopal stages were monitored for growth and survival.

Larval Rearing

Freshly hatched active zoea of *P. pelagicus* and *P. sanguinolentus* were transferred in groups of 10 individuals into clean bowls of 100 ml capacity. The rearing medium was maintained at appropriate salinity (35± 1ppt), temperature (27±1.0°C) and photophase (14/10 hrs L/D), and changed daily during the early hours of the day before feeding and care was taken to remove the dead larvae and exuviae. The early zoeal stages (upto III zoea) were fed with nutrient enriched mixed diet of *B. plicatilis* and *Artemia nauplii* (OSI Brine shrimp eggs, USA) and the later zoeal stages (IV zoea onwards) were fed only with enriched *Artemia nauplii*. Megalopa and I[st]instar crab were fed with 2days old *Artemia nauplii*. The control larvae were sufficiently fed with non-enriched rotifers of *B. plicatilis* and *Artemia nauplii* (Table 9.1). The feed was given twice a day at 8'O clock in the morning and 5'O clock in the evening hours.

Table 9.1: Feeding of Different Crab Larval Stages with Enriched Live Feeds

Larval Stages	*Morning (Pcs/Larvae)*	*Afternoon (Pcs/ml)*
I zoea	*B. plicatilis* 5-15	*B. plicatilis* 5-20
II zoea	*B. plicatilis* 5-15	*Artemia nauplii* 5-20
III zoea	*B. plicatilis* 15-25	*Artemia nauplii* 5-20
IV zoea	*Artemia nauplii* 20-50	*Artemia nauplii* 20-50
Megalpoa	2 days old *Artemia nauplii* 10	2 days old *Artemia nauplii* 50
First crab instar	2 days old *Artemia nauplii* 10	2 days old *Artemia nauplii* >60

Results and Discussion

The intermoult duration and survival in the larval stages of *P. pelagicus* and *P. sanguinolentus* offered, enriched diet showed better results than control. The duration of total larval cycle was also comparatively less in enriched feed fed larvae than in the control, for both crabs' species (Tables 9.2 and 9.3).

Table 9.2: The Percentage of Survival and Intermoult Duration (Days) of Different Larval Satges of *P. pelagicus* Fed with Enriched and Control Diets.

Larval Stages	*Intermoult Duration (Days)*		*Survival (Per cent)*	
	Control	*Enriched*	*Control*	*Enriched*
I zoea to II zoea	4.1±0.29	3.0±0.18	62.0±0.46	72.0±0.26
II zoea to III zoea	4.2±0.18	3.6±1.18	55.0±1.06	64.1±1.11
III zoea to IV zoea	3.3±1.01	3.0±0.00	50.1±2.00	56.5±1.08
IV zoea to Megalopa	3.1±1.04	3.2±0.48	25.1±0.13	28.6±2.12
Megalpoa to crab instar	8.2±0.18	7.1±0.12	9.23±0.12	12.5±1.80
Total	21-24	19-22		

Table 9.3: The Percentage of Survival and Intermoult Duration (Days) of Different Larval Stages of *P. sanguinolentus* Fed with Enriched and Control Diets

Larval Stages	*Intermoult Duration (Days)*		*Survival (Per cent)*	
	Control	*Enriched*	*Control*	*Enriched*
I zoea to II zoea	4.0±0.13	3.58±0.28	62.0±1.08	71.75±1.38
II zoea to III zoea	4.0±0.31	3.42±0.12	40.0±1.02	44.0±1.38
III zoea to IV zoea	3.2±0.21	2.70±0.29	31.0±0.18	34.4±3.12
IV zoea to Megalopa	3.2±0.18	2.58±0.12	8.2±1.08	12.0±2.94
Megalpoa to crab instar	7	5	3.5±1.32	5.20±0.58
Total	18-21	15-18		

The mixed diet of rotifer *B. plicatilis* and *Artemia nauplii* has effectively supported the larval development. But, high mortality rate was observed during the transition period between III and IV zoeal stages, which ultimately resulted in poor survival and growth. Similar results have also been obtained by Frank *et al.* (1975) during the larval development of *R. harrisii*. Levin and Sulkin (1984) reported that *Eurypanopeus depressus* larvae fed with brine shrimp *Artemia nauplii* showed accelerated development in the early stages upto III zoea, but survival was affected during the IV zoeal stage. Similarly, neither blue crab *C. sapidus* (Sulkin, 1978) nor the shore crab *M. mercenaria* has completed zoeal development when they were fed only with rotifer (Sulkin and Van Heukelem, 1980). Sulkin and Norman (1976) illustrated that the mud crabs, *R. harrisii* and *Neopanope texana sayi* could be developed to megalopa on the rotifer diet, but the survival was lower and development time was prolonged as compared to the brine shrimp-fed control. It may be due to poor dietary value of the diet used. Frank *et al.* (1975) have stated reason for higher mortality rate during the transition period between III and IV zoeal stages and explained that the metabolic activity significantly increases which results in the increase in energy requirements, but the normal diets very often fail to provide the required nutrients.

However, in the present study the larval survival improved significantly and the rate of development also increased considerably by enhancing the nutritive value of the *B. plicatilis* as well as *Artemia nauplii* using cuttle fish liver oil. According to Sorgeloos and Leger (1992), the method of bioencapsulation is widely applied at marine fish and crustacean hatcheries all over the world for enhancing the nutritional value of *Artemia* nauplii and rotifer *B. plicatilis*. Further the British, Japanese and Belgium researchers have developed enrichment products and procedures using selected microalgae and/or microencapsulated products, yeasts and/or emulsified preparations, using self-emulsifying concentrates and/or microparticulate products respectively. Apart from the above, the enrichment could be done by exposing *Artemia nauplii* and rotifer to fish oil to get enriched diet. The reason for such enrichment is because of the presence of high levels of PUFA which also provides a better opportunity for large scale enrichment of *Artemia nauplii* and rotifers (Leger *et al.*, 1987; Sorgeloos and Leger, 1992). This view and methodology has been supported by Bengston *et al.* (1991) who have also proved that the highest enrichment levels in *Artemia* as well as *Brachionus* could be obtained by using emulsified concentrations.

The cuttle fish liver oil improved the larval development in the present study which is consistent with the results of other decapods larvae. Bengston *et al.* (1991) showed that lipid – free residue of the short necked clam *Tapes philippinarium* did not promote growth in larvae of the shrimp *P. japonicas*, but the supplement of Pollack liver oil supported development. The results of present study prove that the enrichment of live feed improves both survival and shortening of larval development duration in the two crab species *viz.*, *P. pelagicus* and *P. sanguinolentus.*

Acknowledgments

The finical support provided by UGC is greatly acknowledged.

References

Bengston, D.A., P. Leger and P. Sorgeloos, 1991. Use of *Artemia* as a food source for aquaculture. In: *Artemia biology*. R.A. Browne, P. Sorgeloos and C.N.A. Trotman (Eds), Pp 255-285., CRC Press, Boca Raton, Florida, U.S.A.

Folch, J., M. Lees and G.H. Solane-Stanley, 1956. A simple method for the isolation and purification of total lipids from the animal tissues. *J. Biol.Chem.,* 226: 497-509.

Frank, J.R., S.D. Sulkin and R.P. Morgan, 1975. Biochemical changes during larval development of the Xanthid crab *Rhithropanopeus harrisii* I. Protein, total lipid, alkaline phosphates and glutamic oxaloacetic transaminase. *Mar. Biol.,* 32: 105-111.

Kannupandi. T, A. Veera Ravi and P. Soundarapandian, 2003. Efficacy of enriched diets on the larval development and survival of an edible crab, *Charybdis lucifera* (Fabricius). *Indian J. Fish.,* 50(1): 21-23.

Leger, Ph., E.N. Foucquaert and P. Sorgeloos, 1987. International study on the *Artemia* XXXV. Techniques to manipulate the fatty acid effectiveness for the marine crustacean *Mysdopsis bahia* (M). In: *Artemia research and its applications*, Vol. 3. P. Sorgeloos, D.A. Bengston, W. Decleir and E. Josepers, (Eds.). Universa Press, Wetteren, Belgium, Pp 411-424.

Leger, Ph., D.A. Bengston, K.L. Simpson and P. Sorgeloos, 1986. The use and nutritional value of *Artemia* as a source. *Ocenogr. Mar. Biol. Annual Review*, 24: 521-623.

Levin, D.M. and S.D. Sulkin, 1984. Nutritional significance of long chain polyunsatured fatty acids to the zoeal development of the brachyuran crab, *Eurypanopeus depressus* (Smith). *J. Exp. Mar. Biol. Ecol.,* 81: 211-223.

Sorgeloos, P. and Ph. Leger, 1992. Improved larviculture outputs of marine fish, shrimp and prawn. *J. World Aquacult. Soc.*, 23: 251-264.

Sulkin, S.D., 1978. Nutrition requirements during larval development of the Portunid crab, *Callinectus sapidus* Rathbun. *J. exp. Mar. Biol. Ecol.*, 34: 29-41.

Sulkin, S.D. and K. Norman, 1976. A comparison of two diets in the laboratory culture of the zoeal stages of the brachyuran crabs *Rhithropanopeus harrisii* and *Neopanope* sp. *Helgol. Wiss. Meeresunters.*, 28: 183-190.

Sulkin, S.D., and W.F. Van Heukelem, 1980. Ecological and evolutionary significance of nutritional flexibility in planktonic larvae of the deep sea red crab *Geryon guinquedens* and the stone crab *Menippe mercenaria*. *Mar. Ecol. Prog. Ser.*, 2: 91-95.

Watanabe, T. 1993. Importance of docosahexaenoiic acid in marine larval fish. *J. World Aquacult. Soc.*, 24: 2.

Chapter 10

Estimation of Activity of Some Metabolic Enzymes in the Serum of Fish *Cyprinus carpio* as a Function of their Growth in Juveniles

K. L. Jain, Meenakshi Jindal* and Simmi

Laboratory of Fisheries, Department of Zoology and Aquaculture, C.C.S. Haryana Agricultural University, Hisar – 125 004, India

ABSTRACT

Varying age samples of fish *Cyprinus carpio* were collected to study the activity of some metabolic enzymes in the serum of fish as a function of their growth in juveniles. Growth rate in terms of mass and length gain were found lowest in 40 days old fish (6.4 per cent), but highest in 130 days old fish (16.6 per cent). The activity of succinate dehydrogenase (SDH) and glutamate dehydrogenase (GDH) in serum increased from 17.5 per cent to 106.3 per cent and 11.3 per cent to 41.1 per cent, respectively, in between 70 days old fish to 130 days of growth. The pyruvate dehydrogenase (PDH) activity, however, remained same from 70 to 100 days. All the variables examined, showed a positive correlation with growth rate. Condition factor showed the highest variability with GDH in serum (R^2=0.631). Activities of acid and alkaline phosphatase increased

* Corresponding author: E-mail: meenjind@gmail.com

significantly from 40 days to 100 days. Acid and alkaline phosphatase enzyme activities were strongly correlated with growth in length as well as growth in mass. The levels of dehydrogenases in serum reflect metabolic changes in fish liver on account of not only their growth differences, but also due to the environmental stress, which is counter acted by their energy budget, hence measuring activities of these enzymes could be an dependable estimate of fish age or the stress.

Keywords: *Growth rate, SDH, GDH, PDH, Metabolic enzymes.*

Introduction

Fisheries are most rewarding activity in today's time of socio-economic developments, particularly in developing countries, including India. Development of fisheries in inland water bodies is, however, being increasingly impeded due to various environmental constraints, coupled with residual wastes of inorganic fertilizers and other toxic chemicals such as heavy metals, pesticides, petroleum hydrocarbon, detergents and pulp mills effluents (Ghosh *et al.*, 2000). More seriously, contaminated water destroys aquatic life and reduces its reproductive ability, disturb enzymatic activities in different tissues and produce various physiological changes in aquatic organism including fish (Jain and Mittal, 2004). In fact, deterioration of aquatic ecosystem due to such toxicant has consequent effect on age as well as the early growth stages of fishes. Martinez *et al.* (1999) emphasized upon fish size or growth rate as an important factor determining metabolic enzymes activities.

A strong positive correlation between growth rate and the activity of some glycolytic enzymes like lactate dehydrogenase (LDH), pyruvate kinase (PK) and phosphofructokinase (PFK) was shown by Pelletier *et al.*(1993) in white muscle of Atlantic cod (*Gadus morhua*). In order to assess accurate prediction of various test parameters and to evaluate age dependent measures present studies were aimed to investigate alterations in certain serum metabolic enzymes involved in glycolytic cycle and tissue metabolism in early growth stages of *Cyprinus carpio*. The early growth stages of fish are generally employed in testing animals for various toxicological studies.

Materials and Methods

Analysis of Serum Enzyme Activity

The fish samples of varying age from 40 days to 130 days old (5-10 cm in length) were collected in the months of June to October from a local fish farm around Hisar district. Their length and weight were measured to determine their condition factor and blood was removed in Eppendorf tube. Serum was separated and kept at 4°C for 30 minutes. It was centrifuged at 3000rpm for 10 minutes. The supernatant was removed and used as stock solution for enzymes activity.

The analysis of the activity of the enzymes Succinate dehydrogenase (SDH), Pyruvate deydrogenase (PDH) and Glutamate dehydrogenase (GDH) were carried out following the methods given by Nachlas *et al.* (1960). Reaction mixture was prepared by adding 0.5 ml of sodium succinate, sodium pyruvate and sodium glutamate for SDH, GDH and PDH, respectively. It was treated sequently with phosphate buffer, gelatin and then the respective enzyme and PMS. Absorbance was recorded within 30-60 minutes at 540nm. The blank contained all the reagents except the substrate, which was replaced by sodium fumarate. Varying amounts of INT (0.04 to 0.24 mg) were used for preparing standard curve. DNPH was used as reducing agent. The selected dehydrogenase enzyme activity was expressed in terms of µmoles of formazan formed/mg protein/h.

Assay of phosphatases was carried out by the method of Jones (1969), using p-nitro phenyl phosphate as a substrate, p-nitro phenol and nitro phenol are colorless at acidic or neutral pH, but at alkaline pH of 11 it gives yellowish color with absorbance maxima at 410 nm. Mixture of enzyme extract and substrate solution were incubated and made alkaline NaOH to stop the reaction and OD of yellow colored product *i.e.* *p*-nitro phenol was recorded at 410 nm. A standard curve was prepared using graded concentrations of *p*-nitrophenol phosphate. Enzyme activity was expressed as µmoles of *p*-nitro phenol released per unit time per mg of protein.

Results and Discussion

Age and Growth Parameters

Data regarding length and weight measurements and the

growth rate in terms of percent length gain/day and mass gain/day as a function of growth period is shown in Table 10.1. Maximum growth increment in terms of gain in length was evident in 130 days old fish *i.e.*10.85, as compared to the earlier growth stages. The 40 days old fish showed percent length gain of 7.0 per cent only. Likewise, growth rate in terms of mass gain was also lowest in 40 days old fish (6.4 per cent), but highest in 130 days old fish (16.6 per cent).

Growth in mass, infect is likely to influence metabolic capacity when growing fish restores protein stores including some soluble enzymes, but enzymes which actually increase as a result of growth may not be directly involved in the growth process, although their increase activities may facilitate future growth. The later period of growth is primarily attributed to gain in mass as explained above and requires significant increase in the absorbance of nutrients. Couture *et al.* (1988) stated that muscle LDH stores are depleted in periods of fasting and than restored when food is available and that the resulting high muscles LDH activities favour further growth by enhancing swimming capacity.

Serum Dehydrogenase Enzyme Activity

In serum, SDH showed a rapid increase from 70 days old fish to 100 days old fish *i.e.* 0.242 µg of formazan formed/mg of protein/h in 70 days and 0.346 µg of formazan formed/mg of protein/h in100 days (Table 10.2). But the activity of GDH in serum increased from 11.3 per cent to 41.1 per cent in between 70 days old fish to 130 days of growth. The PDH activity, however, remained same from 70 to 100 days. The mean values were 0.209 µg of formazan formed/mg of protein/h in 40 days and 0.329 µg of formazan formed/mg of protein/h in 130 days old fish, with an increase up to 19.6 per cent in 70 days and 57.4 per cent in 130 days old fish. As per the regression equations (Table 10.3), SDH varied with growth in mass up to 92.6 per cent and in length up to 92 per cent, as compared to condition factor (R^2=0.553). PDH alone showed highest variability of 95.6 per cent with growth in length and 68.9 per cent variability with condition factor. All the variables examined, showed a positive correlation with growth rate. Condition factor showed the highest variability with GDH in serum (R^2=0.631).

Table 10.1: Growth Rate in Length, Mass and Condition Factor of the Different Age Groups of *C. carpio* in the Fish Pond Water

Age (Days)	*Initial Length (cm)*	*Final Length (cm)*	*Initial Mass (g)*	*Final Mass (g)*	*Growth Rate (% length gain day^{-1})*	*Growth Rate (% mass gain day^{-1})*	*Final Condition Factor*
40	4.9* (4.6–5.5)	10.3 (10.0–10.8)	14.0 (10.8–15.8)	27.6 (26.8–30.6)	7.0 (6.8–7.4)	6.4 (6.3–6.6)	9.2 (2.1–2.5)
70	10.3 (10.0–10.8)	14.8 (14.4–15.2)	27.6 (24.4–30.6)	50.9 (46.5–58.9)	8.5 (8.2–8.8)	9.3 (9.1–9.6)	6.1 (1.51–1.57)
100	14.8 (14.4–15.2)	20.3 (19.9–20.8)	50.9 (46.5–58.9)	138.4 (129.2–150.2)	10.3 (10–10.6)	15.1 (14.9–15.2)	6.8 (1.60–1.66)
130	20.3 (19.9–20.8)	25.1 (24.8–25.4)	138.4 (129.2–150.2)	203.9 (168.6–244.4)	10.85 (10.7–11)	16.6 (16.4–16.8)	5.0 (1.1–1.4)

* Values are the mean of 4 observations.

Table 10.2: Mean Values of the Age Related Changes in Succinate Dehydrogenase (SDH), Glutamate Dehydrogenase (GDH), Pyruvate Dehydrogenase (PDH) and the Phosphatase Enzymes in Serum of Fish *Cyprinus carpio*

Age (Days)	*SDH*	*GDH*	*PDH*	*Acid Phosphatase*	*Alkaline Phosphatase*
40	0.21	0.18	0.21	0.35	0.37
70	0.24* (17.5)	0.21 (11.3)	0.25 (19.6)	0.38 (9.3)	0.40 (8.4)
100	0.35 (67.9)	0.23 (24.8)	0.26 (23.9)	0.43 (20.9)	0.45 (21.6)
130	0.43 (106.3)	0.26 (41.08)	0.33 (57.4)	0.46 (30.2)	0.52 (40.1)
CD (5%)	0.21	0.21	0.21	0.01	0.01
F-cal	910.25	1430.60	2475.08	161.33	696.92

* SDH, GDH and PDH are expressed as μg of formazan formed/mg of protein/h, and the phosphatases as μmole of Pi released/mg of protein/h.

Values in parentheses are percent increase in enzyme activity in serum over the 40 days growth stage.

Table 10.3: Linear Regression Between the Activity (units/g wet tissue) of Dehydrogenase and Phosphatase Enzymes and Growth in Length, Mass and Condition Factor in Serum of Fish *Cyprinus carpio* (Level of significance is p<0.0001)

Variables	*Growth in Length (% gain/day)*	*Growth in Mass (% gain/day)*	*Condition Factor*
A. Dehydrogenase			
SDH	Y=-0.21+0.05GL (R^2=0.92)*	Y=0.06+0.02GM (R^2=0.93)	Y=0.58-0.16CF (R^2=0.55)
GDH	Y=0.05+0.02GL R^2=0.91)	Y=0.14+0.01GM (R^2=0.87,)	Y=0.32-0.06CF (R^2=0.63)
PDH	Y=0.001+0.029GL (R^2=0.95)	Y=0.14+0.01GM (R^2=0.92)	Y=0.43-0.09CF (R^2=0.68)
B. Phosphatase			
Acid phosphatase	Y=0.07+0.04GL (R^2=0.95)*	Y=0.07+0.04GM (R^2=0.95)	Y=0.55-0.08CF (R^2=0.63)
Alkaline phosphatase	Y=0.11+0.04GL (R^2=0.87)	Y=0.11+0.01GM (R^2=0.88)	Y=0.62-0.12CF (R^2=0.60)

These studies indicated a change of 106.3, 41.1 and 57.4 per cent, respectively in serum SDH, PDH and GDH enzyme activity. Growth in mass, infect is likely to influence metabolic capacity when growing fish restores protein stores including some soluble enzymes, but enzymes which actually increase as a result of growth may not be directly involved in the growth process, although their increased activities may facilitate future growth.

Serum Phosphatase Enzyme Activity

Activities of acid and alkaline phosphatases were found increased significantly from 40 days to 100 days (Table 10.2). Mean values varied from 0.354 μmole of Pi released/mg of protein/h in 40 days old fish to 0.428 μmole of Pi released/mg of protein/h in 100 days in acid phosphatase and 0.369 μmole of Pi released/mg of protein/h in 40 days to 0.449 μmole of Pi released/mg of protein/h in 100 days in alkaline phosphatase enzyme with an increase of 8.4 per cent to 21.6 per cent. Acid and alkaline phosphatase enzyme activities were strongly correlated with growth in length as well as growth in mass (R^2= 0.952 and 0.879, respectively, Table 10.3) in *C. carpio.* Acid phosphatase alone explained 95 per cent of the variability of growth rate in length and 96 per cent in mass growth. Condition factor showed the highest variability with PDH *i.e.* 69 per cent and with alkaline phosphatase it only explained 60 per cent of the variability.

The Acid and Alkaline phosphatases are groups of enzymes that hydrolysed phosphatemonoesters in the relatively non specific manner with optimal activity in acidic and alkaline enzyme. The increase in the activities of these enzymes with fish age in liver as evident in this study might be associated with the increasing metabolic activities of the liver such as the expenditure activities to meet the high energy need, with a maximum increase after 100 days of growth. The later period of growth is primarily attributed to gain in mass as explained above and requires significant increase in the absorbance of nutrients. Heterogeneity of vascular alkaline phosphatase activities were also correlated with the animal activities by various early workers (Kwan and Ito, 1987).

Ellsaesser and Clem (1987) recorded differences in serum electrolytes, metabolites and enzymes in laboratory acclimated and field pond fish, as a measure to establish accurate values of various serum components for channel catfish, in order to detect stress related changes. They also reported differences in the pond fish collected in the summer and those collected in the winter and these differences were attributed to dietary differences and increase in serum LDH activity is set to indicate depletion of muscles glycogen on account of stress or starvation.

Sullivan and Somero (1983) recorded activities in muscles of the glycolytic enzyme LDH and PK. These activities were increased significantly with increasing body size in sable fish, *Anoplopolma fimbria*. Sharma (1999) has also reported that the treatment of the fish with carbaryl further led to a marked increase in the activities of transaminases (GOT and GPT), phosphatases (acid and alkaline) and lactate dehydrogenase in the fish serum. The magnitude of the effect being dependent on the pesticide concentration and duration of exposure. The increase in lactic acid concentration with subsequent decrease in glucose concentration indicates an enhanced rate of glycolysis due to pesticide stress.

Changes in enzymatic activities as evident in this study could be well correlated with metabolic requirement of the fish as per the growth with age. The relationship between growth rate and the glycolytic enzyme activities such as LDH, PK (pyruvate kinase) and PFK (phosphofructokinase) was further analysed by Pelletier *et al.* (1993) who have shown a strong positive correlation in the white muscle of Atlantic cod.

The activity of acid phosphatase (AP) in liver macrophage aggregates (MA-AP) of different fish species was used as a marker for a pollution-induced modulation of the digestive capacity of phagocytes (Katja, 2003). A significant increase in the phosphatase activity (both acid and alkaline) was also reported in liver of an air breathing fish, *Channa punctatus* by Hota and Pradhan (1994), on account of exposure to sub lethal concentration of mercuric chloride (0.2ppm) for 24, 48, 72 and 96. The increase and decrease in the activity of metabolic enzymes is a clear evidence that changes in the metabolic environment of the fish may be due to age or stress as both these parameters influence dietary intake.

Sastry and Sharma (1980) observed increase in the level of serum enzymes namely glutamate-oxaloacetate transaminase and glutamate-pyruvate transaminase after acute exposure to mercury in *Channa punctatus*. Increased activities of these enzymes may be due to the damage of liver. Age and growth related changes in liver composition and metabolic enzymes activities are interlinked with toxicity in various water resources. Physiological parameters tell us how the various enzymes activities changes in different tissues of vertebrates as according to their age and growth.

Gill *et al*. (1991) observed heterogenetity in the response of alkaline phosphatase in various organs in the freshwater sybarite fish on exposure to Cadmium. They demonstrated that alkaline phosphatase was unaffected in liver and gills, stimulation in kidney and ovary and inhibited in the gill. The increase in the activities of the Acid and Alkaline phosphatase might be associated with the increasing metabolic activities of the liver such as the expenditure activities to meet the high energy need after 100 days of growth.

Heterogeneity of vascular alkaline phosphatase activities were also correlated with the animal activities by various early workers (Kwan and Ito, 1987). Cadmium was also said to show more changes in the alkaline phosphatase, succinate dehydrogenase, glyceraldehydes phosphate dehydrogenase in several tissues.

The levels of dehydrogenases in serum reflect metabolic changes in fish liver on account of not only their growth differences but also due to the environmental stress, which is counter acted by their energy budget, hence measuring activities of these enzymes could be a dependable estimate of fish age or the stress.

References

Couture, P., Dutil, J. D. and Guderley, H. 1998. Biochemical correlates of growth and condition in juvenile Atlantic cod (*Gadus moruha*) from Newfoundland. *Canadian Journal of Fish Aquatic Science*, 55: 1591-1599.

Ellsaesser, C. F. and Clam, L. W. 1987. Blood serum chemistry measurements of normal and acutely stressed Channel catfish. *Comparative Biochemical Physiology*, 88(3): 589-594.

Ghosh, S., Das, A. K. and Vass, K. K. 2000. DDT, HCH and endosulfan residues in the lower stretch of river Ganga. *Geobioscience*, 27: 161-164.

Gill, T., Tewari, H. and Pande, J. 1991. *In vivo* and *in vitro* effect of cadmium on selected enzymes in different organ of the fish *Barbus conchonius* Ham. (Rosy barb). *Comparative Biochemical Physiology*,100C,501.

Hota, A. K. and Pradhan, A. K. 1994. Mercuric chloride toxicity on a freshwater fish, *Channa punctatus*: effect on liver phosphatase activity. *Proceedings of Academic Environmental Biology*, 3(1): 87-91.

Jain, K. L. and Mittal, V. 2004. Heavy metal pollution in surface water bodies and its impact on fishes. Proceedings of the National Workshop on "Rational use of water resources for aquaculture", CCS HAU, Hisar.

Jones, K. C. 1969. Similarities between gibberellins and related compounds in inducing acid phosphatase and reducing sugar released from barley endosperm. *Plant Physiology*, 44: 1695: 1700.

Katja, B. 2003. Acid phosphatase activity in liver macrophage aggregates as a marker for pollution-induced immunomodulation of the non-specific immune response in fish. *Helgoland Marine Research*, 57: 166-175.

Kwan, C. Y. and Ito, H. 1986. Comparative studies of acid and alkaline phosphatase activities in nonvascular smooth muscle membranes. *Comparative Biochemical Physiology*, 86 (3): 483-488.

Martinez, M., Couture, P. and Guderley, H. 1999. Temporal changes in tissue metabolic capacities of wild Atlantic cod (*Gadus moruha*) from Newfoundland. *Fishery Physiology and Biochemistry*, 20: 181-191.

Nachlas, M. N., Margulies, S. I. and Seligman, A. M. 1960. A clorimetric method for the estimation of succinate dehydrogenase activity. *Journal of Biological Chemistry*, 235: 499-503.

Pelletier, D., Gurderely, H. and Dutil, J. D. 1993. Effect of growth rate, temperature and season on glycolytic enzyme activities in white muscle of cod *Gadus morhua*. *Journal of Experimental Zoology*, 265: 477-487.

Sastry, K. V. and Sharma, S. K. 1978. The effect of in vivo exposure of endrin on the activities of acid, alkaline and glucose-6-

phosphatase in liver and kidney of *Opheocephalus* and *Channa punctatus*). *Bulletin of Environmental Contaminated Toxicology*, 20 (4): 456-60.

Sharma, B. 1999. Effect of carbaryl on some biochemical constituents of the blood and liver of *Clarias batrachus*, a fresh-water teleost. *Journal of Toxicological Science*, 24 (3): 157-64.

Sullivan, K. M. and Somero, G. N. 1983. Size- and diet-related variations in enzymatic activity and tissue composition in the sablefish, *Anoplopoma fimbria*. *Biological Bulletin* (Woods Hole), 164: 315-326.

Chapter 11

Use of Supplementary Feed for Sustainable Inland Aqua Farming: A Case Study

Basudev Mandal[1*], *Subasish Mukherjee*[1] *and Bimal Kinkar Chand*[2]

[1]*Department of Aquaculture Management and Technology, Vidyasagar University Midnapore – 721 102, West Bengal, India*

[2]*Directorate of Farm, Research and Extension, West Bengal University of Animal and Fishery Sciences, 68, Kshudiram Bose Sarani, Kolkata – 37, West Bengal, India*

ABSTRACT

Present study deals with the analysis of appropriate inclusion level of feed ingredients in fish diets, methods of feed processing and the economics of using such ingredients in aquaculture operations and also to survey the present form and availability of supplementary aqua feed in West Bengal. Such a study will ensure proper resource combination to improve fish production and thereby to increase the profit. The development of supplementary feeding should, therefore, aim at establishing proper feeding rates and frequencies for maximum efficiency under changing conditions. The outcome

* Corresponding author: E-mail: bdmandal@yahoo.co.in; basudevmondal@vidyasagar.ac.in

feedback based on questionnaires from different groups of people who are directly or indirectly involved in fish culture, revealed that fish production is increased in the polyculture techniques through the use of supplementary feeds.

Keywords: *Supplementary feed, Polyculture, Feed ingredients, Cost benefit.*

Introduction

The aquaculture industry in India, in particular, is embarking on an increasing phase of intensification, driven by multitude of economic forces among which is an increasing need for even the small scale farmers to transform from subsistence aquaculture to a revenue generating enterprise. In the light of increasing competition for primary resources, such as water and land, such economic needs become even more demanding. The increasing intensification of the industry at large however, has to take place within the frame work of certain potential global constraints on the industry such as "fish meal trap" (Wijkstrom and New, 1989). In such a scenario issues related to supplementary feeding are likely to come to the forefront.

In reality, the role of supplementary feeds in the nutrition of the cultured organism is much more intricate and is closely interlinked to the dynamics of the pond ecosystem. In semi-intensive culture there are complex interactions between the natural food organisms and supplementary feeding practices. Farmers usually select a diet on the basis of cost and perceived quality with additional inputs from extras such as feed company backup, computer feed programs etc. The only comparison a farmer can make between diets is historical–that is he can compare a new one with one he used previously by analysis of his farm records (growth rates, FCRs, etc.). When a new diet becomes available, the benefits from it must be obvious in terms of improved growth, FCRs or product quality. If a new diet is more expensive than existing ones, there must be an accompanying decrease in the FCR if feed cost per unit of fish produced is not to go up. Naturally those involved in fish feed formulation need to know the nutritional requirements of a particular stock species and the means by which these can best be met to give maximum growth in a cost-effective manner. The objective of the study is to analyze the satisfaction level of fish production using supplementary feed in West Bengal.

Supplementary Feeding in India

The suitable application of artificial diets along with natural food where the availability of natural food becomes a limiting factor is known as supplementary feed (Jhingran,1997).The traditionally used supplementary feed for carps consists of mixture of rice bran and one of locally available oil cake in the ratio of (1:1). Mitra and Mahapatra (1956) conducted experiments on major carp fry with different quantities of zooplankton; phytoplankton, cow dung and oil cake and found that survival was highest with oil cake and lowest with cow dung. Tripathi *et al.* (1979) obtained a high survival (over 80 per cent)at spawn using a diet comprising a mixture of fish meal, ground nut cake and rice bran in the ratio of (1:1:1). Patra *et al.* (2005) evaluated *Nymphoides cristatum* as a feed for young *Labeo rohita*, and got very much promising results and recommended for in corporation in the diet of *Labeo rohita*. Mandal *et al.* (2005, 2008 and 2009) observed that artificial feed is the only food source of Indian major carps reared in abandoned china clay mines of West Bengal.

A review of available literature on this aspect of fish culture seems to indicate that almost all the work done has been concentrated on feeding the fry at the hatchery or nursery stage. Thereafter, availability of food for the growing fry, fingerlings and yearlings is left to the natural food production of the waters. The obvious reasons for this omission of artificial feeding appear to be the consideration of economy in undertaking this operation in larger waters, as well as lack of sufficient information on suitable supplementary feeds for the various cultured fishes. The common items of feed used in India in nursery and rearing ponds are various types of oil cakes mixed with rice bran. With the recent initiation of intensive fish culture programs in the various states of India, it was felt necessary to develop a more suitable and balanced diet for cultured fish. With this in view, preliminary work commenced in this Research Institute, the main aim of which was to enhance fish production within a minimum rearing period.

While the present work is limited to the observations on survival and growth of fish at the fry stage, it is proposed to continue the investigations to evolve suitable, cheap and effective artificial feed for fingerlings as well as for adult fish. While taking note of the inherent handicaps involved in feeding fish, the authors feel that along with the development of a suitable feed, it may be necessary to evolve and improve the feeding technique in order to derive maximum benefits from artificial feeding. Though work on artificial feeding has been carried out by many other workers, emphasis has been on finding out the vitamin, mineral and other food requirements. The present work is mainly concerned with evolving a practical nursery diet for the fry of Indian carps.

Materials and Methods

The study was conducted in the Midnapore district of West Bengal in India. Though fishing is not a predominant occupation in the district but still it occupies an important place in the district economy. Most of the labour force is engaged in fishing and its allied activities in Paschim and Purba Midnapore. The capture and culture fisheries of Indian major carp and shrimp farming are growing widely in the district.

The study is essentially based on primary data and of both descriptive and analytical data analyses. The study has been carried out at several villages of undivided Midnapore district in West Bengal. To arrive at the six sample villages- namely Norghat, Bhandbhara and Geonkhali in Purba Midnapore and Chaungal, Sirshi, and Rampura are selected and from each village 10 farmers have been selected and were selected by applying proportionate random sampling technique. The different major and minor aqua feed traders have also been interviewed.

To accomplish the stated objectives, a comprehensive questionnaire has been canvassed to the members. Through extensive interface and formal and informal interviews, additional information has been gathered and opinion elicited. Data collection was done using pre tested interview schedule. Two types of schedules were prepared *i.e.* one for the fishermen in local language and the other for traders. The independent variables of the study included age, education, occupation, annual income, family type, experience, social participation, communication asset, innovativeness, scientific orientation and the dependent variables are about the type of feed they use and the culture method. The format of questionnaire is given below. The data generated has been processed and tabulated through statistical tools such as standard deviation with the help of certain software like Ms Excel and in part manually.

Figure 11.1

Questionnaire for Traders

Country________Region________Ingredients________Common name________Generic name________Classification________

Protein_______Energy_______Rhoguage_______Other________

Composition

Average proximate analysis (per cent): Moisture____________

Fibre_________Ca_________Protein_________NFE_________

P_________Lipid_________Ash_________NaCl_________

Toxic factor and growth inhibitor_________

Process

Solvent extractor_________Machine dried_________

Milled_________Expeller_________Sun dried_________

Rolled_________Other_________

Quality

Grade scale: 1 2 3 4 5 Uniformity_________

Major contaminant(s)...........

Usage

Rank:________; Human_________Poultry_________

Livestock_________Fish_________

Methods

Scratch_________Farm mix_________Commercial mill_________

Availability

Harvest Year round availability

Local_________

Regional_________

National_________

Import________

Price: Rs/Kg________________Perishability________________

Packing

Sacked____Jute____Paper____Bulk ____Paper laminated____

Others____

Questionnaires for Farmers

Dear Respondent,

We are conducting a survey of aquatic feed market. We would be grateful if you could answer the following questionnaire in this regard.

(1) General information

*Name*__________________________

Age (___) less than 18 years (___) 18 – 25 years (___) 25 – 35 years

Village______District______Family number______Adult_____

Children: Below 5 (___) Above 5 (___) Educational qualification____

Profession: Govt. Service/Private Service/Student/Business/Any other.

Family income: (___) Less than Rs.36, 000 pa. (___) Rs.36, 000 to 72,000 pa. (___) Above Rs.72, 000 pa.

Household articles: TV/Freeze/Mobile/Motor cycle/Car

(2) Resources

Agricultural land______ha Crop______Fishery land______ha Live stock______

(3) Culture Details

Name of the culture species:__________________

Type of culture: Intensive/semi intensive/super intensive/extensive

Total Number of ponds______

Number of large ponds______/Number of small Ponds______

Pond preparation schedule: followed/not followed

Source of water supply________________________

Stocking material: auto stocking/selective stocking.

Seed source: wild collection/Hatchery seed (PCR tested)

Cost of the seed purchased______Stocking type: One time/in batches

Stock rate______No./ha Seed size:________________________

Culture Duration: 3 Month/4 Month/5 Month/6 Month

Harvesting: 1 Time/Multiple time, Average harvest size: 30 gm/40 gm/50 gm

Total harvest______Sale Price:__________________/kg

Selling type: Farm gate/Agent

(4) About Feed

Do you use supplementary feed? (___) Yes, (___) No

What type of feed: Commercial feed/Farm made feed/Local ingredient

If commercial feed which brand are you using at present and why ?

Please name a few commercial feed you have hard off

(1)_________(2)_________(3)_________(4)_________

FCR of the particular feed you are using:_______________

Purchase type: Cash/Credit. If Credit: By back/after harvest

What type of feed you are using: Pellet/Mesh

Is it species specific: tiger feed/scampi feed/general

In commercial feed you are using: Starter/Grower/Finisher

Level of use: Through out culture period/during limited period nearly to harvest

Technical service available from the company: Yes (___), No (___)

If yes then only about feed/total including water quality management and disease.

Feed purchase: From manufacturer/From Agent/Whole seller/From retailer.

Feed application: Broad casting (hand/well defined pan)/Tray/ Broad casting with check tray.

Feeding time: 5 pm/2 pm/5 am/8 am.

Feeding rate:_____body weight per cent/day, or as per the chart given by the manufacturer or own calculationor based upon feed up take.

Feed frequency:______Feed price:______Rs/kg

Package type: 10 kg/25 kg/50 kg/100 kg

Type of bag: Jute/Plastic/Paper laminated/Inner polythene present or absent

Proximate composition mentioned: Yes (___), No (___)

If yes on the bag/inside

Expiry date: absent/given (after 3 month/6 month/9 month/12 month)

Feed sample collected: Yes (___), No (___)

Thank you

Results

Tabulation of Data

The information obtained by interviewing the 80 farmers is tabulated below:

N = Total number of farmers interviewed.

Personal Details

Age	*N=80*
Less than 18 yrs	Nil
18 – 25 yrs	19
25 – 35 yrs	52
Above 35 yrs	9
Qualification	*N = 80*
> 10th standard	48
10+2	28
Graduate	3
Higher education	1
Profession	*N = 80*
Student	Nil
Government service	Nil
Private service	13
Business	67
Any other	Nil
Economic Condition	*N = 80*
> Rs36,000 pa	14
Rs36,000 – 72,000 pa	32
Above Rs72,000 pa	26
BPL card holders	8
Pond Size	*N = 80*
< 0.4 ha	22
0.4 – 1 ha	30
> 1.0 ha	28

Type of Species Cultured	*N = 80*
Fin fish	Carp culture = 40
Shell fish	Shrimp = 30
Fish cum other	10

Farmers Dependent Upon Culture of Shrimp

There are 30 farmers was depend upon shrimp farming

Type of Species Cultured	*N = 30*
Penaeus monodon	28
Macrobrachium rosenbargii	2
Pond size	*N = 30*
< 0.4 ha	Nil
0.4 – 1 ha	9
> 1.0 ha	21
Source of water	*N = 30*
Ground water	8
Through creek water	22
mixed	Nil
Stocking time	*N = 30*
1 time stocking and 1 time harvesting.	26
Multiple stocking and 1 time harvesting	2
Multiple stocking and multiple harvesting	Nil
1 time stocking and multiple harvesting	2
Seed Stocking	*N = 30*
Nursery management by stocking in happa for 4 days	4
Stocking in Nursery Pond	*13*
Prior to stocking	Nil
Direct stocking	13

Source of Seed	*N = 30*
Auto stocking	2
Selective stocking with natural seed	Nil
Selective stocking with hatchery seed	28
Type of Hatchery Seed	*N = 30*
Certified seeds(PCR tested)	28
Non certified seed	2
Stocking Density	*2,75,000*
Seed size	PL- 16
Price of seed	0.5 p
Pattern on Feed Type Used	*N = 30*
No supplementary feed	Nil
Local ingredients	3
Farm made feed	Nil
Commercial pellet feed	27
Procurement of Feed	*N = 30*
Directly from the company	Nil
From the whole sale dealer	30
From retailer	Nil
Mode of Payment	*N = 30*
Cash payment	18
Credit payment	Nil
Cash and credit	8
Buy- back	4
Mode of Packaging of Feed	*N = 30*
25 kg bag	26
50 kg bag	Nil
> 50 kg bag	Nil
< 25 kg bag	4
Pattern on Feed Application	*N = 30*
100 per cent broad casting	8
broad casting with check tray	22
Tray feeding	Nil

Feeding Frequency	*N = 30*
Once in a day	Nil
Twice in a day	Nil
Thrice in a day	Nil
More than 3 times in a day	30
Feeding Time	*N = 30*
Only during night time	Nil
During day and night time (more during night)	23
Day and night equal	7
Day and night(more during day time)	Nil
Only during day	Nil
Harvest Size	*N = 30*
> 40gm	4
30 – 40gm	26
< 30 gm	Nil
Size of Prawn Harvest	*N = 30*
Almost uniform size (80 per cent similar size)	21
Un uniform size	9
Harvesting Method	*N = 30*
Total harvest by draining water	2
Harvest using net (cast net/drag net)	24
Harvest using trap	4
Type of Feed Bags	*N = 30*
Poly bag	4
Paper laminated poly bag	22
Jute/gunny bags	4
Post Harvesting Handling	*N = 30*
Chill kill	nil
Icing soon after harvest	17
Icing within 2 hrs after harvest	25
Sending to arrat without icing	8

Application of Urea in Relation to N

Available N in Soil (mg/100g soil)	*Urea to be Applied (Kg/ha)*
12.5	100
25	50
50	25

Type of Prawn Sold	*N = 30*
Selling whole prawn (shell on and head on)	7
Selling head less prawn	23
Selling cost	Rs/kg
Size of prawn	
> 40 gm	150
30 – 40 gm	220
< 30 gm	Nil

Mode of Selling	*N = 30*
Selling for processing at farm	7
Selling at local arrat	19
Selling through co-operatives	4

Method of feed calculation	*N = 30*
Based on feeding chart supplied by manufacturer	29
Based on sampling + feeding chart	Nil
Based on feed up take	1

Local ingredients	*Cost of Feed Rs/kg*	*Commercial Pellet Feed*	*Cost of Feed Rs/kg*
Fish meal	Rs20/kg	C.P FEED	Rs55
Soya meal	Rs40/kg	Godrej agrovet	Rs42
		Ms Lakshmi fertilizer	Rs36

Dose of Manure in Carbon Content Soil

Organic Carbon in Soil per cent	Raw Cow Dung (kg/ha)	Dry Chicken Manure (kg/ha)
1	500	175
0.5	1000	350
0.25	2000	700

Application of Super Phosphate in Relation to Phosphorus

Available P in Soil (mg/100 g soil)	Super Phosphate to be Applied (kg/ha)
1.5	100
3	50
6	25

While using inorganic fertilizers, care needed to be taken to avoid over fertilization. The fertilization dosage given may not hold good for all the waters. The best way to regulate the fertilization schedule is through monitoring the algal bloom conditions based on the colour or transparency of the water. Fertilizing the pond is still an art rather than a science.

Farmers Dependent on Culture of Carps

There are 50 farmers surveyed who depend on carp culture and their data are tabulated below:

N= total number.; Farmers using only rice bran: N= 8

Species	Energy Supplement (Rice Bran) Kg/ha/day	Initial Wt. (g)	Final Wt. Without Feed (g)	Final Wt. with Feed (g)	Stocking Density Fish/m^2	Growth Period
Common carp	5–8	13	330	360	0.10	5 month
Silver carp	5	8	250	290	1.50	5 month
Tilapia	1.5–2	13	73	85	2.00	2 month

Farmers using only rice bran+ ground nut oil cake: N= 23

Species	*Energy Supplement (Rice Bran+ Groundnut Oil Cake) Kg/ha/day*	*Initial Wt. (g)*	*Final Wt. Without Feed (g)*	*Final Wt. with Feed (g)*	*Stocking Density Fish/m²*	*Growth Period*
Common carp	5–8	13	330	376	0.10	5 month
Silver carp	5	8	250	322	1.50	5 month
Tilapia	1.5–2	13	73	98	2.00	2 month

Farmers using only commercial feed N=15

Species	*Energy Supplement (Commercial Feed) Kg/ha/day*	*Age of Fish*	*Final Wt. Without Feed (g)*	*Final Wt. with Feed (g)*	*Stocking Density Fish/m²*	*Growth Period*
Common carp	5–8	0+	227	376	0.91	5 month
Silver carp	5	2+	1591	1620	1.50	5 month
Tilapia	1.5–2	1+	369	417	0.15	2 month

Farmers depend only on poultry manure: N= 4

Species	*Energy Supplement (Poultry manures) Kg/ha/day*	*Age of Fish*	*Final Wt. Without Feed (g)*	*Final Wt. with Feed (g)*	*Stocking Density Fish/m²*	*Growth Period*
Common carp	4	0+	227	398	0.91	5 month
Silver carp	5	2+	1591	1876	1.50	5 month
Tilapia	1.5–2	1+	369	425	0.15	2 month

Table shows the yields of fish and ducks (eggs and live weight for meat) from an integrated farming at Banapur, in Paschim

Midnapore, West Bengal, India using a 1.48-ha pond stocked with a total of 9,400 fish (6,340/ha) harvested after 9 and 12-mo growth periods (for Catla alone, a second stocking was made after the 9-mo harvest).

Fish Species	*Average Initial Wt (g)*	*Composition of Stock (%)*	*Growth Period (mo)*	*Average Wt at Harvest (g)*	*Total Wt at Harvest (kg)*	*Survival (%)*	*Contribution to Total Yield (%)*
Catla							
First stocking	43	10	9	904	801.150	96.3	19.2
Second stocking		18	3	498	429.300		
Rohu	28	18	9 12	413 633	346.500 488.650	90.6	13.1
Mrigal	23	28	9 12	495 513	305.800 677.300	72.6	15.4
Silver carp	5	15	9 12	1643 1920	632.000 1,300.300	94.2	30.2
Grass carp	5	10	12	1720	1,092.000	76.8	17.1
Common carp	5	19	12	206	161.200	53.1	2.5
Miscellaneous	–				163.100		2.5
Total yield					Kg 6,397.300		

No. of Duckling Raised	*Average Initial Weight (g)*	*Growth Period (mo)*	*Average Finished Weight (kg)*	*Total Finished Weight (kg)*	*No. of Eggs Produced*
100	250	12	2.500	250	1,835

Total Cost Details

Details of the input costs and returns for fish culture and duck rising are shown in the table. The cost of production for fish was Rs 1.61/kg. The fish were sold at a fixed government price of Rs 5.50/kg as against the prevailing farm gate price of Rs 8.00/kg by deliberate policy to make them available to local consumers. The eggs were also sold at a very low price.

Integration of fish culture with chicken farming has only recently been initiated in India. The results have been tabulated below with input costs and returns for a year's production of fish, duck eggs and ducks sold for meat from an integrated farming experiment at the Benapur Paschim Midnapore, West Bengal, India using a 1.48-ha pond stocked at 6,340 fish/ha (US$1.00 = Rs 8.00).

Input Costs	*Rate/ha*	*Quantity*	*Unit Cost or Value (Rs)*	*Total cost or Value (Rs)*
Mahua cake	250 ppm	6,400 kg	0.38/kg	2,432.00
Lime	250 kg	372.5 kg	0.30/kg	111.75
Fingerlings	6,340	9,400		2,506.75
Insecticide (BHC)	1 kg	1.5 kg	7.50/kg	11.25
Netting and other miscellaneous costs	–			997.51
Labor (security and fish handling)				2,761.54
Pond rental		1.48 ha	1,000.00/ha/yr	1,500.00
Total				10,320.80
Returns Fish sales		6,397 kg	5.50/kg	35,183.00

Ducks Input costs			
Ducklings	140	approx 9.00/head	1,272.00
Poultry feed	1,510 kg	1.00/kg	1,510.00
	Quantity	*Unit Cost or Value (Rs)*	*Total Cost or Value (Rs)*
Medication			97.00
Depreciation cost of floating duck			
House-contribution cost, Rs2,500			500,000
Life expectancy, 5 yr			
Total			3,379.00
Returns			
Egg sales	1,835	0.40/egg	734.00
Ducks for meat	250 kg (live weight)	10.00/kg	2,500.00
Total			3,234.00

Summary	*Rupees (Rs.)*
Total operational cost	= 13,699.80
Interest on working capital at 10 per cent	= 1,369.98
Total variable costs	= 15,069.78
Total returns	= 38,417.00
Net profit	= 23,347.72
Net profit as a per cent of total variable costs	= 155 per cent

Cost Estimation

Total estimated cost is 163100.00. Expected annual return is Rs.90000.00 from sale of 1500 kg fishes. Detailed analyses of the observations are given in the subsequent tables:

Cost Estimate

A)	**Investment cost**			
	Pond construction/0.28ha with 2.5 m average depth	1 no.	45/cu.m	94,500.00
	Sub Total			**94,500.00**
B)	**Cost of implements, equipment and machineries**			
	Hapa	2 nos.	200 each	400.00
	Drag nets	1 no.	5000 each	5,000.00
	Sub total			**5,400.00**
	Total investment cost			**99,900.00**
C)	**Operational cost**			
	Wages/salary	1 no.	2000/month	24,000.00
	Lime	590 kg	5/kg	2,950.00
	Fish seed	1550	5 each	7,750.00
	Feed	2350 kg	10/kg	23,500.00
	POL etc.	Ls	5,000.00	
	Total operational costs			**63,200.00**

Total data collected was 80 among which 30 were the shrimp farmers and the rest were the finfish farmers.

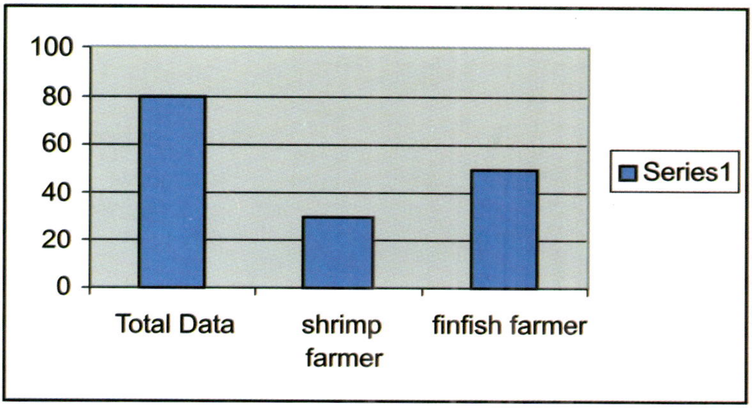

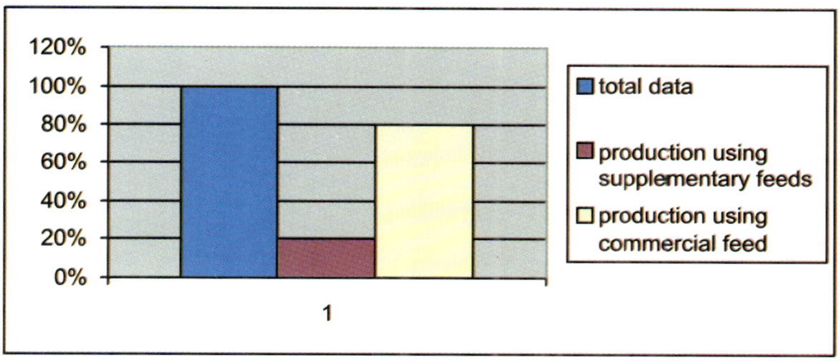

Figure 11.1: Shrimp Farmers Using the Type of Feed

Discussion

The present investigation indicated that the cultured species largely depend on the natural feed. To achieve sustained production, it is essential that supplementary additions of fertilizers should be resorted to. Both organic manures and inorganic fertilizers are generally used for the purpose. The dosage of organic manure to be applied is dependent on the organic carbon content of the soil.

The livestock-fish farming has opened up a new horizon of high animal protein production at very low cost. Apart from providing cheap protein-rich food, integrated farming has proved to be an efficient means of waste disposal and has allowed savings

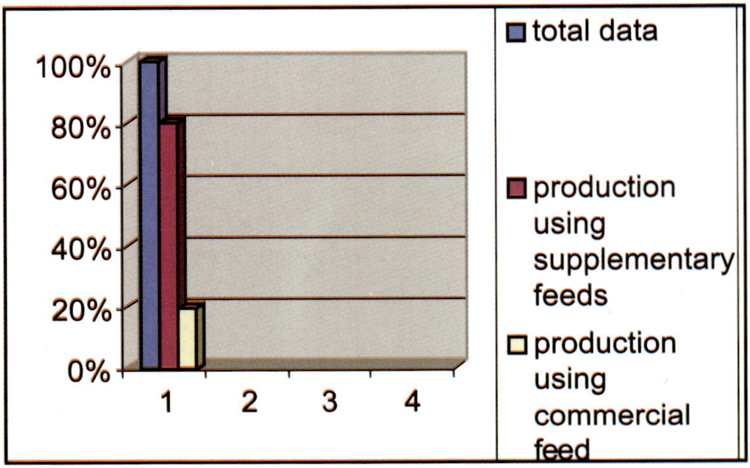

Figure 11.2: Finfish Farmers Using the Type of Feed

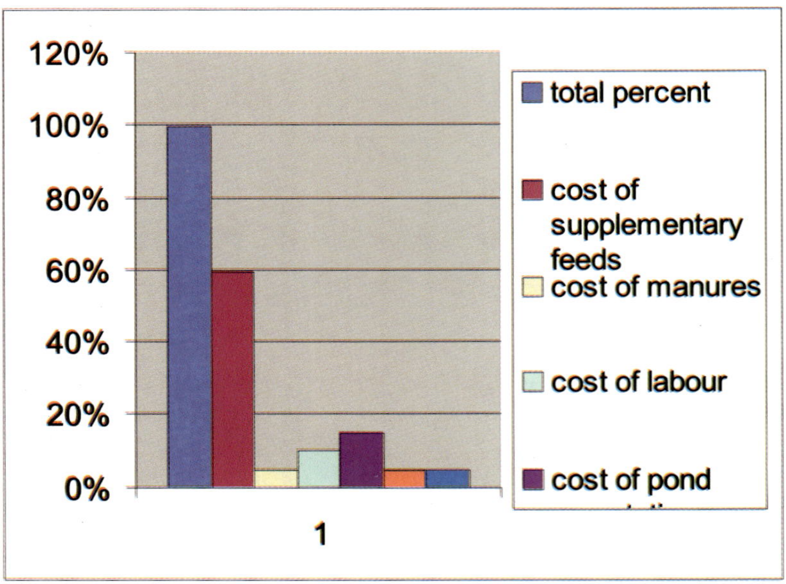

Figure 11.3: Cost of Production in Carp Culture

on the use of inorganic fertilizers and supplemental feeds in fish production.

The fish, besides consuming the large amounts of natural food produced in the pond, were also observed to feed directly (especially the bottom feeders) on duck manure. The fish yield of 4,323 kg/ha/yr from the duck-fish, without the use of any supplemental feed or inorganic fertilizers, is also high compared to the yields of 4,290 kg/ha/yr, obtained from the pond during 1973 and 1974 (Jhingran, 1997; Jhingran and Sharma, 1978). 3,543 kg/ha/yr in Eastern India (Anon, 1998), and 4,258 kg/ha/yr, West Bengal (Murshed *et al.*, 1998), which were achieved with heavy supplemental feeding and fertilization.

The raising of ducks on fish ponds fits very well with the ecological niche concept of polyculture. In conventional polyculture, the major niches are all occupied by various species of fish, except for the water surface niche which is more or less underutilized or unutilized. The ducks consume organisms such as tadpoles, mosquitoes, larvae of dragon flies and other insects, molluscs and aquatic weeds, and thus do not compete significantly with the fish for any food items commonly found in conventional polyculture of carp. Their manure fertilizes the pond, and their disturbance of the substrate, while feeding, helps to release nutrient from the soil and further increases fish production. By day, the ducks distribute their manure over the whole pond area.

The polyculture ponds described here for integrated farming since lack any aquatic vegetation and the ducks therefore depend mostly on feeding by man. The number of ducks it is recommended, should be kept to an easily manageable number, just adequate for sake of pond fertilization. In the present experiment, 100 ducks provided approximately 10,000 kg of manure for over 12 months. The results and experiences obtained suggest that 100 to 150 ducks can give adequate fertilization of 1 ha of water. This is comparable to the recommendation of Behrendt (2001): *i.e.* 300/ha average and 100/ha where natural food is limited.

Ducks are likely to eat small fish and should be excluded from nursery ponds. Fingerlings of 10 cm or over are recommended for stocking in duck-fish systems. Integrated livestock-fish farming not only increases fish production but also cuts down the cost of fish culture operations considerably. The average cost of production in conventional polyculture with supplemental feeding and inorganic fertilization was Rs 2.93/kg in Eastern India (Anon, 1976). Murshed *et al.* (1998) have also recorded Rs2.67/kg as the cost of fish production by conventional methods in District Nadia, West Bengal. The present costs of production of Rs1.07/kg (pig-fish) and Rs1.61/kg (duck-fish) are much lower.

The annual profits made from duck farming are not very impressive, but the total income from an integrated farming system must be considered to assess economic viability. The waste materials, *i.e.* duck droppings, act as a substitute for supplemental fish feed and fertilizers, which in conventional fish culture account for 58.6 per cent of the total input cost (Anon, 1998). The gap between the demand and supply of inorganic fertilizers is increasing every day due to intensive cropping of the high yielding varieties of cereals. The recycling of organic wastes, through integrated farming systems, can help solve this problem. The cattle fodder, vegetables, and fruit crops grow on the terraced embankments of the pond, which are not normally utilized in the fish culture operations, and it provided fodder for the grass carp and pigs and, therefore, extra income to the farmer. The organic detritus removed from the pond bottom can also serve as an efficient fertilizer for growing land crops. This opens up possibilities of combining horticulture and fish culture.

While integration of livestock farming with fish culture gives high yields at low cost, the systems require effective management. One of the problems is the difficulty in combining and balancing the expertise needed in fish and animal husbandry. Over concentration on one may work to the detriment of the other. Monitoring the DO and BOD of the pond water is absolutely essential as a thick organic sediment settles at the pond bottom, which hastens the depletion of dissolved oxygen and enhances the production of toxic gases, and can result in fish kills. The application of manure should be regulated according to the pond water DO. The pond should also be desalted when a significant amount of detritus has accumulated at the bottom.

Animal excreta are a potential source of infection for various parasites and diseases. Studies are being made to investigate the possible human health hazards from integrated livestock-fish farming systems. Obviously, the livestock should be maintained in good health and under hygienic conditions. Since food and feeding costs generally constitute the highest operating cost item of semi-intensive fish and shrimp grow-out farming operations, it is essential that the feed is formulated and presented in such a manner so as to provide maximum production efficiency at a minimum cost. Commercial fish and shrimp farmers are concerned therefore with converting fish or shrimp feed into fish or shrimp flesh as quickly and as efficiently as possible.

However, the relative importance of growth rate and feed conversion efficiency or survival (in the case of larvae/juveniles) will depend upon the cost of the feed in relation to the market value of the farmed product. When fish/shrimp stocking density and standing crop is such that the natural productivity of the water body alone cannot sustain adequate fish or shrimp growth, an exogenous supplementary diet can also be fed as a direct 'supplementary' source of dietary nutrients for the cultured fish or shrimp; the dietary nutrient requirements of the cultured fish or shrimp species being supplied by a combination of natural live food organisms and supplementary diet feeding. Supplementary feeds usually consist of low-cost agricultural/animal by-products, and may involve the use of a single food item in its fresh or unground state (*i.e.* such as mill sweepings, beer waste or rice bran) or the use of a combination of different feed items in the form of a feed mash or pellet. Although supplementary feeds are used as a direct source of dietary nutrients for the farmed species, when used in excess these products may also exert a fertilization effect on the water body. With this feeding strategy higher fish and shrimp stocking densities are possible and consequently higher fish/shrimp yield per unit area. This feeding strategy is typical of a semi-intensive farming system.

References

Anon, 1998. Report, National Commission on Agriculture. Part VII. 531 p.

Anon, 2000. First 128 case studies of composite fish culture in India. Bull. Cent. Inland Fish. Res. Inst., Barrackpore 23. 152 p.

Behrendt, A. 2001. Profit at all levels of the carp pond. *Fish Farmer,* 1(4): 42.

Jhingran, V.G. and Sharma, B.K. 1978. Operational research project on aquaculture in West Bengal. Bull. Cent. Inland Fish. Res. Inst. Barrackpore. 276 p.

Jhingran, V.G. 1997. Fish and Fisheries of India. Hindustan Publishing Corporation, New Delhi, India.

Mandal, B and Patra, B.C., 2005. Application of biotechnological tools for aquaculture practices in the pita (Khadan) of abandoned china clay mines. Biotechnology in Environmental Management. Edited by T.K. Ghosh, T. Chakraborty and G. Tripathi. APH Publisher, New Delhi. pp 313-324.

Mandal, B., Patra, B.C. and Chand, B.K., 2008. Possibilities of sustainable aquaculture practice in abandoned china clay mines of West Bengal, India. *Research Journal of Fisheries and Hydrobiology* (INSINET Publication), 3 (2): 36-40.

Mandal, B., Patra, B.C. and Chand, B.K., 2009. Restoration of soil and water quality of abandoned china clay mines of West Bengal for fish production. *Proceedings of the International Conference* (Food Security and Environmental Sustainability). Published by Agriculture and Food Engineering Department, IIT, Kharagpur, West Bengal, India. SFA: 11-18.

Mitra, G.N. and Mahapatra, P., 1956. On the role of Zooplankton in the nutrition of carp fry. *Indian J. Fish*, 3: 299-310.

Murshed, S. M., S. N. Roy, D. Chakraborty, M. Randhir, and V. G. Jhingran. 1998. "Potential and Problems of Composite Fish Culture Technology in West Bengal." Bulletin of Central Inland Fishery Research Institute (Barrackpore, India) 25: 11.

Patra, B.C., Maity, J., Debnath, J. and Patra, S., 2005. Making aquatic weed V: Nutritional evaluation of *Nymphoides cristatum* as a non-conventional feed ingredients for an Indian major carps *Cirrhinus mrigala* (Ham). *Journal of Biological Science*, 11: 40-54.

Sharma, B.K., A. Mukherjee, J. Chander, D. Kumar, M.K. Das and S.R. Das. 1978. Operational Research Project on Composite Fish Culture, Krishnagar, Nadia, West Bengal (Report). First Annual Workshop on Operational Research Projects of the Indian Council of Agricultural Research, held at Karnal, India. 17 p.

Sinha, V.R.P. and B.K. Sharma. 1976. Composite fish culture in large sheets of water. *Indian Farming*, 26(2): 30–31.

Tripathi, S.D., Dutta, A., Sen Gupta, K.K. and Patra, S., 1979. High density rearing of Rahu apawns in village ponds. In:

Symposium of Inland Aquacult. (Abstracts) February, 12 – 13, 1979. CIFRI, Barrackpore, pp: 14.

Wijkström, U.N. and New, M.B. 1989. Fish for feed: a help or a hindrance to aquaculture in 2000. *INFOFISH International*, 6/89: 48-52.

Chapter 12

Influence of Different Feeding Regimes on the Growth Performance and Nutrient Retention of the Walking Catfish, *Clarias batrachus* Fingerlings

Meenakshi Jindal

Department of Zoology and Aquaculture,
CCS Haryana Agricultural University,
Hisar – 125 004, Haryana, India

ABSTRACT

Clarias batrachus fingerlings were maintained under laboratory conditions and fed on a 40 per cent fishmeal (FM) based protein diet on a circadian pattern selecting six different time intervals, *i.e.*, 08^{00}h, 12^{00}h, 16^{00}h, 20^{00}h, 24^{00}h and 04^{00}h O'clock. The results indicated that body weight gain (live weight gain and specific growth rate) in *C. batrachus* on a time restricted meal is optimal when the food is made available at mid-night *i.e.*, 24^{00}h. Better accumulation of protein with less excretion of ammonia in fish body were also reported at this time of feeding. These results concluded that the growth rates of fish fed during night time (mid-night) follow their feed demand significantly

* Corresponding author: E-mail: meenjind@gmail.com

higher with lowest food conversion ratios and food wastage, thus less polluting the water. Therefore, it is beneficial to feed *C. batrachus* at mid-night. This will also alleviate pollution problems in the intensive aquacultural systems.

Keywords: *Ammonia, Circadian pattern, Growth, Live weight gain, Mid-night.*

Introduction

A rhythm reveals it's exogenous (under the influence of an external periodic factor) or endogenous character (which originate from within the organism) under constant conditions. The exogenous rhythm immediately loses its periodicity, while the endogenous rhythm is sustained with a period close to 24h.

Most organisms do not feed constantly, but exhibit a "circadian-like" pattern. Meal timings can interact with food utilization and because the cost of feed is one of the factors that most influences the production cost of intensively cultured fish. Parker (1984) suggested the need to consider feeding rhythms of domesticated fish species as a means to improve the production efficiency. A better knowledge of feeding rhythms in fishes is needed before feeding modification can be applied to aquacultural practice.

Many workers have studied the diel rhythms of feeding in a number of fish species (See Boujard and Leatherland, 1992 for references), however, a lot of work has been done on Indian major carp species but much less work on catfishes.

Among catfishes, air-breathing and non-piscivorous, *C. batrachus* (Indian magur/desi magur/walking catfish) holds an important place. Some of the advantages regarding magur culture are that its flesh is tasty and has less spines. It has medicinal and therapeutic value. Moreover, its flesh is more digestible, for which it is also recommended to patients. It can be cultured in small and shallow ponds (where carp culture is not possible), as it can gulp atmospheric air due to the presence of air breathing organs. Further, it can be cultured in monoculture or in polyculture with carps due to its non-piscivorous nature. It can fetch high market price than carps. Therefore, in the present study, an effort was done to study the circadian rhythms of feeding activity in this fish under controlled conditions. Effect of feeding was examined on the growth, water quality parameters and also on the nutrient retention in the test fish.

Materials and Methods

Specimens of *C. batrachus* were procured from Sultan Singh Fish Farm, Nilokheri, Haryana, India. For a week prior to their experimental use, the fish were standardized in the Fisheries Laboratory, Department of Zoology and Aquaculture, CCS Haryana Agricultural University, Hisar, Haryana, maintained at a temperature of 25 ± 1°C. The fingerlings were fed on fishmeal (FM) based diet (animal protein diet) containing 40 per cent protein already prepared and tested by our laboratory on *H. fossilis* and *C. batrachus* fingerlings while studying their optimum protein requirement. The ingredient and proximate composition of the feed is presented in Table 12.1.

Table 12.1: Ingredient Composition and Proximate Analysis (per cent dry weight basis) of the Experimental Diet when Fishmeal was Used as the Main Protein Source

Ingredients	*(Per cent)*
Groundnut Oil Cake (GNOC)	60.0
Rice Bran (RB)	5.0
Fishmeal (FM)	28.0
Binder[a]	5
Chromic Oxide[b] (Cr_2O_3)	1
Mineral premix[c] and amino acids (MPA)	1
Proximate Analysis (per cent)	
Crude Protein	40.25
Crude Fat	7.00
Crude Fibre	3.50
Ash	7.30
Nitrogen Free Extract (NFE)	42.95
Gross Energy KJg^{-1}	19.65

a: Used is Carboxyl methyl cellulose to make the diets water stable

b: Used as an external indigestible marker for estimating apparent digestibility.

c: Added to supplement the diets with minerals and amino acids. Each Kg contains Copper–312mg; Cobalt–45mg; Magnesium–2.114g; Iron -979mg; Zinc–2.13g; Iodine 156mg; DL-Methionine–1.92g; L-lysine mono hydrochloride–4.4g; Calcium 30 per cent and Phosphorus–8.25 per cent.

The fish learned to eat this feed in a couple of days. The amount consumed was not controlled. Some fish took a pellet and then dropped it. They or others then consumed it, or the food pellet remained at the bottom of the aquarium. Moreover, the food seen to be suspended in the water, as fine particles that eventually may have lost some solutes and may have disintegrated. It appeared that the fish ate at different rates.

Following 7 treatments were maintained and fingerlings in a group of 10 fish (mean weight 7.04 ± 0.13g) for each treatment were fed daily @ 5 per cent body weight (BW) daily for the whole experimental duration of 120 days.

Fish were bulk weighed at every 20th day interval and the feeding rates were adjusted accordingly. Fish were exposed to their respective diet for 4 hr. during each ration, thereafter; the uneaten feed was siphoned out, stored and dried separately for calculating the daily feed consumption.

Sl.No.	*Treatment/Feeding Time*
Group 1	Control (feeding *ad libitum*)
Group 2	Feeding at 08^{00}h
Group 3	Feeding at 12^{00}h
Group 4	Feeding at 16^{00}h
Group 5	Feeding at 20^{00}h
Group 6	Feeding at 24^{00}h
Group 7	Feeding at 04^{00}h

The faecal matter voided by the fish was also collected from each aquaria separately and dried at 60°C for subsequent determination of digestibility. At the termination of experiment, fish from all the treatments were weighed individually and processed for carcass composition.

Analytical Techniques

The feed ingredients, experimental diets, faecal samples and fish carcass were analysed according to AOAC, 2000. Chromic oxide levels in the diets as well as in the faecal samples were estimated spectrophotometrically (Spyridakes *et al.*, 1989). Live weight gain (g), growth percent gain, specific growth rate (SGR, per cent d^{-1}), protein efficiency ratio (PER) and gross conversion efficiency (GCE) were calculated using standard methods (Steffens, 1989). Apparent nutrient digestibility (APD) of the diets were calculated (Cho *et al.*, 1982).

$$APD = 100 - \frac{100 \times \text{per cent } Cr_2O_3 \text{ in diet} \times \% \text{ nutrient in faeces}}{\% \ Cr_2O_3 \text{ in faeces} \times \% \text{ nutrient in diet}}$$

Water Quality Parameters

Water samples for the determination of water quality parameters were obtained in replicates from each treatment at 20 days interval. Water temperature (°C) was recorded daily using digital thermometer and pH, conductivity and dissolved oxygen were monitored using multiline F-set-3 (E. Merck Ltd., Germany). All other parameters were determined (APHA, 1998).

At the end of feeding trials, offer the same feed to the fish in sufficient quantity so that the same is consumed, wait for 2 hours. Maintain a fixed level of water in each aquarium (say 30-40 L). Remove the excess of feed. Water samples from each aquarium were collected at 2h interval for the estimation of excretory levels of ammonical nitrogen (NH_4-N) and ortho-phosphate (*o*-PO_4) following APHA (1998), to see the influence of compounded feeds on pollution status of receiving water in the aquaria and calculated as follows:

$$\text{Ammonical nitrogen excretion } (mgg^{-1}BWd^{-1}) = \frac{NH_4\text{-N } (mgl^{-1}) \text{ in water}}{\text{fish biomass (g) per L of water}}$$

$$\text{Ortho-phosphate production } (mgg^{-1}BWd^{-1}) = \frac{o\text{–}PO_4 \ (mgl^{-1}) \text{ in water}}{\text{fish biomass (g) per L of water}}$$

Statistical Analysis

Data were analysed by One-Way analysis of variance (ANOVA) and Duncan's Multiple Range Test.

Results and Discussions

Water Quality Parameters

Highest value of DO was reported in the fish group fed at midnight (24^{00}h). This indicates that when catfishes were fed at mid-night, then feed is better utilized and less deteriorate the water quality. These results are in agreement with those of Jindal *et al.* (2007a,b). pH remained alkaline. Water temperature fluctuated between 25 to 25.6°C. All the other parameters remained in the optimal range (Table 12.2).

Significantly ($P< 0.05$) lowest values of NH_4-N and *o*-PO_4 were observed in the fish fed at mid night (24^{00}h) as shown in the Table 12.2. After mid-night, these values started increasing. This showed that fishes are active at mid-night and utilized feed better at this time with lesser excretion of ammonia and phosphorous (Jindal *et al.*, 2007b and 2009).

Although water temperature, DO and free CO_2 contents are examples of abiotic factors that might influence the pattern of feeding activity, but the main daily environmental rhythmic Zeitgeber is probably the periodicity of the light/dark alternation. This is in agreement with field studies in which diel rhythms appear to be directly related to the day/night alteration (Boujard and Leatherland, 1992).

Growth and Digestibility Parameters

Fish survival in different dietary treatments were observed high and varied between 92 to 98 per cent (Table 12.3). A significantly ($P< 0.05$) high live weight gain, per cent weight gain and SGR were observed in fingerlings fed at mid-night with lowest FCR and food wastage. The results showed clearly negative correlation between FCR values and live weight gain.

The results should be viewed in the light of data documenting the fact that food offered at different times in the circadian cycle encounters temporally different biologic systems. It has been reported that if food availability is restricted daily to one or another fraction (*eg.*1 or 4 hours) of daily routine such as part of the light span versus part of dark span for a nocturnally active fish species like the Indian catfish, *H. fossilis* or the evening or morning for diurnally active animals such as carps among fishes or human beings, a number of circadian rhythms are shifted in timing and what is critical, not all the same extent (Noeske *et al.*, 1981, Sundararaj *et al.*, 1982, Hossain *et al.*, 2001 and Jindal *et al.*, 2009).

Table 12.2: Water Quality Parameters of Different Treatments when Fish *C. batrachus* Fingerlings was Fed with 40 per cent Fishmeal Based Diets

Parameters	*Treatments (40 per cent FM based diets)*						
	Control	*08^{00} h*	*12^{00} h*	*16^{00} h*	*20^{00} h*	*24^{00} h*	*04^{00} h*
Dissolved oxygen (DO) mg/l	5.8±0.000	5.6±0.200	5.4±0.000	5.8±0.200	6.4±0.600	6.6±0.200	5.4±0.200
pH	7.75±0.005	7.65±0.005	7.45±0.005	7.70±0.000	7.45±0.005	7.50±0.014	7.55±0.005
Water temperature (°C)	25.5±0.100	25.2±0.150	25.5±0.005	25.4±0.100	25.5±0.001	25.4±0.300	25.3±0.010
Conductivity micro (μ) mhos cm^{-1}	0.53±0.001	0.53±0.000	0.52±0.001	0.49±0.001	0.46±0.000	0.53±0.000	0.49±0.000
Free Carbon dioxide (Free CO_2) mg/l	16.6±0.200	16.7±0.000	16.6±0.000	16.3±0.010	16.5±0.100	16.8±0.000	16.5±0.141
Total alkalinity (mg/l)	218±2.000	217±1.000	215±1.000	211±1.000	212±0.000	209±1.000	216±2.000
Total hardness (mg/l)	222±2.000	229±1.000	222±2.000	224±4.000	217±1.000	219±1.000	217±1.000
NH_4-N excretion (mg/100g BW of fish)	2.147±0.055	2.149±0.070	1.974±0.040	1.589±0.080	1.388±0.020	1.278±0.070	1.604±0.035
o-PO_4 excretion (mg/100g BW of fish)	1.158±0.050	1.037±0.012	1.013±0.020	0.901±0.020	0.685±0.012	0.559±0.065	0.788±0.030

*: All values are mean±S.E. of mean of 3 observations; * BW = Body Weight.

Table 12.3: Effect of Feeding 40 per cent FM Based Diets on Growth Performance, Digestibility and Nutrient Retention in *C. batrachus* Fingerlings

Parameters	*Treatments (40 per cent FM based diets)*						
	Control	*08^{00} h*	*12^{00} h*	*16^{00} h*	*20^{00} h*	*24^{00} h*	*04^{00} h*
Survival (per cent)	92	96	98	94	97	96	94
Live Weight gain (g)	$7.37^{A}\pm$ 0.003	$8.28^{B}\pm$ 0.004	$8.56^{C}\pm$ 0.004	$9.64^{D}\pm$ 0.002	$12.80^{E}\pm$ 0.001	$14.95^{F}\pm$ 0.004	$8.57^{A}\pm$ 0.015
Fish length gain (cm)	$3.10^{A}\pm$ 0.000	$3.15^{B}\pm$ 0.001	$3.30^{B}\pm$ 0.005	$3.35^{B}\pm$ 0.005	$3.75^{C}\pm$ 0.005	$4.15^{D}\pm$ 0.005	$3.30^{B}\pm$ 0.005
Growth per cent gain in BW	$158.94^{AB}\pm$ 3.800	$169.05^{AB}\pm$ 2.715	$180.97^{B}\pm$ 6.405	$201.05^{C}\pm$ 1.050	$256.01^{D}\pm$ 1.850	$310.93^{E}\pm$ 5.715	$180.54^{B}\pm$ 2.978
Growth/day per cent BW	$0.888A^{B}\pm$ 0.090	$0.916^{AB}\pm$ 0.080	$0.949^{BC}\pm$ 0.017	$1.002^{D}\pm$ 0.025	$1.123^{E}\pm$ 0.035	$1.217^{F}\pm$ 0.015	$0.950^{BC}\pm$ 0.010
Specific growth rate SGR (per cent d^{-1})	$0.413^{A}\pm$ 0.011	$0.429^{B}\pm$ 0.014	$0.449^{C}\pm$ 0.015	$0.478^{D}\pm$ 0.020	$0.552^{E}\pm$ 0.010	$0.614^{F}\pm$ 0.010	$0.388^{G}\pm$ 0.050
Food Consumption/ day per cent BW	$4.730^{A}\pm$ 0.001	$4.765^{B}\pm$ 0.000	$4.770^{B}\pm$ 0.002	$4.770^{B}\pm$ 0.001	$4.270^{C}\pm$ 0.001	$3.935^{D}\pm$ 0.004	$4.700^{B}\pm$ 0.003

Contd...

Table 12.3–Contd...

Parameters	*Treatments (40 per cent FM based diets)*						
	Control	*08^{00} h*	*12^{00} h*	*16^{00} h*	*20^{00} h*	*24^{00} h*	*04^{00} h*
Feed Conversion ratio (FCR)	5.345^{A}± 0.006	5.205^{BC}± 0.004	5.025^{C}± 0.115	4.760^{D}± 0.002	4.270^{E}± 0.001	3.935^{F}± 0.004	4.750± 0.009
Protein efficiency ratio (PER)	0.184^{A}± 0.000	0.214^{B}± 0.001	0.207^{C}± 0.001	0.242^{D}± 0.000	0.319^{E}± 0.000	0.374^{F}± 0.001	0.259^{G}± 0.004
Gross conversion ratio (GCE)	0.187^{A}± 0.002	0.192^{AB}± 0.002	0.199^{B}± 0.005	0.210^{C}± 0.001	0.233^{D}± 0.000	0.254^{E}± 0.003	0.185^{AB}± 0.002
Apparent protein digestibility (APD) per cent	82.72^{A}± 0.000	84.83^{B}± 0.005	85.16^{C}± 0.001	85.98^{C}± 0.00	86.63^{D}± 0.004	87.65^{E}± 0.006	85.78^{C}± 0.003

* BW = Body Weight; * Duration of experiment = 120 days.

* Mean with the same letter in the same row are not significantly ($P>0.05$) different.

* All values are Mean ± S.E. of mean of three observations.

* Data was analyzed by Duncan Multiple Range test.

Table 12.4: Effect of Feeding 40 per cent FM Based Diets Under Different Treatments on Proximate Carcass Composition (per cent wet weight) in *C. batrachus* Fingerlings

Composition Carcass	*Treatments (40 per cent FM based diets)*						
	Control	*08^{00} h*	*12^{00} h*	*16^{00} h*	*20^{00} h*	*24^{00} h*	*04^{00} h*
Moisture	$66.78^{A}\pm0.035$	$64.89^{B}\pm0.007$	$65.28^{C}\pm0.005$	$65.02^{C}\pm0.003$	$64.11^{B}\pm0.004$	$63.88^{D}\pm0.007$	$65.02^{C}\pm0.005$
Crude protein	$19.53^{A}\pm0.003$	$20.89^{B}\pm0.003$	$21.09^{C}\pm0.004$	$21.27^{C}\pm0.003$	$22.46^{D}\pm0.003$	$23.03^{E}\pm0.005$	$21.13^{C}\pm0.004$
Crude Fat	$3.69^{A}\pm0.001$	$3.93^{B}\pm0.002$	$4.09^{C}\pm0.004$	$4.23^{C}\pm0.002$	$4.53^{D}\pm0.007$	$4.91^{E}\pm0.008$	$4.17^{C}\pm0.002$
Total Ash	$3.34^{A}\pm0.002$	$3.64^{B}\pm0.015$	$3.64^{B}\pm0.015$	$3.73^{B}\pm0.000$	$3.82^{C}\pm0.001$	$4.07^{D}\pm0.001$	$3.68^{B}\pm0.000$
Nitrogen Free extract (NFE)	$6.65^{A}\pm0.001$	$6.49^{B}\pm0.004$	$6.19^{C}\pm0.007$	$5.74^{D}\pm0.008$	$5.08^{E}\pm0.004$	$4.11^{F}\pm0.005$	$6.00^{C}\pm0.005$
Gross Energy (KJg^{-1})	$7.218^{A}\pm0.002$	$7.642^{B}\pm0.002$	$7.603^{B}\pm0.005$	$7.687^{B}\pm0.002$	$7.971^{C}\pm0.004$	$8.088^{D}\pm0.002$	$7.673^{B}\pm0.005$

* Mean with the same letter in the same row are not significantly ($P>0.05$) different.

* All values are Mean ± S.E. of mean of three observations.

* Data was analyzed by Duncan Multiple Range test.

Table 12.5: Relationship Between Water Quality and Growth Parameters in *C. batrachus* Fingerlings Fed on 40.25 per cent FM Based Diets Under Different Treatments/Feeding Timings

Diets	*DO mg/l*	*Excretory Wastes*		*Growth Parameters*			*Fish Carcass Protein (Per cent)*	*Fish carcass Fat (Per cent)*
		NH_4-N (mg/100g BW of fish)	*o-PO_4 (mg/100g BW of fish)*	*Wt. gain (g)*	*FCR*	*APD*		
Control (*ad libitum*)*	6.4	2.147	1.158	7.37	5.34	82.72	19.53	3.69
08^{00}h FM	6.6	2.149	1.037	8.28	5.21	84.83	20.89	3.93
12^{00}h FM	6.4	1.974	1.013	8.56	5.03	85.16	21.09	4.09
16^{00}h FM	5.8	1.589	0.901	9.64	4.76	85.98	21.27	4.23
20^{00}h FM	5.4	1.388	0.685	12.80	4.27	86.63	22.46	4.53
24^{00}h FM	5.4	1.278	0.559	14.95	3.94	87.65	23.03	4.91
04^{00}h FM	6.2	1.604	0.788	8.57	4.75	85.78	21.13	4.17

Fish Carcass Composition

The data on carcass composition (Table 12.4) indicates that the fish fed at mid-night had high retention of protein and fat. Further, accumulation of carcass protein paralleled with that of weight gain, indicated that growth was mainly due to the accumulation of protein. These results are supported by Kalla (2001) and Jindal *et al.* (2007a, b).

The growth and digestibility parameters, protein accumulation and fat deposition in the fish carcass were found to be negatively correlated with NH_4-N excretion and *o*-PO_4 production (Table 12.5). These results are in agreement with those of Mazid *et al.*(1994) and Jindal *et al.* (2007 a, b; 2009).

These results clearly indicate that meal timings play an important role in food utilization and growth in *C. batrachus*. When feed was given at different timings of light/dark cycle, then the group of fish which was fed at mid-night shows better growth performance paralleled with better accumulation of carcass protein in fish body. These results concluded that the growth rates of fish fed during night time (mid-night) follow their feed demand significantly higher with lowest food conversion ratios and food wastage. Therefore, it is beneficial to feed *C. batrachus* at mid-night.

This type of study will also be of interest as regards to the general biology of the species. In addition, an understanding of the natural feeding rhythms of this commercially important fish will be mandatory in order to permit the development of feeding regimes that optimize growth. Therefore, it will be preferable to exploit such information in "farming inland waters".

Acknowledgements

The authors acknowledge funding received under the scheme "Women Scientist Scholarship Scheme for Societal Programmes (WOS-B), Department of Science and Technology, Government of India" for carrying out this research. Mr. Sultan Singh, owner Sultan Fish Seed Farm, Nilokheri, Haryana, India is fully acknowledged for providing *C. batrachus* fingerlings, free of cost, for carrying out the research work.

References

AOAC (Association of Official Analytical Chemists). 2000. Official methods of analysis. Assoc. Off. Anal. Chem. Washington, Sc, USA.

APHA (American Public Health Association). 1998. Standard methods for the examination of water and wastewater. APHA, AWWA, EPFC, 20th Ed., New York.

Boujard, T. and Leatherland, J.F.. 1992. Demand feeding behaviour and diel patterns of feeding activity in *Oncorhynchus mykiss* held under different photoperiod regimes. *Journal of fish biology*, 38: 130-142.

Cho, C.Y., Slonger, S.J. and Bayley, H.S. 1982. Bioenergetics of salmonid fishes. Energy intake, expenditure and productivity. *Comp. Biochem. Physiol.*, 73B: 25-41.

Hossain Mostafa, A.R., Haylor Graham, S. And Beveridge Malcolm, C.M. 2001. Effect of feeding time and frequency on the growth and feed utilization of african catfish, *Clarias gariepinus* (Burchell 1822) fingerlings. *Aquaculture Research*, 32 (12): 999.

Jindal M., Garg, S.K., Yadava, N.K. and Gupta, R.K. 2007a. Effect of replacement of fishmeal with processed soybean on growth performance and nutrient retention in *Channa punctatus* (Bloch.) fingerlings. *Livestock Research for Rural Development, USA*, Volume 19, Article #165. Retrieved from http://www.lrrd. Org./ lrrd19/11/jind19165.htm

Jindal, M., Garg, S.K. and Yadava, N.K. 2009. Effect of feeding defatted canola on daily excretion of ammonical nitrogen (NH_4-N) and ortho-phosphate (*o*-PO_4) in *Channa punctatus* (Bloch). *Livestock Research for Rural Development*, Volume 21, Article # 35 Retrieved from http://www.Irrd.org./lrrd21/3/jind21035.htm

Jindal, Meenakshi, Garg, S.K. and Yadava, N.K. 2007b. Effect of replacement of fishmeal with dietary protein sources of plant origin on the growth performance and nutrient retention in the fingerlings of *Channa punctatus* (Bloch.) for sustainable aquaculture. *Pb. Univ. Res. J. (Sci.)*., 57: 133-138.

Kalla, A. 2001. Effect of Supplementary feeding in some teleosts on growth, digestibility and water quality parameters in intensive fish culture system. Ph.D. Thesis. Guru Jambheshwar university, Hisar, India.

Mazid, M.A., S. Sultan, M. Kamal, M.A. Hossain and S. Gheyasuddin. 1994. Preparation of feed from indigenous sources for the optimum growth of tilapia (*Oreochromis niloticus*). The Third Asian Fisheries Forum Asian Fisheries Society, Manila, Philippines.

Noeske, T.A., Erickson, D. and Spieler, R.E. 1981. The time of day goldfish receive a single daily meal affects growth. *J. World maricult. Soc.*, 12: 73-77.

Parker, N.C. 1984. Chronobiologic approach to aquaculture. *Trans. Amer. Fish. Soc.*, 113: 545-552.

Spyridakes P., Metailler, R., Gabandan, J. and Riaza, A. 1989. Studies on nutrient digestibility in European sea bass *Dicentrarchus labrax* I. Methodical aspects concerning faeces collection. *Aquaculture*, 77: 61-70.

Steffens, W. 1989. Principles of fish nutrition. Ellis Horwood, Chichester.

Sundararaj, B.I., Nath, P. and Halberg, F. 1982. Circadian meal timing in relation to lighting schedule optimizes catfish body weight gain. *J. Nutr.*, 112: 1085-1097.

Chapter 13

Amelioration of Anti-oxidant Vitamins on Ciprofloxacin Induced Toxic Changes in Blood

B. Kalaivani[1]*, *K.M. Priya Therciny*[2], *and G. Vanithakumari*[2]

[1]*Sri Sarada College for Women, (Autonomous), Salem – 636 016, Tamil Nadu, India*

[2]*Department of Zoology, Bharathiar University, Coimbatore – 641 046, Tamil Nadu, India*

ABSTRACT

The aim of the present study was to investigate the impact of the combined administration of vitamin A, and C on ciprofloxacin induced toxic changes in blood. Adult male Wistar strain weighing about 150 – 250g was subdivided into seven groups. The rats in the control group were received 0.9 per cent saline orally. The other group received ciprofloxacin at a low dose of 4.2 mg. The third group received ciprofloxacin at a dose of 6.6mg/100g body weight/day. The fourth group received 6.6 mg of vitamin A along with high dose. The fifth group received 8.3mg of vitamin C along with high dose. Final group rats received ciprofloxacin at a dosage of 6.6mg/100g

* Corresponding author: E-mail: kalai03win@yahoo.co.in

body weight/day respectively for 7 days and the animals were kept for further seven days without giving any drug a withdrawal group. Haematological parameters like RBC, WBC and differential counts were studied. The body weight of the animals significantly increased with the number of days of treatment and was dose dependent. RBC, WBC, neutrophil, lymphocytes were reduced, basophil and monocytes were raised after ciprofloxacin treatment. Vitamin A and C supplementation to high dose group showed marked improvement of RBC, WBC and differential counts. In particular vitamin A improves RBC and WBC. These experimental results strongly indicate the protective effect of vitamin A and C against toxic effects of ciprofloxacin on blood.

Keywords: *Ciprofloxacin, RBC, WBC, Basophil, Neutrophil, Monocytes, Lymphocytes, Vitamin supplementation.*

Introduction

Antimicrobial drugs are the greatest contribution of the present day therapeutics. Fluroquinolone antimicrobials have one or more fluorine substitutions. The more recent introduction of fluorinated – 4 quinolones are the ciprofloxacin (CIPRO) which represents a particularly important therapeutic advance, since these agents have a broad antimicrobial activity (Hooper *et al.*, 2000). Relatively few side effects appear to accompany with the use of these fluoroquinolones and resistance to their action does not develop rapidly (Hooper *et al.*, 2000). Clinical experience has indicated that they have some undesirable side effects including cutaneous reactions like phototoxicity, juvenile cartilage toxicity, and although the incidence is very low, adverse central nervous system reactions including epileptogenic convolutions (Kucers *et al.*, 1997).

Ciprofloxacin administration increases the number of granulocytes, macrophage, colony forming cells, while the white blood cells decrease in numbers. Colony forming factor production by cultured mice spleen cells in the presence of poke weed mitogen is increased three fold by the addition of ciprofloxacin at a dose of 5 – 10mg/litre (Kletter *et al.*, 1991). Consuming 3 to 6 mg of β-carotene daily will maintain plasma β-carotene blood levels in the range associated with a lower risk of chronic diseases (Institute of Medicine, 2001). Red blood cells, like all blood cells are derived from precursor cells called stem cells. These stem cells are dependent on retinoids for normal differentiation into red blood cells. Additionally vitamin A appears to facilitate the mobilization of iron from storage sites to the developing red blood cells for incorporation into Hb, the oxygen carrier in blood cells (Ross, 1999). The supplementation of vitamin C have improved anemia. Thus the present study is an attempt to investigate the probable adverse effect of ciprofloxacin on haematological parameters.

Materials and Methods

Ciprofloxacin hydrochloride (Neocip–500) was purchased from OKASA Ltd, Goa, India. Vitamin A from VSV Ltd, Mumbai, India. Vitamin C (celin–500) from Galaxo Smithkline Pharmaceutical Company Ltd., Sisco Research Laboratories, Pvt Ltd, India and Loba Chemie, Austria.

Animals

The experiments were performed in Wistar strain rats (200g) provided by Tamil Nadu Agricultural University, Coimbatore. They were housed in a clean and well ventilated animal house with a constant 12 hrs light and darkness schedule, They were provided with free access to standard rat pellets and water throughout the 2 weeks acclimatization periods. They were then carefully monitored until the end of the treatment. The experiments were approved by the Animals Care Committee of the Institute (722/02/a/CPCSEA).

Animal Treatment

Experimental Protocol

The animals were weighed and divided into three groups of five animals each.

Group I: Control

The healthy rats were selected and treated as control and they received saline orally. A separate batch of five rats was maintained for vitamin supplementation groups and received gingelly oil orally.

Group II: Short Duration

The animals selected for short duration treatment were treated with ciprofloxacin at twelve hours interval for seven days.

Group III: Long Duration

Here the animals were treated with ciprofloxacin at twelve hours interval for thirty consecutive days.

Group II and Group III was further sub-divided into six groups, each group consisting of five animals. The animals received the following regimen of treatment and all the treatments were designed on the basis of adult human dosage prescribed by the physicians and interpolated to the body weights of rats.

(*a*) Low dose

The animals selected for low dose treatment received 250mg. of ciprofloxacin/60kg body weight as an oral dose.

(*b*) High Dose

The animals received 400mg of ciprofloxacin/60kg body weight as an oral dose.

(*c*) High Dose + Vitamins A

The animals received 400mg. of ciprofloxacin followed by 7.5mg of vitamin A/60kg body weight, as an oral dose.

(*d*) High dose + Vitamin C

The animals received 400mg of ciprofloxacin followed by 500mg of vitamin C/60kg body weight as an oral dose.

(*e*) High dose + Vitamin E

The animals received 400mg of ciprofloxacin followed by 600mg of vitamin E/60kg body weight, as an oral dose.

(*f*) High dose withdrawal

The experimental animals received 400mg of ciprofloxacin as an oral dose and were allowed a withdrawal period of the drug for further seven days for short duration and one month for long duration.

Suitable controls were maintained for each duration of treatment. However, as there was no difference in any parameter among control group, a common control was employed in the present study.

Sample Collection

The rats selected for experiments were weighed before and after treatment. Twenty four hours after the last treatment schedule the rats were sacrificed by decapitation method. Blood was collected and was kept at room temperature. Then serum was separated immediately by centrifugation of blood at 3000xg for 30 minutes and stored at 20°C until utilized for hormonal assays. Few ml of blood were collected in a separate vial containing a drop of EDTA or Heparin to prevent coagulation. This blood was used for estimating haematological parameters.

Blood Analysis

The haemoglobin concentration of blood was estimated by the method using Shali's haemoglobinometer (Samuel, 1986). Erythrocytes/Red Blood Cells and leukocytes/White Blood Cells were counted by the method of Samuel (1986) using haemocytometer. Differential count is to estimate the percentage distribution of white blood corpuscles in a blood smear (Merck Veterinary Manual, 1979).

Statistical Analysis

The data were statistically analysed and expressed as mean ± standard error of mean (SEM). The data were further compared by Duncan's Multiple Range Test (DMRT), Irristat version 3/93, Biometrics Unit, International Rice Research Institute, P.O.Box 933, Manila, Philippines (Alder and Roessler, 1977).

Results

Effect on Red Blood Corpuscles (RBC) Count (Table 13.1)

In short duration, a dose dependent 34 per cent to 27 per cent reduction in red blood count was seen after ciprofloxacin administration for a short duration. Vitamins A and E were more effective (95 per cent and 93 per cent respectively) in preventing the drug induced adverse effect than vitamin C (79 per cent). Withdrawal of drug could bring complete restoration of the RBC counts. Whereas, in long duration of ciprofloxacin treatment simulated the short durational effect on RBC counts. But the percent reduction in RBC count was higher in low dose (31 per cent) and high dose (49 per cent) treatment groups. The vitamin A and E supplementations proved to be better in restoring the RBC counts than vitamin C and the withdrawal of drug restored this parameter to normalcy better than short duration group.

Table 13.1: Effect of Ciprofloxacin Alone and in Co-administration with Vitamin Supplements on Red Blood Cell Counts of Rats

Sl.No.	Treatments	RBC Count (In millions/mm³ of blood)			
		Short Duration	Percentage (per cent)	Long Duration	Percentage (per cent)
1.	Control	7.33±0.09	–	7.33±0.09	–
2.	Low dose	5.22±0.1**f	–29	4.85±0.1**e	–34
3.	High dose	4.81±0.1**g	–9	3.87±0.06**f	–47
4.	High dose + vitamin A supplementation	6.82±0.07**c	–19	6.97±0.02**b	–5
5.	High dose + vitamin C supplementation	5.96±0.001**e	–7	6.07±0.09**d	–17
6.	High dose + vitamin E supplementation	6.66±0.1**d	–34	6.82±0.05**c	–7
7.	High dose withdrawal	7.06±0.06b	–29	7.28±0.07a	–0.7

Effect on White Blood Corpuscles (WBC) Counts (Table 13.2)

In short duration treatment, like the RBC counts, the WBC counts were also decreased at both low dose (22 per cent) and high dose (34 per cent), respectively of ciprofloxacin treatment. All the three vitamins namely A, C and E could prevent the decrease caused by the high dose of the drug by 50 per cent. However, drug withdrawal was effective in restoring these parameters to normalcy. However, in long duration, about 23 per cent to 29 per cent reduction in WBC counts was observed after low and high doses of ciprofloxacin when given for a longer period. The vitamins A, C and E were less effective than short duration groups in preventing the high dose drug induced reduction of counts. Withdrawal of the drug was again effective in complete restoration of the WBC count to control level.

Effect on Differential Count

Effect on Neutrophil Counts (Table 13.3)

In short duration, only the high dose drug treatment induced 22 per cent decrease in the neutrophil count at this duration of treatment. Supplementations with vitamins C and E were only partially effective in raising the neutrophil counts whereas, vitamin A effectively restored these parameters to normalcy. Similarly, drug withdrawal also effectively restored the neutrophil count. Unlike short duration group a dose dependent decrease in neutrophil count was seen in low dose (11 per cent) and high dose (18 per cent) drug treatment in long duration.

Table 13.2: Effect of Ciprofloxacin Alone and in Co-administration with Vitamin Supplements on White Blood Cell Counts of Rats

Sl.No.	Treatments	WBC Count (In Thousands/mm³ of blood)			
		Short Duration	Percentage (per cent)	Long Duration	Percentage (per cent)
1.	Control	8.88±0.05[a]	–	8.88±0.05[a]	–
2.	Low dose	6.4±0.17**[d]	–10	6.2±0.2**[d]	–30
3.	High dose	5.13±0.2**[e]	–24	5.5±0.2**[c]	–38
4.	High dose + vitamin A supplementation	6.81±0.03**[c]	–24	6.2±0.07**[d]	–30
5.	High dose + vitamin C supplementation	6.71±0.04**[c]	–23	6.5±0.04**[c]	–27
6.	High dose + vitamin E supplementation	6.97±0.03**[c]	–42	6.3±0.4**[cd]	–29
7.	High dose withdrawal	7.98±0.04**[b]	–28	8.3±0.2[b]	–7

Table 13.3: Effect of Ciprofloxacin Alone and in Co-administration with Vitamin Supplements on Differential Count-Neutrophil Counts of Rats

Sl.No.	Treatments	Neutrophil Counts (In per cent)			
		Short Duration	Percentage (per cent)	Long Duration	Percentage (per cent)
1.	Control	73.3±1.8[a]	–	73.3±1.8[a]	–
2.	Low dose	70.3±1.0**[ab]	–4	65.6±1.9**[bc]	–10.5
3.	High dose	57.0±1.7**[c]	–22	60.0±0.8**[c]	–18
4.	High dose + vitamin A supplementation	73.3±2.1[a]	0	54.3±2.4**[d]	–26
5.	High dose + vitamin C supplementation	62.6±1.5**[e]	–15	53.7±4.4**[d]	–27
6.	High dose + vitamin E supplementation	64.6±2.4**[d]	–12	56.0±2.9**[d]	–24
7.	High dose withdrawal	71.6±0.7[b]	–2	66.7±1.4**[b]	–9

Effect on Eosinophil Counts (Table 13.4)

In short duration treatment, the eosinophil counts showed a decrease in both low (60 per cent) and high (30 per cent) dose ciprofloxacin treatment groups, while the vitamin supplementations effectively restored the eosinophil count to near normalcy, drug withdrawal could exert such an effect only by 87 per cent. Whereas in long duration treatment, ciprofloxacin caused a similar decrease in the eosinophil counts like the short duration group, only at low doses. However, the drug had raised the eosinophil counts when given at higher doses. The vitamin supplementations had maintained the drug-induced response, though to a lesser degree. Vitamin E was more effective than vitamin A and C. Drug withdrawal in these high dose groups was not able to prevent the decrease in this parameter.

Table 13.4: Effect of Ciprofloxacin Alone and in Co-administration with Vitamin Supplements on Differential Count-Eosinophil Counts of Rats

Sl.No.	*Treatments*	*Eosinophil Counts (In per cent)*			
		Short Duration	*Percentage (per cent)*	*Long Duration*	*Percentage (per cent)*
1.	Control	8.7±0.3[c]	–	8.7±0.3[c]	–
2.	Low dose	7.3±0.5**[b]	–16	7.3±0.3**[d]	–16
3.	High dose	5.6±0.3**[a]	–36	10.7±0.7**[a]	23
4.	High dose + vitamin A supplementation	8.3±0.9[b]	–5	9.7±1.7**[b]	11.5
5.	High dose + vitamin C supplementation	8.6±1.4[b]	–1	9.7±0.7**[b]	11.5
6.	High dose + vitamin E supplementation	8.0±0.9[a]	–8	10.7±0.9**[a]	23
7.	High dose withdrawal	7.6±0.2**[d]	–16	7.3±1.4**[d]	–16

Effect on Basophil Counts (Table 13.5)

In short duration treatment, the basophils were decreased in a dose dependent manner. While the low dose administration caused 41 per cent, the high dose caused a 78 per cent decrease. Vitamin supplementations brought about a 50 per cent restoration of the basophilic counts, when given to the high dose drug treated animals.

Similarly, drug withdrawal also could partially restore the basophilic counts. In case of long duration treatment, unlike short duration groups ciprofloxacin at low dose only could induce a decrease in basophilic counts by 37 per cent, whereas the high dose of the drug caused only a 15 per cent of decrease of the basophil counts. Neither vitamins supplementations nor drug withdrawal could prevent the drug induced decrease in the basophil counts. Among the vitamin groups, vitamin C appears to exert a more adverse effect.

Table 13.5: Effect of Ciprofloxacin Alone and in Co-administration with Vitamin Supplements on Differential Count-Basophil Counts of Rats

Sl.No.	*Treatments*	*Basophil Counts (In per cent)*			
		Short Duration	*Percentage (per cent)*	*Long Duration*	*Percentage (per cent)*
1.	Control	2.7±0.7[d]	–	2.7±0.7[a]	–
2.	Low dose	1.6±0.3**[b]	–41	1.7±0.3**[d]	–37
3.	High dose	0.6±0.1**[a]	–78	2.3±0.3**[b]	–14
4.	High dose + vitamin A supplementation	1.6±0.5**[b]	–41	1.7±0.8**[d]	–37
5.	High dose + vitamin C supplementation	1.6±0.2**[b]	–41	1.3±0.3**[e]	–52
6.	High dose + vitamin E supplementation	1.6±0.8**[b]	–41	2.0±0.9[c]	–26
7.	High dose withdrawal	2.0±0.4[c]	–26	1.7±0.3**[d]	–37

Effect on Lymphocyte Counts (Table 13.6)

In short duration treatment, nearly 16 per cent decrease in lymphocytes was observed after low dose ciprofloxacin administration for a shorter duration. High dose of the drug had decreased the same by 32 per cent. Vitamins supplementations as well as drug withdrawal could partially (50 per cent) raise the lymphocyte counts like control. In long duration treatment, both low and high doses of drug had brought similar reduction in lymphocyte counts. However, the three vitamins were unable to exert an impact on the lymphocytes when given along with the high dose. Whereas drug withdrawal was able to partially restore the lymphocyte counts.

Table 13.6: Effect of Ciprofloxacin Alone and in Co–administration with Vitamin Supplements on Differential Count-lymphocyte Counts of Rats

Sl.No.	*Treatments*	*Lymphocyte Counts (In per cent)*			
		Short Duration	*Percentage (per cent)*	*Long Duration*	*Percentage (per cent)*
1.	Control	25.3±0.5[a]	–	25.3±0.5[a]	–
2.	Low dose	21.3±1.1**[bc]	–16	21.3±1.0**[b]	–16
3.	High dose	17.3±1.4**[d]	–32	16.3±0.7**[e]	–36
4.	High dose + vitamin A supplementation	21.0±1.2**[bc]	–17	16.3±1.8**[c]	–36
5.	High dose + vitamin C supplementation	20.6±1.1**[c]	–19	15.7±0.5**[d]	–38
6.	High dose + vitamin E supplementation	22.0±1.2**[b]	–13	16.3±0.7**[c]	–36
7.	High dose withdrawal	21.3±1.0**[bc]	–16	18.7±3.0**[b]	–27

Table 13.7: Effect of Ciprofloxacin Alone and in Co-administration with Vitamin Supplements on Differential Count-Monocyte Counts of Rats

Sl.No.	*Treatments*	*Monocyte Counts (In per cent)*			
		Short Duration	*Percentage (per cent)*	*Long Duration*	*Percentage (per cent)*
1.	Control	6.3±0.5[b]	–	6.3±1.0[c]	–
2.	Low dose	5.3±0.5**[cd]	–16	5.3±1.4**[d]	70
3.	High dose	6.0±1.4[bc]	–5	10.7±0.5**[a]	70
4.	High dose + vitamin A supplementation	6.6±1.4[ab]	–4.8	10.7±1.4**[a]	70
5.	High dose + vitamin C supplementation	7.3±0.5**[a]	16	9.7±0.3**[b]	54
6.	High dose + vitamin E supplementation	4.6±0.7**[e]	–27	10.7±1.4**[a]	70
7.	High dose withdrawal	5.0±0.4**[d]	–16	4.7±0.5**[e]	–25

Each value is the mean ± SE of five animals.

** Control vs Treatment significant at 1 per cent level by ANOVA.

Mean± SE followed by a common letter are not significantly different at the 5 per cent level by DMRT (a, b, c etc).

Effect on Monocyte Counts (Table 13.7)

In short duration treatment, the low dose drug treatment brought a 16 per cent decrease in monocyte counts, whereas, high dose treatment had no marked effect on the monocyte counts. Among the three vitamins, vitamin C was very effective in restoring the monocyte counts (16 per cent) than vitamin A, whereas, vitamin E, supplementation and withdrawal of the drug had brought a decrease in the monocyte count (21 per cent and 27 per cent, respectively) compared to the high dose drug treated group. Whereas in Long duration treatment, the low dose drug treatment as well as drug withdrawal in high dose drug treated animal, had brought a significant 70 per cent decrease in the monocyte counts. On the contrary high dose drug treatment and vitamins supplementations had induced a marked increase in the monocyte counts, except for vitamin C where, only a 54 per cent increase was observed.

Discussion

Effect on Red Blood Corpuscles (RBC)

Ciprofloxacin had brought about a 35 per cent–50 per cent reductions in red blood cell count when administered at both low and high doses in the present study. Ciprofloxacin was shown to affect the bone marrow's capacity to produce red cells by Caraman (1993), or impair erythropoiesis. Ciprofloxacin at even smaller doses than the present study was seen to decrease the erythrocyte count (Kalpana, 2002). Excess blood loss, more erythrocyte destruction and enzyme deficiency can also cause anemia. Similar effect has been noted with the antibiotic chloramphenicol (Keller and Follath, 1998) and tetracycline which produce aplastic anemia. Abnormal rigidity of the erythrocyte leads to sickle-cell anemia and results in premature destruction (Dow *et al.*, 1996). Decreased RBC counts and decreased haematocrit and haemoglobin levels were observed in dogs when benfluralin was given orally to dogs in varying doses, similar to ciprofloxacin (USEPA/OPP, 1994). Butafenacil administration in mice caused lower mean values of erythrocyte count, haemoglobin concentration and haematocrit with an increase in neutrophil and monocyte count. In carfentrazone-ethyl administrated mice, the mean corpuscular haemoglobin (MCH) and mean corpuscular volume (MCV) were reduced but RBCs were increased in short term studies and a reverse condition was noticed in long term studies in a dose dependent manner. Diisopropyl flurophosphate (DFP) administration caused a fall in the RBC, plasma and whole blood fractions (Gordan *et al.*, 1991), and dichlofluanid significantly reduced the eosinophils and monocytes without reducing the red blood cell counts and white blood cell counts. Diflubenzuron decreased haematocrit haemoglobin concentration and RBCs. Tilmicosin, drug used to treat respiratory disorders in animals, caused reduction in RBC and WBC counts in mice (Yazer *et al.*, 2004), and in rabbits. Azithromycin and clarithromycin caused the same effect in humans (McEvay and Litvak, 2001). Indomethacin, a NSAID reduced the packed cell volume (PCV), RBC, MCV and MCH in rats (Adedapo and Aiyelotan, 2001). By chloramphenicol administration the PCV, MCH and neutrophils were reduced in a study conducted in rabbits (Saba *et al.*, 2002). Epogen is an injectable drug that stimulates the production of RBC cell.

Vitamins A and E supplemented to ciprofloxacin treated animals caused an 85 per cent–90 per cent restoration of RBC count in the present study. This effect of vitamin A may be due to improved mobilization of iron from body stores into the circulation and into hematopoietic tissue as vitamin A is essential for normal hematopoiesis. In the absence of vitamin A, increased iron absorption increased tissue iron concentration and possibly was associated with impaired erythropoiesis (Roodenburg. *et al.*, 1994). Layrisse *et al.* (1998) has reported that the inhibitory effect of polyphenols and phyrates on iron absorption was reduced by vitamin A. The improvement of iron mobilization from the body stores into the circulation and the increase of erythropoiesis with vitamin A supplementation were indicated by a sharp decline in serum ferritin concentration. Vitamin C supplementation has also proved effective in restoring the drug induced decrease in RBC counts, when given with high doses of ciprofloxacin.

A favorable effect of ascorbic acid on the acceleration of blood clotting may be due to vitamin C's facilitative effect on iron absorption on activation of vitamin A (The Merck Manual, 2004). Vitamin E has also the same effect as that of vitamins A and C in ciprofloxacin treated rats. It was proved by Pauling mega – vitamins/amino acid therapy that the vitamins such as A, C and E strengthen and heal blood vessels. Vitamin E can prevent haemolysis of RBCs, and protect blood vessels from oxidation and damage (Merck Manual, 2004). Other findings have corroborated with the present data and have proved that vitamin A supplementations have brought about major reductions in the number of anemic cases and improve anemia and growth (Mwanri *et al.*, 2000). Dreyfuss *et al.* (2000) have observed vitamin A deficiency to lead to anemia by inhibiting erythropoiesis directly (West and Roodenburg, 1992) or through inhibition of iron mobilization (Bloem *et al.*, 1989; Wolde – Gebriel *et al.*, 1993). Supplementation of vitamin A improves haemoglobin concentration and erythropoiesis (Muslimatun *et al.*, 2001). Kalpana (2002) also observed a similar result with ciprofloxacin administration in a short-term study in rats.

Effect on White Blood Corpuscles (WBC)

White blood cell count (WBC) is the count of leucocytes, which make the immune system and defend the body against infection. A high WBC means that the immune system has been activated for some reason or the body is infected. A low WBC count might be concerned with problem in the bone marrow or may be due to the side effects of various drugs. In the present study, a 20 per cent – 35 per cent reduction in leucocytes count was noticed with low and high doses of ciprofloxacin showing similar to RBC count. By the withdrawal of the drug, about 90 per cent restoration of leukocyte was noticed. Davidson and Macleod (1971) observed that the reduction in agranulocytes is due to poisoning by a variety of drugs. Avoiding NSAIDs, as well as aspirin is better, because these drugs interfere with blood platelets and cause prolonged bleeding and low WBC count may indicate the existence of bone marrow diseases or an enlarged spleen (Barrick and Vogel, 1996). An elevated WBC count only shows some type of inflammation by infectious agents like virus, bacteria etc. Disorders of blood production can result in elevated white cell counts, accompanied by blasts, and low white cell counts leads to several syndromes. Isolated cases of leukopenia and eosinophilia are known to occur with the use of ciprofloxacin (Wang *et al.*, 1986). High concentration of ciprofloxacin inhibited lymphocyte proliferation and this was proved by a study on mice by Valera *et al.* (1995). It will be right to affrim based on this study that ciprofloxacin had some haematotoxic effects.

Antioxidant vitamins A, C and E reduced the drug effect by about 77 per cent–80 per cent in both short and long durations of treatment. Of the three major antioxidants, vitamin C is effective in cell mediated immune function and leucocytes are effective in recycling ascorbate (Rumrey and Levine, 1998). *In vitro* evidence indicates that vitamin C may modulate leukocyte's phagocyte action, blastogenesis, immunoglobin production, chemotaxis and adhesiveness (Jariwalla and Harakeh, 1996). Vitamin C is important for its antioxidant effects, direct antimicrobial action and/or effects on various immune system modulators such as histamine and prostaglandins (Jacob and Sotoudeh, 2002). Vitamin E appears to increase the hypoprothrom-binemic response to oral anticoagulants. While clinical evidence of the interactions of vitamin A and E with oral anti coagulants is still limited, existing case reports warrant precaution as vitamin C inhibits warfarin's absorption and reduces its anticoagulant effect. Diclofenac and piroxicam – induce bone marrow lymphopoiesis as well as their respective interactions with µ–tocopherol are associated with parallel changes in monoamines levels, especially nor- epinephrine and serotonin (Fuller *et al.*, 2000).

Effect on Differential Count

Differential count sub classifies each component of the white cell population. High population can be a sign of infection which, also indicate certain types of leukemia and low population indicates bone marrow diseases or an enlarged spleen (bloodbook.com). Low WBC could be a sign of the result of various chronic diseases and also be a side effect of various drugs, particularly chemotherapeutic drugs for cancer treatment (NHL Cyberfamily, 2005).

Effect on Neutrophils

Neutrophils act as first line of defense against invading pathogens by destroying them and by stopping them from multiplying in the body (Mike Gleeson, 2004). In the present study, there was an 18 to 22 per cent decrease in neutrophil count after high dose of ciprofloxacin was administered at both the duration of its treatments. The withdrawal group of short duration brought effective recovery than that of long duration group. Chloramphenicol reduced the neutrophil count (Saba *et al.*, 2002). Similar to ciprofloxacin some drugs (*e.g.* carbimazole; suiphonamides) may increase the neutrophil counts and they might be destroyed by neutrophil antibodies and reduced vitamins B_{12} manufacture and finally bone marrow failure (CWS, 2002). Butafenacil and carfentrazone-ethyl, increased the neutrophil count in male mouse. A person with HIV infection usually has a normal neutrophil count but some drugs given to fight the HIV infection might cause a drop in neutrophil count (CWS, 2002). Vitamin E is a good supplement for older people to maintain a healthy immune system and vitamin A and retinoic acid play a vital role in the development and differentiations of WBC, that play critical roles in the immuno response (Merck Manual, 2004). Vitamin C is excellent for its immune support and have a positive influence in fighting and preventing many diseases (Elson, 2004) but in the present study all the vitamins supplemented A, C and E could not prevent the ciprofloxacin induced decrease in the neutrophil counts.

Effect on Eosinophils

Eosinophils protect the body against allergic reactions and parasites. Elevated levels indicate an allergic response and low count is normal (CWS, 2002). The WBC count can be decreased by certain medications. Azido de oxythymidine (AZT) is an example for haematotoxicity, which was observed in a study with mice (NTP Toxicology and carcinogenesis studies of AZT, 1999). 1,1-Difluoroethane reduced eosinophil in rats (http://www.fluoride action.org/pesticides/1,1-difluoroethane.epa.iris. htm).

In the present study, all treatment groups of rats in long duration treatment have shown an elevated eosinophil count except low dose drug treated and withdrawal group. This indicated a development of allergic response to ciprofloxacin by long-term administration of this drug and the condition was reversed in short duration treatment groups. The drug withdrawal also brought only partial recovery, suggestive of persistent effect of ciprofloxacin on the eosinophils.

Effect on Basophils

Basophils are the least common of the WBCs and their count at zero is quite normal. They are involved in fighting bacteria and their increased number reflects a possibility of parasitic activity in the body (NHL Cyberfamily, 2005). In the present study, the basophil count was seen to be raised in the high dose ciprofloxacin treated groups at long duration only and the same was very low in the short duration treatment. This may be due to the development of allergy by the body against ciprofloxacin when used continuously for a longer period. Vitamins especially A and C supplements to this drug treated rats helped the body to fight against the allergic reaction of the drug when given for a longer duration. However, all the three vitamins acted effectively and equally in the short duration treatment period by lowering the basophil counts.

Effect on Lymphocytes

Lymphocytes are involved in protecting the body from viral infections and an elevated level of lymphocytes indicated an active viral infection and a depressed level indicated an exhausted immune system (CWS, 2002). The NSAIDs like diclofenac and piroxicam were shown to induce bone marrow lymphopoiesis as well as they interacted with vitamin E and causes parallel changes in monoamines level especially those of norepinephrine and serotonin (CWS, 2002). Diclofenac alone and its combination with vitamin E did not produce any consistent effect on whole blood (Fuller, 2000). In the present investigation, the lymphocyte count was observed to decrease in both durational studies with ciprofloxacin. The antioxidant vitamins also could not prevent the decrease of lymphocytes. The drug withdrawal could also bring no recovery indicating a permanent immune exhaust caused by ciprofloxacin for a prolonged period.

Effect on Monocytes

Monocyte cells are helpful in fighting severe infections and are considered the body's second line of defense against infection. Their elevated levels are seen in tissue breakdown and indicate chronic infections. Their low level are indicative of a good state of health (CWS, 2002). Butafenacil increased the monocyte count in male mouse. 1,1-Difluoroethane reduced monocytes in rats.

In the present investigation, the monocyte count was raised markedly by the high dose ciprofloxacin administration for long duration with vitamins supplementation. However a reverse process was observed, in short-term study with this drug. So by long term administration of ciprofloxacin there might be some tissue breakdown and this might in turn lead to elevated monocyte counts in the experimental animals. Neither the supplemented vitamins nor the withdrawal of drug in long term study groups could stop the tissue damage caused by the drug effectively.

Thus, the above data obtained in this study suggests that RBC, WBC, neutrophil, lymphocytes were reduced, basophil and monocytes were increased after the ciprofloxacin treatment. In particular vitamin A improves RBC and WBC.

References

Adedapo, A.A. and Aiyelotan, O., 2001. Effect on chronic administration of indomethacin on haematological parameters in rats. *Afr J. Biomed Res.*, 4; 159-160.

Alder,H.L. and Roessler, E.B., 1977. The analysis of variance. In: Introduction to probability and statistics, Freeman (ed. San Fransisco, USA, pp.319 – 385.

Barrick, B., and Vogel, S., 1996. Applications of Laboratory Diagnostics in HIV Nursing. In: Grady C, Buenning – Betchel, C Boland M (eds): Nursing clinic of N. America. Philadelphia; W.B.Saunders, pp. 41-46.

Bloem, M.W., Wedel.M. and Aeagger, R.J., 1989. Iron metabolism and Vitamin A deficiency in children in Northest Thailand. *Am. J. Clin. Nutr.*, 50: 332-338.

Carman, R.H., 1993. Haematology. In: Hand book of medical laboratory technology.2nd edn., Christain medical association of India. Banglore. pp. 76 – 155.

CWS, 2002.CellMate Wellness System ™ Blood Chemistry Definitions, Haematology. webmaster@carbonbased.com.

Davidson, S.S. and Mecleod, J., 1971. Disorders of blood and blood forming organs. In: The principles and practice of medicine. 10th edn., Churchill Livingstone, Edinburgh. pp. 828 – 891.

Dow, J., Lindsay,G. and Morrison, J., 1996. Blood molecular functions of cells and plasma. In: Biochemistry-Molecules, Cells and the body. Addison – Wesley publishing company, Inc, England. pp. 355 – 378.

Dreyfuss, M.L., Stoltzfus, R.J., Shrestha, J.B., Pradhan, e.k., Leclerq, S.C, Khartry, S.K., Shrestha, S.R., Katz, J., Albonico, M. and West, K.P., 2000. Hookworms, Malaria and Vitamin A deficiency contribute to anemia and iron deficiency among pregnant women in the plains of Nepal. *Journal of Nutrition,* (130): 2527-2536.

Fuller, C.J., May, M.A. and Martin, K.J., 2000. The effect of vitamin C and vitamin E supplementation on LDL oxidizability and blood cells in young smokers. *J.Am.Coll. Nutr.*, 19(3): 361-369.

Gordan, C.J., Fogelson, L., Richards, J.and Highfill, J., 1991. The National Technical Information Service. Report No. NTIS/PB92-158658.

Hooper, D.C., Wolfson, J.S., Ng, E.V. and Swartz, M.N., 2000. Mechanism of action and resistance to ciprofloxacin. *AM J Med.*, 82 (supply 4A): 12-20.

Institute of Medicine, Food and Nutrition Board, 2001. Dietary Reference Intakes. Vitamin A, Vitamin E, Arsenic, Boron, Chromium, Copper, Iodine, Iron, Manganese, Molybdenum, Nichol, Silicon, Vanadium and Zinc. National Academy press, Washington, USA

Jacob, R.A., and Sotoudch, G., 2002. Vitamin C function and status in chronic disease. *Nutr. Clin. care., 5* (2): 66 – 74.

Jariwala, R.J. and Harakeh, S., 1996. Antiviral and immunomodulatory activities of vitamin C. *Sub Cell Biochem.*, (25): 213-231.

Kalpana, S., 2002. Ciprofloxacin induced haematological and biochemical changes and the role of vitamin A and C as rescue in albino rats. M.phil dissertation Bharathiar University, Coimbatore.

Keller, H. and Follath, F., 1988. Misscellaneous antibiotics, In: Mayler's side effects of drugs.Dukes, M.N.G.(eds).11th ed. Elsevier Science Publishers, NewYork. pp. 546-556.

Kletter, Y., Riklis, I.and Fabian, I., 1991. Enhanced repopulation of murine hematopoietic organs in sublethally irradiated mice after treatment with ciprofloxacin. *Blood,* 78: 1685 – 1691.

Layrisse, M., Garcia- Casal, M.N., Solano, L., Baron, M.A., Arguello, F., Llovera, D., Ramirez, J., Leets, I. and Tropper, E., 1988. Vitamin A reduces the inhibition of iron absorption by phytates and polyphenols. *Food Nutr. Bull.*,(19): 3-5.

Mc Evay, G.K. and Litvak, K., 2001. AHFS Drug 2001 information, American Soeciety of Healthy System Pharmacists, USA.

Mike Gleeson., 2004. What Blood Tests can tell you about possible causes of under performance. www.medicdirect sport.com.

Muslimatun,S., Schmidt, M.K., Schultink, W., West, C.E., Hautvast, J.G.A.J., Gross, R. and Muhilai, 2001. Weekly supplementation with iron and vitamin A during pregnancy increases haemoglobin concentration but decreases serum ferritin concentration in Indonesian pregnant women, *J. Nutr.*, (131): 85-90.

Mwanri, L., Worsley, A., Ryan, P. and Masika, J., 2000. Supplemental vitamin A Improves anemia and growth in anemic school children in Tanzania. *journal of Nutrition*, (130): 2691-2696.

Ross A.C., 1999. Vitamin A and Retinoids. In: Shils, M. (eds.). Nutrition in health and disease, 9th edn., Baltimore Williams and Wilkins, pp. 305-327.

Rumrey, S.C. and Levine, M., 1998. Absorption, transport and deposition of vitamin C in humans. *J. Nutr. Biochem.*, (9): 116-130.

Saba, A.B., Awe, O.E., Akinloye, A.K. and Oladele, G.M., 2002. Haematologicl changes accompanying prolonged ocular chloramphenicol administration in laboratory rabbits. *Arf. J. Biomed. Res*: 5: 131-135.

Samuel, R.M., 1986. Haematology In: Notes on Clinical lab techniques 4th ed. Tyyer and Sons, Madras, pp. 23-27.

The Merck Manual, 2004. Sec. 1 nutritional disorder, chap S. Vitamin E Deficiency, dependency and toxicity, Merck and Co. INC.

USEPA/OPP, 1994. Support Document for the addition of chemicals from Federal Insecticide, Fungicide, Rodenticide Act (FIFRA. Active ingredients to EPCRA section 313. U.S. Ennvironmental Protection Agency, Washington D.C.

Wang, C., Sabbai, J. and Corrado, M., 1986. World wide clinical experience with norfloxacin: efficacy and safety. Scand. *J. Infect. Dis*, 48: (supply): 81.

Wolde- Gabriel, Z., West, C.E., Haile Gebru, Tadesse, A.S., Fisseha, T., Gabre, P., Ayana, G. and Hautvast, J.G.A.G., 1993. Interrelationship between vitamin A, iodine and iron status in school children in Shoa Region, Central Ethiopia. *Br. J. Nutr.*, (70): 593-607.

Yazer, E., Oztekin,E., Sivrikaya, R.,Col,R., Elmas, M. and Bas, A.L., 2004. Effects of different doses of tilmicosin on malondialdehyde and glutathione concentrations in mice. *Acta Vet Brno*; 73: 69-72.

Chapter 14

Impact of Chromium on Haematological Parameters in Freshwater Fish *Channa punctatus*

Deepali Saxena and Madhu Tripathi*

Aquatic Toxicology Research Laboratory, Department of Zoology, University of Lucknow, Lucknow – 226 007, U.P., India

ABSTRACT

Chromium is well known pollutant and found to be in every part of environment. In water it is discharged from the various industries. The aim was to observe the effect of exposure against two concentrations of hexavalent chromium (2.5 mg/L and 5 mg/L) on haematological parameters of a freshwater fish *Channa punctatus* under laboratory conditions. Significant decrease in Hb per cent, Total erythrocyte count, PCV per cent, MCHC and Lymphocytes per cent as well as significant increase in Clotting time, Total leucocyte count, ESR, MCH, MCV, Neutrophils per cent, Basophils per cent, Eosinophils per cent and Monocytes per cent were observed in experimental groups in comparison to control. This study reflects the extent of the toxic effect of hexavalent chromium

* Corresponding author: E-mail: drmtripathi@gmail.com

and metal induced cumulative deleterious effects at various functional levels in fish.

Keywords: *Chromium, Channa punctatus, Haematological parameters, Chromium toxicity.*

Introduction

Rapid industrialization has left behind various effluents which are added to the aquatic system. Metallic compounds in these effluents accumulate in various parts of animal or plant body (Barman and Lal, 1994) causing huge mortification of inhabiting biota including palatable fish. Heavy metals due to their bio-accumulative and non-biodegradable properties constitute a core group of aquatic pollutants. The direct or indirect bioaccumulation of metals in organisms is of great significance because it affects the food chain (Vagas *et al.*, 2001). Chromium is an element of subgroup VI. It is naturally occurring element found in rocks, animals, plants, soil and in volcanic dust and gases. The most common forms of chromium are chromium(0), trivalent (or chromium(III)) and hexavalent (or chromium(VI)) but trivalent and hexavalent chromium compounds predominate in use having various industrial applications (Durgan, 1972; Zey, 1984). Most of the chromium present in nature is result of industrial emissions (Steven *et al.*, 1976). Its particulate enters in aquatic medium through effluent discharge from industries, such as tanneries, textiles, electroplating, mining, dyeing and printing industries, photographic and pharmaceutical industries. Emissions from burning coal and oil, steel production, stainless steel welding, chemical manufacturing, waste streams from electroplating, leather tanning and textile industries as well as those that make dyes and pigments can increase chromium(III) and chromium(VI) level in air as well as into waterways. Hexavalent chromium compound is mostly present in natural water (Vutukuru, 2005) and is estimated to be 100 times more toxic than trivalent form (Schroeder, 1961 and Dell and Campbell, 1971). The chromate ion $(Cr_4)^{+2}$, the dominant form of chromium(VI) in neutral aqueous solutions, can readily cross cellular membranes via non-specific anion carriers (Danielsson *et al.*, 1982), while chromium(III) is poorly transported across membranes. The differences in membrane transport can explain the differences in the abilities of these two oxidation states of chromium to produce tissue damage. In animal body Cr(VI) is quickly converted to a stable form Cr(III) by natural defense mechanism (ASTDR, 1999).

Toxicity of metals has been reviewed by various workers (Subramanian and Varadraj, 1993; NTP, 2007, 2008a and Vinodhini and Narayanan, 2009). Haematological studies in some fish species have also been reported under chromium stress (Singh, 1995 and Vutukuru, 2005). However, studies on exposure to chromium is concerned are limited on freshwater palatable fish.

In the present study an attempt has been made to observe the effect of chromium on parameters of fish blood.

Materials and Methods

Experimental fish, *Channa punctatus* weighing 40 ± 10 gm and 14 ± 4 cm length were collected from local resources. They were treated with 0.1 per cent $KMnO_4$ solution and acclimated with proper aeration and food in glass aquaria, 15 days prior to start of the experiment. Water was changed every alternate day after feeding. The physico-chemical properties of water were determined by standard protocols of APHA *et al.* (2005) which are as follows:

Table 14.1

Sl.No.	Parameter	Value
1.	Temperature (°C)	26 ± 2
2.	pH	6.9 ± 0.20
3.	Dissolved Oxygen (DO) mg/L	8.4 ± 0.20
4.	Alkalinity (mg/L)	90-100
5.	Water hardness (mg/L) as $CaCO_3$	118-120

Water was changed every alternate day after feeding. For experiment, fish were divided into three groups, each group containing 6 fish. Group I served as control, while the group II and group III were exposed to chromium. Group II was treated with 2.5 mg/L (low concentration) and group III was treated with 5 mg/L (high concentration) of hexavalent chromium. The source of hexavalent chromium is Potassium di-chromate ($K_2Cr_2O_7$, Ranbaxy Laboratories Limited, Punjab, India). Experiment was conducted for 60 days. After the completion of exposure period blood sample from both the exposed and control group fish were collected by serving the caudle peduncle into vials containing heparin as 4.5 units/ml blood. Total erythrocyte count (TEC) and Total leucocyte count (TLC) were done by Neubauer's haemacytometer using physiological saline (9.0 g NaCl/L) and turk's solution as diluting fluid respectively. Hemoglobin per cent was measured by Sahli's Hemometer, Erythrocyte sedimentation rate (ESR) and PCV by Wintrob's tube method, Clotting time (CT) by capillary tube and Mean corpuscular hemoglobin (MCH), Mean corpuscular hemoglobin content (MCHC) and Mean corpuscular volume (MCV) were determined by methods of Dacie and Lewis (1977). Differential leucocyte count (DLC) was performed on methanol fixed Giemsa stained blood smear. The mean values of control and exposed fishes were compared following't' test.

Results and Discussion

Comparative data of haematological parameters of fish after low and high concentration chromium treatment along with control is summarized in Table 14.2. It is clear from the table that there is a significant decrease in HB per cent, Total erythrocyte count, PCV per cent, MCHC, Lymphocytes per cent and also significant increase in Clotting time, Total leucoocyte count, ESR, MCV, MCH, Neutrophils per cent, Basophils per cent, Eosinophils per cent and Monocytes per cent under stress of both low and high concentration of chromium(VI) exposure.

Blood is a pathophysiological reflector of whole body and therefore, blood parameters are an asset in diagnosing the structural and functional status of body organs exposed to toxicants (Sampath *et al.*, 1998). Metallic ions present in river water enter the fish body through permeable surfaces and they get accumulated in various organs (Al-Mohanna, 1994). These metallic ions are the main probable cause of the pathophysiological anomalies in fish.

The fall in HB per cent, Total erythrocyte count, PCV per cent, MCHC, Lymphocytes per cent upon treatment with chromium, shows occurrence of acute anemia in toxic conditions. The count of erythrocytes is quite a stable index and the fish body tries to maintain this count within the limits of physiological standards, using various physiological mechanisms of compensation, especially under stress. The hemoglobin concentrations reflect the supply of an organism

Table 14.1: Hematological Alterations in Fish *Channa punctatus* Exposed to Chromium (VI)

Parameters	*Control Group*	*Experimental Groups*	
		Low Concentration	*High Concentration*
Hemoglobin per cent	13.65±0.75	10.98±0.93**	8.97±0.86****
Clotting time (sec.)	77.19±1.04	82.83±2.62**	87.58±3.75***
Total erythrocyte count (x 10^6/cumm)	3.90±. 0.58	2.89±0.16*	2.44±0.15***
Total leucoocyte count (x 10^6/cumm)	19.01±0.64	20.67±0.52**	23.37±0.84****
ESR (mm/hr)	0.50±0.06	0.63±0.04*	0.71±0.03***
PCV (per cent)	56.69±1.45	48.74±2.61***	43.83±2.75****
MCV (mm^3)	158.58±1.78	161.55±1.17*	166.30±1.62****
MCH (g per cent)	34.66±0.45	35.38±0.84*	35.64±1.13***
MCHC (per cent)	22.14±0.56	21.21±0.48	19.49±0.83***
Lymphocytes per cent	64.79±1.15	60.56±1.35***	57.34±1.75****
Neutrophils per cent	24.63±0.53	25.97±0.68	26.31±0.73*
Eosinophils per cent	0.64±0.08	0.72±0.05*	0.97±0.11****
Basophils per cent	1.64±0.04	1.90±0.16	2.10±0.25
Monocytes per cent	8.38±0.20	10.88±0.96***	13.37±1.20****

The observations are mean±SE values.

*P<0.05, **P<0.02, ***P<0.01 and ****P<0.001 (in comparison to control).

with oxygen and the organism itself tries to maintain them as much stable as possible (Vutukuru, 2005). Heavy metals might alter the properties of hemoglobin by decreasing their affinity towards oxygen binding capacity rendering the erythrocytes more fragile and permeable, which probably results in cell swelling, deformation and damage (Witeska and Kosciuk, 2003). Earlier works also reported a fall in Hb per cent, PCV, MCHC, RBC count in freshwater fishes exposed to metals like chromium, cadmium, copper, zinc and nickel indicating anemia, erythropenia and leucopoiesis (Christensen *et al.*, 1972; Sen *et al.*, 1992; Singh, 1995; Nanda and Behera, 1996; Vincent *et al.*, 1996 and Vutukuru, 2005). The Total erythrocyte count, Hb per cent and MCH were appreciably declined in *Labeo rohita* exposed chromium reflecting the anemic state of fish which was due to iron deficiency and its consequent decreased utilization for hemoglobin synthesis. This is in accordance with a similar study on *Labeo rohita* which also reported hypo chromic microlytic anemia under lead chloride stress (Reddy *et al.*, 1998). Anemia in fish is an early manifestation of acute and chronic intoxication of chromium. Further, a significant decrease in TEC, hemoglobin per cent, MCH and hematocrit were also reported in *Channa punctatus* exposed to both copper and chromium and this decrease is more pronounced in fishes exposed to chromium suggesting that the metal induces acute anemia under toxic conditions (Singh, 1995). In the present study, the anemia could be probably due to structural alterations of heme leading to disturbed hemoglobin synthesis and also the inhibitory effect of chromium on the enzyme system in the synthesis of hemoglobin cannot be ruled out as suggested in earlier studies (Johansson and Larrsson, 1979 and Vutukuru, 2005).

The increase in Clotting time, Total leucoocyte count, ESR, MCV, MCH, Neutrophils per cent, Basophils per cent, Eosinophils per cent and Monocytes per cent were reported by many workers (Lynch *et al.*, 1969; Singh, M.; Saxena *et al.*, 2001; Gupta *et al.*, 2002; Gupta, 2003; Lethtinum *et al.*, 1990 and Karuppasamy *et al.*, 2005), is the sign of macrocytic anemia. This type of anemia may reflect possible chronic liver disease (Dacie and Lewis, 1977). The possible reason for increase in blood clotting time may be decreased plasma proteins and imbalance in ionic composition. Chromium is known to decrease protein content in blood serum (Kumar and Barthwal, 1991 and Ruhela *et al.*, 2006).

The chromium produced leukocytosis (increased leucocytes) probably due to the tissue damage and subsequent removal of cell debris (Mc Leay and Brown, 1974). The increase in MCH and MCV values with decrease in MCHC perhaps is due to toxic substances in the medium causing differences in the haemopoietic activity (Lethtinum *et al.*, 1980 and Karuppasamy *et al.*, 2005). This study shows that chromium is toxic to palatable fish.

Acknowledgements

The authors are thankful to Head of Zoology Department, University of Lucknow, Lucknow for providing necessary laboratory facilities to carry out this work.

References

Al-Mohanna, M.M., 1994. Residues of some heavy metals in fishes collected from (Red Sea Coast) Jizan, Saudi Arabia. *J. Environ. Biol.*, 15(2): 149-257.

APHA, AWWA and WEF, 2005. In: *Standard methods for the examination of water and wastewater*, 21th Ed. American Public Health Association, Washington, DC.

ASTDR, 1999. Toxicological profile for chromium (update. U.S. Department of health and human service, Public Health Services.

Barman, S.C. and Lal, M.M., 1994. Accumulation of Heavy Metal (Zn, Cu, Cd and Pb) in soil and cultivated vegetables and weeds grown in industrially polluted fields. *J. Environ. Biol.*, 15(2): 107-115.

Christensen, G.M.; Mc Kin, G.M.; Brinos, W.A. and Hunt, E.P., 1972. Changes in the blood of the brown bull head (*Ictalurus nebulosus*) following short term and long term exposure to copper(II. *Toxicol. Appl. Pharmacol.*, 23: 417-427.

Dacie, J.V. and Lewis, S.M., 1977. Practical Hematology, ELBS, Churchill, Livingston.

Danielsson, B.R.G., Hassoun, E.and Dencker, L., 1982. Embryo toxicity of chromium. Distribution in pregnant mice and effects on embryonic cells *in vitro*. *Arch. Toxicol.*, 51: 233-245.

Dell, B.L.O' and Campbell, B.J., 1971. Trace elements metabolism and metabolic function, in M. Florkin and E. H. Stotz (Eds.), Comprehensive Biochemistry, Vol. 21, Elsevier, Amsterdam, 21: 179-266.

Durgan, P.R., 1972. Biochemical Ecology of Water Pollution, Plenum Press, New York.

Gupta, R., 2003. Pathophysiological consequences to freshwater fish *Channa punctatus* induced by fluoride. Ph.D. Thesis, University of Lucknow, Lucknow.

Gupta, R.; Gopal, K.; Tripathi, M. and Sharma, U.D., 2002. Haematological changes in freshwater fish, *Labeo rohita* and *Channa punctatus* induced by fluoride. Prof S.B. Singh Commomoration Volume *Zoological Society of India*. 1: 105-114.

Johansson-Sjobeck, M.L. and Larrsson, A., 1979. Effect of inorganic lead on α-dehydratase activity and haematological variables in *Salmo gairdnerii*. *Arch. Environ. Contam. Toxicol.*, 8: 419-431.

Karuppasamy, R.; Subathra, S. and Puvaneswari, S., 2005. Haematological responses to exposure to sublethal concentration of cadmium in air breathing fish, *Channa punctatus* (Bloch). *J. Environ. Biol.*, 26(1): 12-128.

Kumar, A. and Barthwal, R., 1991. Hexavalent chromium effects on haematological indices in rats. Bull. Environ. *Contam. Toxicol.*, 46: 761-768.

Lethtinum, K.J.; Kierkegard, A.; Jakobson and Wandell, A., 1990. Physiological effect in fish exposed to effluents from mills with six different bleaching process. *Ecotoxicol. Environ. Saf.*, 19: 33-46.

Lynch, M.J.; Raphael, S.S.; Mellor, L.D.; Sapre, P.D. and Inwood, M.J.H., 1969. Medical laboratory technology and clinical pathology. W. B. Saunders Co., Philadelphia.

McLeay and Brown, D.J., 1974. Growth stimulations and biochemical changes in juvenile Coho Salmon (*Onchorhynchus kistuch*) exposed to bleached Kraft pulp mill effluent for 200 days. *J. Fish Res. Bd. Can.*, 31: 1043-1049.

Nanda, P.and Behera, M.N., 1996. Nickel induced changes in some hemato-biochemical parameters of catfish, *Heteropneustes fossilis*. *Environ. and Ecol.*, 14 (1): 82-85.

NTP, 2007. NTP technical report on the toxicity studies of sodium dichromate dihydrate (CAS No. 7789-12-0) administered in drinking water to male and female F344/N rats and B6C3F1 mice and male BALB/c and *am*3-C57BL/6 mice. Washington, DC: National Toxicology Program. Toxicity Report Series Number 72.

NTP, 2008a. NTP technical report on the toxicology and carcinogenesis studies of sodium dichromate dihydrate (CAS No. 7789-12-0) in F344/N rats and B6C3F1 mice (drinking water studies. Washington, DC: National Toxicology Program. NTP TR 546.

Reddy, J.; Kalarani, S.; Tharakanadha, B.; Reddy, D.C. and Ramamurthi, R., 1998. Changes in energy metabolism of the fish *Labeo rohita* in relation to prolonged lead exposure and recovery. *J. Ecotoxicol. Environ. Monit.*, 8(1): 43-53.

Ruhela, S.; Pandey, A.K. and Khare, A.K., 2006. Effect of experimental *Procamallanus* infrction on certain blood parameters of *Clarias batrachus* (Teleostei: Siluroidae: Clariidae). *J. Ecophysiol. Occup. Hlth.*, 6: 73-76.

Sampath, K.; James, R.; Ali Akbar, K.M., 1998. Effect of copper and zinc of blood parameters and prediction of their recovery in *Oreochromis mossambicus* (Pisces: Cichlidae). *Indian J. Fish*, 45: 129-139.

Saxena, R.; Gupta, R.; Tripathi, M. and Gopal, K., 2001. Fluoride induced haematological alterations in the freshwater fish *Channa punctatus*. *J. Ecophysiol. Occup. Hlth.*, 1: 139-146.

Schroeder, H.A.; Balassa J.J. and Tipton, I.H., 1961. Abnormal trace metals in mean: Chromium. *J. Chron. Dis.*, 15: 941-964.

Sen, G.; Behera, M.K. and Patel, P., 1992. Effect of zinc on hemato biochemical parameters of *Channa punctatus*. *J. Ecotoxicol. Environ. Monit.*, 2(2): 89-92.

Singh, M., 1995. Haematological responses in a freshwater teleost *Channa punctatus* to experimental copper and chromium poisoning. *J. Environ. Biol.*, 16(4): 339-341.

Steven, J.D.; Davies, L.J.; Stanley, E.K.; Abbott, R.A.; Ihnat, M.; Bidstrup, L. and Jaworski, J.F., 1976. Effects of chromium in the Canadian environment. Nat. Res. Coun. Canada, NRCC No. 15017. 168 pp. Avail. from Publications, NRCC/CNRC, Ottawa, Canada, K1A OR6.

Subramanian, M.A. and Varadraj, G., 1993. Impact of tannery effluents on biochemical constituents in the haemolymph of the water prawn *Macrobrachium idella* Heller. *J. Environ. Biol.*, 14(3): 255-259.

Vagas, V.M.F.; Migliavacca, S.B.; DeMelo, A.C.; Horn, R.C.; Guidobono, R.R.; Ferreira, I.C.F.S. and Pestana, M.H.D., 2001. Genotoxicity assessment in aquatic environments under the influence of heavy metals and organic contaminants. *Mutat. Res.*, 490: 141-158.

Vincent, S.; Ambrose, T.; Cyril, A.K.L. and Selvanayagan, M., 1996. Heavy metal cadmium influenced anemia in *Catla catla*. *J. Environ. Monit.*, 2: 89-92.

Vinodhini, R. and Narayanan, M., 2009. The impact of toxic heavy metals on the hematological parameters in common carp (*Cyprinus carpio* L.). *Iran. J. Environ. Health Sci. Eng.*, 6(1): 23-2.

Vutukuru, S.S., 2005. Acute effects of hexavalent chromium on survival, oxygen consumption, haematological parameters and some biochemical profiles of the Indian major carp, *Labeo rohita*. *Int J. Environ. Res. Public Health.*, 2(3): 456-462.

Witeska, M. and Kosciuk, B., 2003. Changes in common carp blood after short-term zinc exposure. *Environ. Sci. Pollut. Res*, 3: 15 – 24.

Zey, J.N. and AW, T.C., 1984. American transportation crop. (Health hazard evaluation report No. 82-025-1413) Cincinnati, OH. National Institute of Occupational Safety and Health.

Chapter 15

Behavioural Alterations in Freshwater Stinging Catfish, *Heteropneustes fossilis* (Bloch) after Fluoride Toxicity

*Sandeep Bajpai, Anand Kumar, Nalini Tripathi, Seema Tewari and Madhu Tripathi**

Aquatic Toxicology Research Laboratory, Department of Zoology, University of Lucknow, Lucknow – 226 007, U.P.. India

ABSTRACT

Static bioassay test was conducted to determine the short-term toxicity of chemical contaminant fluoride on freshwater catfish, *Heteropneustes fossilis*. Fishes were exposed to various concentrations of fluoride upto 96 hours and the percent mortality was recorded. The LC_{50} value (96th hour) was found to be 386.56 mg F/L. During experimentation, behavioural patterns were critically observed after 24, 48, 72 and 96 hours. The major behavioural alterations marked till 96 hours of exposure include loss of schooling, irregular opercular movement, disturbed surfacing, loss of balance, change in orientation, erratic swimming, hyperactivity, jerky swimming, vertical hanging, profuse mucus secretion and loss of

* Corresponding author: E-mail: drmtripathi@gmail.com

equilibrium. The results indicate that fluoride is toxic at high concentration producing dose-dependent increase in mortality and abnormality in behaviour. The fish specie is therefore recommended as good bioindicators for the risk assessment of aquatic environment due to fluoride.

Keywords: *Acute toxicity, Behavioural alterations, Catfish, Fluoride toxicity, Heteropneustes fossilis.*

Introduction

Increasing population gives rise to increasing demands for basic needs such as food and water. To fulfill these requirements with limited resources has become a problem without solution. The rapid industrial growth throughout the world as well as in India to meet the demands of ever-increasing population has caused tremendous environmental degradation (Beg and Ali, 2008; Arner *et al.*, 2009). Contamination of environment has created serious threat to all kinds of life in the form of pollution, which has now become a global problem.

Among all natural resources water is the most precious one. It cannot be created by any technical advancement we make but, at the same time serves as a reservoir for man-made pollutions. Primary source of water pollution are industrial effluents and agrochemicals that are posing threat to the health and productivity of aquatic animals particularly fish which is continuously exposed to the contaminated media (Joshi *et al.*, 2002; Camargo, 2003). Level of fluoride, which is included among chemical contaminants (Sharma, 2002) is gradually increasing in water bodies due to its widespread use in several products by industries (EPA, 1997) and unscientific disposal of industrial wastes into freshwater bodies (Shankar *et al.*, 2007). Fluoride is accumulated in bones causing skeletal deformities (Tripathi *et al.*, 2006) and various soft tissues of animals altering the normal physiology (Chinoy, 1991; Gupta *et al.*, 2001; Bhatnagar *et al.*, 2007). It is also known to affect enzymes activity and other metabolic processes (Chitra *et al.*, 1983; Kumar *et al.*, 2007) and induce behavioural alterations (Tripathi *et al.*, 2004). Chemical agents are known to induce behavioural toxicity affecting both the instinctive and learned type of behaviour in aquatic organisms (Sprague, 1971).

Most of the studies regarding fluoride toxicity have been carried out on mammalian models and human population. There are very scanty information available regarding its effect on fishes, despite their great economic and nutritional value. Keeping all these views into consideration, the present investigation has been designed to evaluate the acute toxicity of fluoride and its impact on behavioural responses of freshwater catfish *Heteropneustes fossilis,* a common edible fish, found almost in every part in India.

Materials and Methods

Live and healthy specimens of adult *Heteropneustes fossilis* of both sexes of 16.0±0.5 cm length and 30.0±2.0 gm weight were bought from the local fish market from nearby areas of Lucknow. Fishes were examined for injury and rinsed in 0.1 per cent $KMnO_4$ solution. They were acclimatized to standard laboratory conditions (APHA, 2005) in dechlorinated tap water in aquaria (60x30x45cm) with continuous aeration (by the help of aerator and stone diffusers) for a period of about 20 days. The physico-chemical parameters were maintained as listed in Table 15.1. Water of aquaria was renewed on alternate days and fishes were fed with goat liver and dried prawn pieces. Feeding was stopped two days before commencement of the experiment to reduce the excretory products and to avoid any possible change in toxicity of fluoride.

Table 15.1: Physico-chemical Properties of the Test Water during Experimentation

Parameters	*Values with Units*
Temperature	27±1.5°C
pH	7.1±0.2
Dissolved Oxygen	7.7±1.5 mg/L
Alkalinity as $CaCO_3$	225–231 mg/L
Hardness as $CaCO_3$	290–300 mg/L
Phosphate	0.30–0.33 mg/L
Nitrite	0.10–0.11 mg/L
Nitrate	7.0–7.1 mg/L
Fluoride	0.1 mg/L

The source of fluoride used in the study was in the form of Sodium Fluoride (NaF), MW 41.99 (Qualigens Fine Chemicals Ltd. Mumbai, India). A stock solution was prepared using sodium fluoride by dissolving measured quantity of toxicant in double distilled water to give a concentration of 100mg F/ml. This stock solution was further diluted to prepare the desired concentrations with chlorine free tap water. The physico-chemical properties of holding water during exposure of fish were maintained according to standard methods of (APHA, 2005).

A static bioassay test was performed according to standard methods (APHA, 2005) to determine the 96 hour median lethal concentration (LC_{50}) of fluoride. Initial range finding experiments were performed to derive the suitable concentrations for LC_{50} determination by using five different concentrations of fluoride along with the control group. In each concentration, ten adult fish (both male and female) were introduced. Definitive tests were subsequently performed using various concentrations of fluoride solution. The mortality was recorded at 24, 48, 72 and 96 hours. The mortality data was statistically analyzed by Trimmed Spearman Karber Method (Hamilton *et al.*, 1977) using computerized package for calculation of LC_{50} value.

During experiment, various behavioural changes such as hyperactivity, swimming patterns, responsiveness, food grasping, schooling, colouration, surfacing and opercular movements were observed at different durations. For opercular beating and surfacing, animals were watched carefully with naked eye with the help of stopwatch for one minute and fifteen minutes respectively. Observations were recorded for six individuals and replicated thrice.

Results

The LC_{50} value for 96 hr of exposure was estimated as 386.56 mg F/L. Control group of fishes showed normal behavioural pattern and compact schooling behaviour (Plate 15.1a). The chief behavioural alterations noticed in *H. fossilis* during acute toxicity exposure were hyperactivity, fast swimming just after the introduction of test solution. The fish tried to avoid the toxic medium which was revealed by their jumping activity. There was an increase in the heart beat rate revealed by the movements of operculum (Table 15.2; Figure 15.1). Erratic swimming, frequent surfacing (Figure 15.2), disturbed schooling and aggregation at the corners of aquaria (Plate

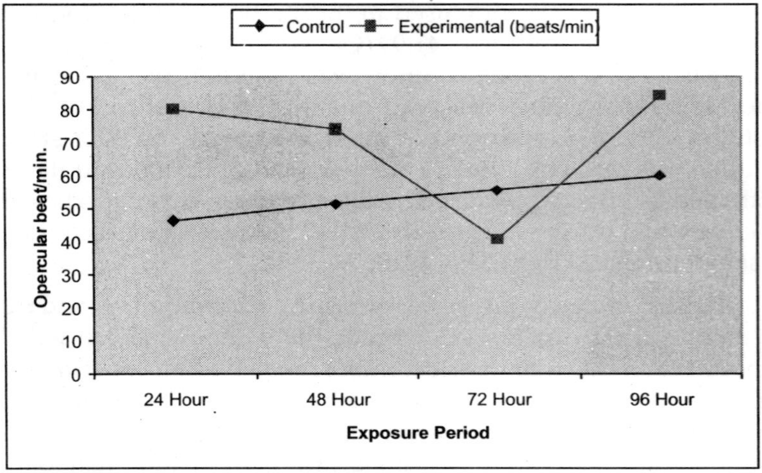

Figure 15.1: Effect of Fluoride on Opercular Beats of *Heteropneustes fossilis* After Acute Exposure

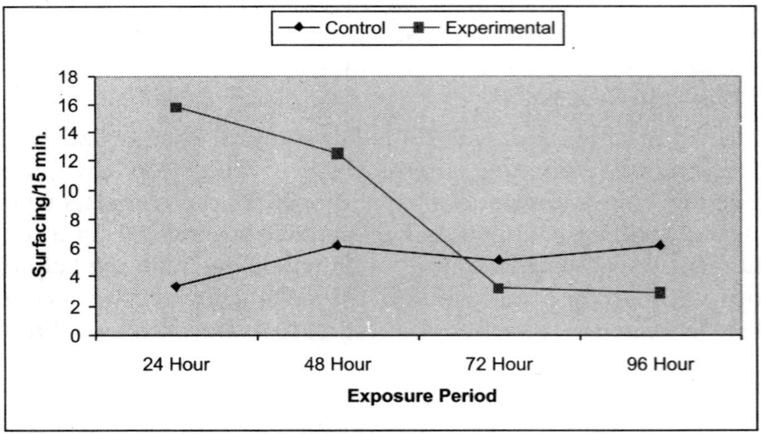

Figure 15.2: Effect of Fluoride on Surfacing Activity of *Heteropneustes fossilis* After Acute Exposure

15.1b) were prominent behavioural changes after 24 hours of exposure. After 48 hours of exposure, loss of balance, vertical hanging in the middle of the water, shocking movements, sudden and frequent turnings, loss of orientation, loss of schooling (Plates 15.1c and d),

Table 15.2: Alterations in Behavioural Responses after Acute Exposure of Fluoride to Freshwater Catfish, *Heteropneustes fossilis*

Sl.No.	*Group*	*Exposure Period (Hours)*	*Behavioural Responses*									
			A	*B*	*C*	*D*	*E*	*F*	*G*	*H*	*I*	*J*
1	Control	24	–	–	–	–	–	–	–	–	–	–
		48	–	–	–	–	+	–	–	–	–	–
		72	–	+	–	–	–	–	–	–	–	–
		96	–	–	–	–	–	–	–	–	–	–
2	Exposed	24	++	+++	+++	–	+++	–	–	+	–	+
		48	++	++	+	–	+++	+	+	++	+	+
		72	+++	++	+	++	++	++	++	+++	++	++
		96	+++	+	+	+++	+	++	+++	++–	+++	++

A: Mucus secretion; B: Jumping; C: Fast movement; D: Erratic swimming; E: Air gulping; F: Jerky movements; G: Loss of equilibrium; H: Loss of schooling; I: Vertical hanging; J: Loss of colouration.

+: Less; ++: More; +++: Prominent; –: Nil.

increased opercular movements, surfacing activity (Figures 15.1 and 15.2) and profuse mucus secretion were chiefly marked (Table 15.2).

Plate 15.1

Plate 15.1a: Control Group Fishes Showing Compact Schooling Behaviour (Arrow)

Plate 15.1b: Showing Disturbed Schooling Behaviour and Aggregation at the Corners of Aquaria After 24 Hours of Fluoride Exposure (Arrow)

Contd...

Plate 15.1–Contd...

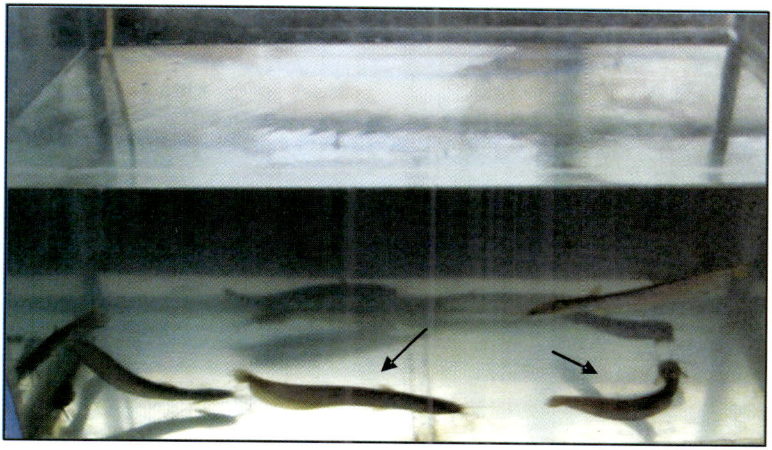

Plate 15.1c&d: Showing Loss of Schooling and Equilibrium After 48 Hours of Fluoride Exposure (Arrow)

Contd...

After 72 hour of exposure, the fish showed less movement, became sluggish and showed loss of equilibrium revealed by vertical hanging with mouth facing the surface and tail towards bottom (Plate 15.1e). Thick mucus coating was seen on the whole body. The

Plate 15.1–Contd...

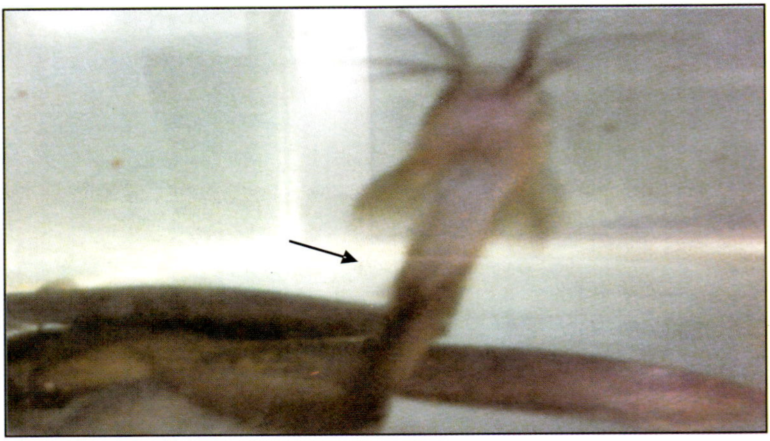

Plate 15.1e: Showing Vertical Hanging After 72 Hours of Fluoride Exposure (Arrow)

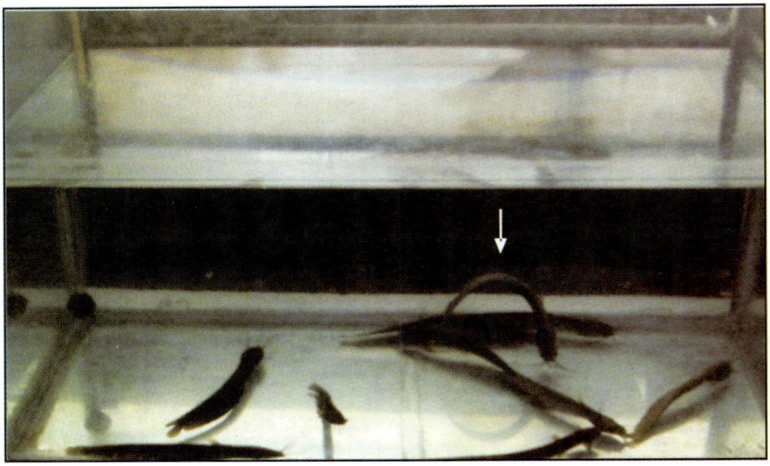

Plate 15.1f: Showing Bending of the Body and Complete Loss of Equilibrium After 96 Hours of Fluoride Exposure (Arrow)

opercular movements and surfacing was reduced (Figures 15.1 and 15.2) and there was marked change in body colouration in comparison to control (Table 15.2). After 96 hours of exposure, the fish showed loss of schooling, complete loss of equilibrium, bending of the body (Plate 15.1f), increased opercular movements and decreased surfacing (Figures 15.1 and 15.2). They do not reacted to gentle tapping. Finally, they were remaining motionless at the bottom of the aquaria and after sometime their bellies turned up and the fish died.

Discussion

Behavioural changes are considered as an outcome of the complex physiological responses and have often been considered as sensitive indicator of stress (Eisler, 1977). The fishes exposed to lethal concentration of fluoride showed hyperactivity, rapid swimming, jumping, increased opercular movement and rapid surfacing due to fluoride toxicity, these observations are in accordance with the findings of Gupta, 2003; Tripathi *et al.*, 2004; Kumar, 2005 and Tripathi, 2007. Similar observations are reported by many other workers with different toxicants (Tripathi and Shukla, 1988; Lata *et al.*, 2001; Prashanth *et al.*, 2005).

The increased swimming and jumping activity after fluoride exposure in initial hours revealed contact avoidance for toxic chemicals in fishes (Devi, 2003). The increased opercular movement and rapid surfacing activity recorded in this study was in accordance with the result of earlier workers (Gupta, 2003; Tripathi *et al.*, 2004). Behavioural alterations in fish observed during experimentation may be due to the alterations in central nervous system during exposure to fluoride and requirement of more oxygen for energy expenditure to cope up with stress condition. Profuse mucus secretion in the fish exposed to fluoride may be regarded as adaptive response for providing additional protection against test chemical and to avoid absorption of the toxicant by general body surface (Devi, 2003; Choudhary and Jha, 2009). Change in body colouration may also be due to deposition of ample amount of mucus on body and alteration in the chromatophore pattern. Similar findings have also been reported in *Channa punctatus* after fluoride exposure by Gupta (2003) and Tripathi *et al.* (2005).

Acute toxicity test play an essential role in assessing the relative toxicity of a chemical to aquatic animals. Toxicity caused by pollutant may induce change in normal behaviour and various physiological activity of organism which may inevitably leads to ecological crisis. Change in the physiology and behaviour of aquatic organisms such as respiration, feeding and swimming performance could be used as rapid and sensitive indicator of behavioural toxicity. Altered behavioural pattern which reveals adverse effects of contaminant exhibits the environmental perturbation. These behavioural patterns in such conditions might explain other consequences such as reduced survival, growth or reproduction. Thus behavioural responses after standardization can serve as biomarkers of environmental stress and fish as primary bio-indicator of degrading water quality particularly due to chemical contaminants.

Acknowledgements

The authors are thankful to Prof. Minakshi Shrivastava, Head, Department of Zoology, University of Lucknow, for providing necessary laboratory facilities to carry out this study.

References

APHA, AWWA and WEF (2005). Standard methods for the examination of water and wastewater, 21st Edition. American Public Health Association. Washington D.C.

Arner, J., Vila, P. and Plautz, C.Z. (2009). Effects of local water contaminants on the development of aquatic organisms. *Sujur.*, 1: 12-26.

Beg, K.R. and Ali, S. (2008). Chemical contaminants and toxicity of Ganga river sediment from up and down stream area at Kanpur. *Amer. J. Environ. Sci.*, 4(4): 362-366.

Bhatnagar, C., Bhatnagar, M. and Regar, B.C. (2007). Fluoride-induced histopathological changes in gill, kidney and intestine of freshwater teleost, *Labeo rohita*. *Fluoride*, 40(1): 55-61.

Camargo, J.A. (2003). Fluoride toxicity to aquatic organisms: A review. *Chemosphere*, 50(3): 251-261.

Chinoy, N.J. (1991). Effects of fluoride on physiology of animals and human beings. *Indian J. Env. Toxicol.*, 1(1): 17-32.

Chitra, T., Reddy, M.M. and Rao, J.V.R. (1983). Levels of muscle and liver tissue enzymes in *Channa punctatus* (Bloch) exposed to NaF. *Fluoride*, 16: 48-51.

Chowdhary, M. and Jha, M.M. (2009). Acute toxicity and behavioural responses of nickel sulphate to the fish *Heteropneustes fossilis*. *Aquacult.*, 10(1): 143-145.

Devi, S. (2003). Behavioural changes in *Oreochromis mossambicus* exposed to endosulfan. *J. Ecobiology,* 15(6): 425-430.

Eisler, R. (1977). Behavioural response of marine poikilotherms to pollutants. *Phil. Trans. R. Soc.* (*Sec B*)., 286: 507-521.

EPA (1997). Public Health Global for Fluoride in drinking water. Pesticide and Environmental Toxicology. Section office of Environmental Health Hazard Assessment, California. Environmental Protection Agency, December, 1997.

Gupta, R. (2003). Pathophysiological consequences to freshwater fish, *Channa punctatus* induced by fluoride. Ph.D. Thesis, University of Lucknow, Lucknow.

Gupta, R., Saroj, Tripathi, M. and Sharma, U.D. (2001). Fluoride induced haematological and histopathological changes in freshwater fish *Labeo rohita. Biological Memoirs,* 26(2): 27-31.

Hamilton, M.A., Russo, R.C. and Thurston, R.V. (1977). Trimmed spearman karber method for estimating median lethal concentration. *Environ. Sci. Technol.,* 11: 714-719.

Joshi, P.K., Bose, M. and Harish, D. (2002). Hematological changes in the blood of *Clarias batrachus* exposed to mercuric chloride. *Ecotoxicol. Environ. Monit.,* 12: 119-122.

Kumar, A. (2005). Evaluation of fluoride toxicity on reproductive system of freshwater fish. Ph.D. Thesis, University of Lucknow, Lucknow.

Kumar, A., Tripathi, N. and Tripathi, M. (2007). Fluoride-induced biochemical changes in freshwater catfish (*Clarias batrachus,* Linn.). *Fluoride,* 40(1): 37-41.

Lata, S., Gopal, K. and Singh, N.N. (2001). Toxicological evaluation and morphological studies in a catfish *Clarias batrachus* exposed to carbaryl and carbofuran. *J. Ecophysiol. Occup. Hlth.,* 1: 121-130.

Prashanth, M.S., David, M. and Mathed, S.G. (2005). Behavioural changes in freshwater fish, *Cirrhinus mrigala* (Ham) exposed to cypermethrin. *J. Environ. Biol.,* 26(1): 141-144.

Shankar, B.S., Reddy, M.T.M. and Balasubramanya, N. (2007). Incidence of fluoride contamination in the ground waters of an industrial area- a case study. *Our Earth,* 4: 19-23.

Sprague, J.B. (1971). Measurement of pollutant toxicity to fish. III sublethal effects and safe concentrations. *Water Res.*, 5: 245-266.

Tripathi, A., Kumar, A. and Tripathi, M. (2006). Effect of fluoride on vertebral column of a freshwater fish, *Channa punctatus*. *J. Appl. Biosci.*, 32(2): 164-167.

Tripathi, A., Kumar, A., Rani, A. and Tripathi, M. (2004). Fluoride induced morphological and behavioural changes in freshwater fish, *Channa punctatus*. *J. Ecophysiol. Occup. Hlth.*, 4: 83-88.

Tripathi, G. and Shukla, S.P. (1988). Toxicity bioassay of technical and commercial formulation of carbaryl to freshwater catfish, *Clarias batrachus*. *Ecotoxicol. Environ. Safe*, 15: 227-281.

Tripathi, M., Tripathi, A. and Gopal, K. (2005). Impact of fluoride on pigmentation of a freshwater fish, *Channa punctatus*. *J. Appl. Biosci.*, 31(1): 35-38.

Tripathi, N. (2007). Evaluation of fluoride toxicity on reproduction in freshwater fish. Ph.D. Thesis, University of Lucknow, Lucknow.

Chapter 16

Adverse Effect of Ampicilin on Brain Lipids in Albino Rats

B. Kalaivani[*1], R. Dhamotharan[2], and G. Vanitha Kumari[2]*

[1]*Department of Zoology, Sri Sarada College for Women, Salem – 636 016, Tamil Nadu, India*

[2]*Division of Endocrinology, Department of Zoology, Bharathiar University, Coimbatore – 641 046, Tamil Nadu, India*

ABSTRACT

Ampicilin is one of the antimicrobial and β-lactum antibiotic of penicillin derivative drug which acts on gram positive and gram negative bacteria by blocking their cell wall synthesis. The present study investigates the probable effect of ampicilin on brain in terms of total lipids, total cholesterol, total glyceride glycerol and total phospholipids. The rats were divided into four groups. Group I: Control rats were given normal saline. Whereas, group II and III rats received ampicilin (80mg/kg body weight) for 5 and 10 days, respectively. Group IV rats received ampicilin for 10 consecutive days and were allowed for a withdrawal period of 5 days. In the present study it was observed that in rat brain there was a marked decrease in the concentrations of total lipids, total glyceride glycerol and total phospholipids, while there was an increase in the total

* Corresponding author: E-mail: kalai03win@yahoo.co.in

cholesterol concentrations. Thus ampicilin has an adverse and inhibitory effect on lipid metabolism.

Keywords: *Ampicilin, Lipid metabolism, Total lipids, Total cholesterol, Total glyceride glycerol, Total phospholipids.*

Introduction

In the present day the usage of drugs for the treatment of many types of ailments has become very common. Even though these drugs have some therapeutic effects on the ailments for which they are taken, they also seem to affect other organ systems to a great extent. These effects may be reversible or irreversible. Pharmacological agents of all sorts affect the central nervous system function and these effects are usually associated with negative personality traits and in extreme cases, severe personality disorders (Goodman and Gilman, 1991).

The CNS stimulants are drugs which primarily act by stimulating the CNS mostly producing a generalized action which may, at high doses, result in convulsions (Tripathi, 1988). Aminoglycosides, to which class ampicilin belongs, are known to exert cytotoxic effects on the nerve cells (Dretchen *et al.*, 1973 and Oshima *et al.*, 1989). Ampicilin is one of the antimicrobial and β – lactum antibiotic of penicillin derivative drug which acts on gram positive and gram negative bacteria by blocking their cell wall synthesis. Based on cytotoxic effects on the nervous system, the present investigation was attempted in order to understand the possible effect of the drug ampicilin on lipid metabolism of brain.

Materials and Methods

All the chemicals and reagents were of analytical grade, obtained from british Drug House (BDH, England and India), E. Merck (Germany and India), Sigma chemicals Company (USA); Lobe Chemie (Austria) and Sarabai M Chemicals (India). The drug Ampicillin sodium was procured from Ranbaxy Laboratory Limited.

Adult female albino rats weighing about 150 – 200 g bodyweight are selected for this work. They were kept in a well ventilated animal house with constant 12 hours of darkness and 12 hours of light schedule. They were fed with rat pellet (Hindustan Lever Limited, India). The rats were divided in to four groups.

Group – I

Control rats received normal saline (0.9 per cent), intramuscularly.

Group – II

Rats received ampicilin (80 mg/kg body weight) for 5 consecutive days

Group – III

Rats received ampicilin (80 mg/kg body weight) for 10 consecutive days

Group – IV

Rats received ampicilin (80 mg/kg body weight) for 10 consecutive days, im, and animals were allowed for a withdrawal period of 5 days.

The body weight of each animal was recorded before and after treatment schedule. Twenty four hours after the last injection the animals were sacrificed by decapitation. The brain tissues were washed in saline and blotted on a filter paper and weighed accurately. The weights were expressed in terms of mg/100g body weight. They were kept frozen until they were subjected to biochemical estimations.

The method of Folch *et al.* (1957) was employed for the extraction of lipids from the tissue. The total lipid was estimated by the method of Frings *et al.* (1972). The tissue total cholesterol was determined by Pareh and Junk (1970). The colorimetric method described by Van Handel and Zilversmith (1957) was used for the estimation of total glyceride glycerol. The modified method of Fiske and Subbarow (1925) as per Marinetti (1962) was employed for phospholipids quantification.

Results and Discussion

Effect of Ampicillin on Body Weight of the Adult Rats

Table 16.1 showed a marked decrease in the body weights of all the experimental groups except withdrawal group after ampicilin treatment given to rats. The withdrawal group attained normalcy compared to other groups. The decreased body weight under ampicilin influence may be due to reduced food and water intake.

Many reviews are available on influence of under-nutrition on lipids of the brain (Dobbing, 1968a; Krob and Winick, 1969).

Table 16.1: Effect of Ampicilin Treatment and its Withdrawal on Bodyweight of the Adult Rats

Sl.No.	Control	5 Days	10 Days	10 Days Withdrawal
1.	100 ± 0.000	96.760±0.926***	94.397±1.375**	06.584±5.122

The body weights are expressed in 100 g.

Each value is mean ± SEM of 5 animals.

Control and Experimental groups as described in the text.

* P < 0.05; ** P < 0.01 and *** P < 0.01.

Effect of Ampicillin on Lipid and Lipids Classes in Different Regions of Brain of Adult Rats

Effect on Total Lipids Concentration

Table 16.2 shows the effect of ampicilin on the total lipid concentration in prosencephalon, mesencephalon and myelencephalon of rats at 5 days and 10 days treatment as well as the drug withdrawal period. In all the brain regions the total lipid concentration was markedly decreased in all the experimental groups. In the brain of control rats, myelencephalon had the highest total lipid concentration followed by procencephalon and mesencephalon. Regional variation of total lipids, total phospholipid, total cholesterol and total glycerol in the different regions of control rats has been observed.

Ampicillin administration caused a significant reduction in total lipid and lipid classes in the different regions of brain. There was a substantial variation in the type of lipid classes that contributed for decrement of the total lipids in the nervous tissues. The effect appears to be persistent, for even after withdrawal of the drug for more than ten days the drug was ineffective in restoring the total lipids to normalcy suggesting the adverse effect of ampicillin on the lipid synthesising machinery of prosencephalon, mesencephalon and myelencephalon. These may probably be due to the drugs cytotoxic effect (Oshima *et al.*, 1989). A significant fall in saturated fatty acid contents and lipid levels were observed by many

Table 16.2: Effect of Ampicilin Treatment and its Withdrawal on Total Lipids Concentration in the Brain of Adult Albino Rats

Parts of Brain	*Control*	*5 Days*	*10 Days*	*10 Days Withdrawal*
Procencephalon	9.091±0.000	8.072±0.003***	8.880±0.003***	6.67±0.003***
Mesencephalon	9.366±0.03	8.817±0.003***	7.859±0.003***	6.445±0.003***
Myelencephalon	9.478±0.003	8.325±0.003***	7.942±0.003***	5.106±0.001***

The values are expressed as mg/g tissue.

Each value is mean±SEM of 5 animals.

Control and Experimental groups as described in the text.

* $P < 0.05$; ** $P < 0.01$ and *** $P < 0.01$ control Vs other groups.

workers in injured brain (Nakahara *et al.*, 1991). The change in lipid classes of the central auditory pathway due to streptomycin, another aminoglycoside, administration has been studied by Faraq and Khana (1986) in rats. A significant fall in saturated and unsaturated fatty acid contents and lipid levels were observed by many workers in injured brain (Nakahara *et al.*, 1991)

Effect on Cholesterol Concentration

Table 16.3 shows that in the control rat brain, cholesterol concentration was found in higher amount in both mesencephalan and prosencephalan compared to myelencephalan. Like the total lipid concentration, the total cholesterol concentration also exhibited a decrease at 5 days of ampicillin treatment in prosencephalon. In mesencephalan and myelencephalan areas of the brain the same was increased significantly. The cholesterol concentration has been reported in the brain of rats by Bharghawa *et al.* (1991) after corticosteroid administration. Similarly, Howard *et al.* (1965)reported an increase in cholesterol concentration in brain of steroid implanted rats. Kaur *et al.* (1989) have shown tetramethazone administration treatment to increase cholesterol synthesis in brain of rats. However in withdrawal groups, there was marked reduction in the total cholesterol in all the brain regions.

Effect on Total Phospholipids Concentration

The prosencephalan and myelencephalan regions showed decreased total phospholipids concentration both at 5 and 10 days of ampicillin treatment (Table 16.4). Whereas, such on effect was observed only after 10 days of treatment in the phospholipid concentration. The above observed decrease in total phospholipids might suggest a decrease in phospholipid biosynthesis or an accelerated catabolism. Ampicillin might only induce an enhanced phospholipid degradation since the total lipids also decreased markedly in the brain tissues.

In the present study, the total phospholipids degradation may be due to the triggering of the activities of phospholipase enzyme or may be due to lipid peroxidation. Incases of brain injury of various mechanism an impaired energy metabolism has been documented (Goto *et al.*, 1988). Mitochondrial membranes, organelles that are essential to cellular metabolism, also contain phospholipids and this ischemic degradation by destruction of these phospholipids

Table 16.3: Effect of Ampicilin Treatment and its Withdrawal on Total Cholesterol Concentration in the Brain of Adult Albino Rats

Parts of Brain	*Control*	*5 Days*	*10 Days*	*10 Days Withdrawal*
Procencephalon	3.871±0.003	3.041±0.006***	4.190±0.002***	3.151±0.003***
Mesencephalon	3.951±0.003	4.412±0.003***	4.337±0.003***	3.289±0.003***
Myelencephalon	3.217±0.003	4.375±0.003***	4.557±0.003***	2.551±0.003***

The values are expressed as mg/g tissue.

Each value is mean±SEM of 5 animals.

Control and Experimental groups as described in the text.

* $P < 0.05$; ** $P < 0.01$ and *** $P < 0.01$ control Vs other groups.

Table 16.4: Effect of Ampicilin Treatment and its Withdrawal on Total Phospholipid Concentration in the Brain of Adult Albino Rats

Parts of Brain	*Control*	*5 Days*	*10 Days*	*10 Days Withdrawal*
Procencephalon	2.810±0.002	2.657±0.003***	2.632±0.003***	1.838±0.003***
Mesencephalon	2.389±0.003	2.375±0.003***	2.228±0.003***	1.579±0.003***
Myelencephalon	3.217±0.003	2.625±0.003***	1.823±0.003***	1.531±0.003***

The values are expressed as mg/g tissue.

Each value is mean±SEM of 5 animals.

Control and Experimental groups as described in the text.

* $P < 0.05$; ** $P < 0.01$ and *** $P < 0.01$ control Vs other groups.

Table 16.5: Effect of Ampicilin Treatment and its Withdrawal on Total Glycerol Concentration in the Brain of Adult Albino Rats

Parts of Brain	*Control*	*5 Days*	*10 Days*	*10 Days Withdrawal*
Procencephalon	2.407±0.00	2.378±0.003	2.058±0.003***	1.681±0.003***
Mesencephalon	3.014±0.002	2.036±0.003***	1.290±0.003***	1.578±0.003***
Myelencephalon	3.033±0.003	1.325±0.003***	1.562±0.003***	1.020±0.003***

The values are expressed as mg/g tissue.

Each value is mean±SEM of 5 animals.

Control and Experimental groups as described in the text.

* $P < 0.05$; ** $P < 0.01$ and *** $P < 0.01$ control Vs other groups.

has been shown by several investigators (Majewska *et al.*, 1978 and Nakahara *et al.*, 1991). A decrease in phospholipids in cell injury has been shown to cause a further deterioration of energy metabolism of the brain (Siesjo, 1984) and contribute significantly for loss of oxidative phosphorylation (Smith *et al.*, 1980).

Effect on Total Glyceride Hlycerol Concentration

Both mesencephalan and myelencephalon regions showed a significant decrease in total glycerol concentration in all the experimental groups studied. However prosencephalan exhibited such a reduction in total glycerol concentration only in 10 days ampicillin treated groups as well as in withdrawal group (Table 16.5). Hawk (1965) has reported very low levels of cholesterol and its esters and triglycerides in the nervous tissues. In the present study, the total glyceride glycerol concentration was comparatively lower than the phospholipids in the prosencephalan and myelencephalan regions of control rats.

Thus the above data obtained in this study suggests an adverse effect of ampicilin in prosencephalon, mesencephalon and myelencephalon. Further, the altered lipid metabolism as indicated by changes in total lipids, cholesterol, neutral lipids and to some extent phospholipids suggests an altered cell permeability, cytotoxicity and probable cell injury.

References

Bhargawa, H.K., Lalitha Tenneti and Telans, S.D., 1991. Corticosteroid administration and lipid metabolism in brain regions during development. *Ind. Biochem. Biophys.*, 28: 214 – 218.

Cumings, J.N., Grundt, I.K. and Yanigahara, T., 1968. *J. Neyrol. Neyrosurg. Psychiat.*, 31: 334 – 348.

Dobbing, J., 1968a. In "Psychopharmacology, Dimensions and Perspectives".(C.R.B. Joce, ed.), p 345. Lippincott, Philadelphia, Pennsylvania.

Dretchen, K.C., Sokoll, M.S., Gergis, S.D. and Long, J.P., 1973. Relative effects of streptomycin on motor nerve terminal end peptide. *Eur. J. Pharmacol.*, 22: 10.

Faruq, N.A. and Khana, L., 1986. Lipid changes in central auditory pathway following streptomycin intoxication in rats. *Indian. J. Med. Res.*, 83: 318.

Fiske, C.H. and Subbarow, Y., 1925. The calorimetric determination of phosphorous. *J Biol Chem.*, 66: 375 – 400.

Folch, J., Lees, M. and Sloans – Stanely, G.H., 1957. A simple method for the isolation and purification of lipids from animal tissues. J. *Biol. Chem.*, 226: 497 – 509.

Frings, C. S., Fendley, T.W., Dunn, R.T. and Qwen, C.A., 1972. Improved determination of total lipids by sulphovanilin reaction. *Clin. Chem.*, 18: 673 – 674.

Goodman, L.S. and Gilman, A.C., 1991. In: The pharmacologic basis of therapeutics, 8th edition, Pergamon Press, Macmillon Publishing corporation, New Yark. pp: 639– 650.

Goto, Y., Okamoto, S. and Yonekawa, Y., 1988. Degradation of phospholipids molecular species during experimental cerebral ischemia in rats. *Stroke.*, 19: 728 – 735.

Kaur, N.N. Sharma, A.K. and Gupta, A.K., 1989. *Indian. J. Biochem. Biophys.*, 26: 371 – 376.

Krob, T.M. and Winick, M., 1969. Pediatria (Santiago). 12, 151.

Majewska, M.D., Strosznajder, J. and Lazarewicz, J., 1978. Effect of ischemic anoxia and barbifurate anesthesia on free radical oxidation of mitochondrial phospholipids. *Brain Res.*, 158: 423–434.

Marinetti, G.V., 1962. Chromatographic separation, identification and analysis of phophatides. *J. Lipid. Res.*, 3: 1 – 20.

Nakahara, I., Kikuch, H. and Taki, W., 1991. Degredation of mitochondrial phospholipids during experimental cerebral ischemia in rats. *J. Neurochem.*, 57: 839 – 844.

Oshima Meiko, Misuko Hashiguchi, Miyuki Nakasuji, Noboru Shindo and Seiichi Shikata, 1989. Biochemical mechanisms of aminoglycoside cell toxicity. I. Accumalation of phospholipids during myeloid body formation and histological studies on myeloid bodies using 12 aminoglycoside antibodies. *J. Biochem.*, 106 (5): 794.

Parekh, A.C. and Jung, D. H., 1970. Cholesterol with ferric acetate-uranyl acetate and sulfuric acid-ferrous sulpahate reagents. *Anal Chem.*, 42: 1423–1427.

Sisejo, B.K., 1984. Cerebral circulation and metabolism. *J. Neurosurg.*, 60: 883–908.

Smith, M.W., Yrjocollan, Myongwon Kahng and Benjamin, F. Tramp., 1980. Changes in mitochondrial lipids of rat kidney during icchemia. *Biochemica et Biophysica acta.*, 618: 192–201.

Tripathi, K.D., 1988. In: Essentials of medical pharmacology, 2nd edition. Jaypee Brothers, Medical Publishers, New Delhi.

Van Handel, E. and Zilversmith, D.B., 1957. Micromethod for the direct determination of triglycerides. *J. Lab. Clin. Med.*, 50: 152 – 157.

Chapter 17

Effect of Ivermectin on Some Biochemical Constituents of the Fish *Channa punctatus*

Kamal Kumar Saxena and Namita Verma*

Pest and Parasite Research Laboratory,
Department of Zoology, Bareilly College, Bareilly – 243 005, India

ABSTRACT

Ivermectin is widely used both as an antiparasitic compound and as an insecticide but it also creates serious threat to non-target organisms mainly aquatic animals. In the present study effect of ivermectin was studied on some biochemical parameters of the fish *Channa punctatus* up to the exposure perriod of 30 days. LC_{50} of ivermectin to this fish was determined and then three sublethal concentrations (0.0005, 0.001 and 0.005 ppm) were selected for the experiment. The activities of acid phosphatase, alkaline phosphatase and amylase were determined in the muscles of *Channa punctatus* at an interval of 5 days up to 30 days. Significant reduction in the activity of acid phosphatases and amylase and enhancement in the activity of alkaline phosphatases were recorded during exposure period. The results obtained also showed a great

* Corresponding author: E-mail: namitaverma38@yahoo.com

degree of fluctuations in the activity of Acid phosphatase (ACP), Alkaline phosphatase (ALP) and amylase due to exposure to sublethal concentrations of ivermectin.

Keywords: *Ivermectin, Acid phosphatase (ACP), Alkaline phosphatase (ALP), Amylase, Channa punctatus.*

Introduction

Appreciation of fisheries and aquatic systems has been accompanied by increasing concern about the effects of growing human populations and human activity on aquatic life and water quality. Pesticides are a group of toxic compounds linked to human use that have a profound effect on aquatic life and water quality.

Aquatic pollution is significant to fisheries and aquaculture industries. The changes of physical, chemical, and biological parameters of water alter the behavior of fishes besides causing mortality. The presence of pesticides within aquatic environment can induce physiological and behavioral changes in aquatic organisms. Pesticides leave residues in water and soil even after several days of the spray in the crop fields. Avermectins used in crop field are highly toxic to the aquatic organism including fish.

Ivermectin is a member of avermectins which is a group of chemicals discovered in 1975, isolated from culture of the actenomycete *Streptomyces avermitilis* (Campbell, 1989; Fisher and mrozik, 1992). Ivermectin is being used for domestic and wild animals to control their ecto and endo parasites. Recently these compounds are also introduced as insecticides. Most of the insecticides are not biodegradable and tend to persist for years together in soil and water. They get transported into the bodies of higher organisms through the food chain.

There are many reports (Bradbury and coats, 1989; Mestress and Mestress, 1992; Moore and Waring, 2001; Saxena and Seth, 2002) related to toxicity of synthetic chemicals on different fish species but there is no information on the metabolism of *Channa punctatus*. So the present study was designed to study the changes in some biochemical parameters of *Channa punctatus* treated with ivermectin.

Material and Methods

Channa punctatus were obtained from local water bodies and acclimated in the laboratory for a maximum period of 30 days and then divided into four experimental and one control groups. Fish then transferred to different aquaria containing various lethal and sub lethal doses of Ivermectin were maintained up to 30 days. During this period fish were sacrificed after 5, 10, 15, 20, 25, and 30 days for the study of the changes in biochemical constituents due to exposure to ivermectin. For the determination of biochemical constituents muscles were collected, weighted and homogenized in glass homogenizer and centrifuged in Zanetzki K-24 refrigerated. The supernatants were separated and used for measuring the activity of enzymes. The activities of acid and alkaline phosphates and amylase were measured by the methods given in Bergmeyer (1971). The experiment was repeated at least four times. The data obtained was analyzed by ANOVA to test its significance.

Results

The pattern of changes in activities of acid phosphatase, alkaline phosphatase and amylase are given in Tables 17.1–17.3 respectively.

Effect of Ivermectin on the Mortality of *Channa punctatus*

When *Channa punctatus* were exposed to ivermectin at the concentration of 0.006 ppm and above mortality was observed. At 0.014 ppm fish died within few minutes and the period was increased with decrease an the concentration of ivermectin. When the concentration was less than 0.006 ppm fish survived for 30 days or more. So three concentrations (0.0005, 0.001 and 0.005 ppm) which were below 0.006 ppm were selected as sub lethal concentration for studying the effect of ivermectin on fish metabolism.

Effect of Ivermectin on Biochemical Activities in *Channa punctatus*

Acid Phosphatase Activity

The fishes exposed to sublethal concentrations of ivermectin showed the significant alterations ($p < 0.05$) in the activity of acid phosphate. Maximum decrease was 14 per cent after exposure to 0.0005 ppm and 0.001ppm of ivermectin for 10 days, while minimum decrease was 45 per cent after exposure to 0.001 ppm and.005ppm of ivermectin for 30 days.

Table 17.1: Effect of Ivermectin on ACP Activity (μ moles pnp/mg protein/hours) in Muscles of *Channa punctatus*

Concentration	*Exposure Period*						
	0 Day	*5 Days*	*10 Days*	*15 Days*	*20 Days*	*25 Days*	*30 Days*
Control	0.30	0.30	0.28	0.30	0.33	0.28	0.33
0.0005 ppm		0.24 – 20%	0.24 – 14%	0.23 – 23%	0.24 – 27%	0.22 – 21%	0.19 – 42%
0.0005 ppm		0.25 – 17%	0.24 – 14%	0.23 – 23%	0.22 – 33%	0.19 – 32%	0.18 – 45%
0.005 ppm		0.24 – 20%	0.22 – 21%	0.23 – 23%	0.19 – 42%	0.19 – 32%	0.18 – 45%

* % change from control value.

Table 17.2: Effect of Ivrmectin on ALP Activity (µ moles pnp/mg protein/hours) in Muscles of *Channa punctatus*

Concentration	*Exposure Period*						
	0 Day	*5 Days*	*10 Days*	*15 Days*	*20 Days*	*25 Days*	*30 Days*
Control	0.24	0.24	0.23	0.21	0.24	0.26	0.26
0.0005 ppm		0.24 0%	0.23 0%	0.23 + 10%	0.23 – 4%	0.24 – 8%	0.24 – 8%
0.0005 ppm		0.23 – 4%	0.22 – 4%	0.23 + 10%	0.26 + 8%	0.24 – 8%	0.26 0%
0.005 ppm		0.26 + 8%	0.24 + 4%	0.24 + 14%	0.23 – 4%	0.24 – 8%	0.23 – 12%

* % change from control value.

Table 17.3: Effect of Ivermectin on Amylase Activity (street-close unit/100ml) in Muscles of *Channa punctatus*

Concentration	*Exposure Period*						
	0 Day	*5 Days*	*10 Days*	*15 Days*	*20 Days*	*25 Days*	*30 Days*
Control	28.57	30.47	30.95	29.98	30.15	29.19	30.12
0.0005 ppm		10.12 – 67%	10.18 – 67%	11.11 – 63%	12.53 – 58%	13.11 – 55%	14.10 – 53%
0.001 ppm		11.11 – 64%	11.26 – 64%	12.10 – 60%	13.50 – 55%	13.56 – 54%	13.89 – 54%
0.005 ppm		12.23 – 60%	12.26 – 60%	13.16 – 56%	14.10 – 53%	14.56 – 50%	14.59 – 52%

* % change from control value.

Alkaline Phosphatase Activity

The fish exposed to sublethal concentrations of ivermectin showed the significant increase when compared the control average value ($p < 0.05$). Maximum increase was 14 per cent after exposure to 0.005 ppm of ivermectin for 15 days and maximum decrease was 12 per cent after exposure to 0.005 ppm of ivermectin for 30 days.

Amylase Activity

The fishes exposed to sublethal concentrations of ivermectin showed the significant ($p < 0.05$) reduction when compared with the control. Maximum decrease was 67 per cent when *Channa punctatus* were exposed to 0.0005 ppm of ivermectin for 5 to 10 days.

Structural and Behavioral Changes in *Channa punctatus*

Structural and behavioural changes were observed in *Channa punctatus*. During exposure period fish remained restless. They showed erratic and uncoordinated swimming. Sometimes they tried to jump out of the aquarium. An important structural change was the formation of a hole on the lower side between pectoral girdles. The diameter of hole increased gradually and in some fishes heart was visible through the hole. The color of body also faded during exposure period.

Discussion

Ivermectin is a known anthelmintic which is recently introduced as insecticide also. The biochemical changes observed in present studies in *Channa punctatus* are due to exposure to sublethal concentrations of Ivermectin. These findings are in accordance with the findings of previous workers (Saxena and Seth, 2002; Saxena and Gupta, 2003; Saxena and Gupta, 2005; Sirohi and Saxena, 2006; Saxena and Sirohi, 2007; Maheswaran *et al.*, 2008). Pesticides are known to cause biochemical toxicological changes in fishes. Avermectins, which is a known antiparasitic compound (Singh and Kaushal, 1995) are also being used as pesticide and is becoming a wide spectrum compound against a variety of pests in agriculture and other purposes.

In the present studies ivermectin caused mortality in *Channa punctatus* at very low concentrations *i.e.* 0.014 ppm. Exposure to ivermectin caused significant alterations in the activities of acid phosphatase, alkaline phosphatase and amylase.

Acid and alkaline phosphatases are reported (Bhatnagar *et al.*, 1995; Rana *et al.*, 2003) to be associated with the transport of metabolites, with metabolism of phospholipids, phosphoproteins, nucleotides and carbohydrates and synthesis of protein. Acid phosphates are responsible for the hydrolysis of ester linkage of phosphates which helps in the autolysis of the cell after its death. The alterations in the activity of acid and alkaline phosphates in present investigations indicates the necrotic changes in the muscles of *Channa punctatus* and the interference of Ivermectin in phosphate and protein metabolism.

The activity of amylase is directly related with the carbohydrate metabolism of the animal. Alterations in the activity of amylase clearly indicate the interference of ivermectinin carbohydrate metabolism of fish. Saxena and Sirohi (2007) have also reported similar changes in *Channa punctatus* due to exposure to pyrethroides pesticide.

In the present studies doramectin/Ivermectin at sublethal concentrations caused structural and behavioural changes in *Channa punctatus* and also altered the activities of acid phosphates, alkaline phosphates and amylase. It is an indicator of the toxicity caused by this compound in this fish at the level. Which may be harmful for its survival and growth.

Acknowledgements

We are thankful to Dr. K. Singh, Head, Department of Biotechnology and Dr. (Mrs.) Sunita Sharma, Reader, Department of Zoology, Bareilly College, Bareilly for their cooperation.

References

Bergmeyer, H.U., 1971. *Methods of Enzymatic Analysis*. Academic Press, New York.

Bradbury, S.P. and Coats J.R., 1989. Comparative toxicology of the pyrethroid insecticides; *Rev. of Environ. Contam Toxicol.*, 108: 133-177.

Bhatnagar, M.C., Tyagi, M. and Tamata, S., 1995. Pyrethroid induced toxicity to phosphatases in *Clarias batrachus*. *J. Environ. Biol.*, 16(1): 11-14.

Campbell W.C., 1989. *Ivermectin and Abamectin*. Springer Verlag, N.Y.

Fisher, M.H. and Mrozik, H., 1992. The chemistry and pharmacology of avermectins. *Annu. Rev. Pharmacol. Toxicol.*, 32: 537–553.

Mestres, R and Mestres, G., 1992. Deltamethrin uses and environmental safety. *Rev. Environ. Contam. Toxicol.*, 124:1-18

Moore, Andrew and Waring, Calin. P., 2001. The effects of a synthetic pyrethroids pesticide on some aspects of reproduction in Atlantic Salmon (*Salmosalar C.*) *Aquat. Toxicol.*, Amsterdam, 52: 1-12

Maheswaran, R., Devapaul, A., Muralidharan, S.,Velmurugan, B., Iguacimuthu, S., 2008. Haematological studies of freshwater fish, *Clarias butrachus* (L.) exposed to mercuric chloride. *IJIB.* 2(1): 49-54

Saxena K.K. and Gupta, P. 2003. Effect of permethrin on the activities of acid and alkaline phosphomonoesterases in a freshwater fish *Channa punctatus. Proc. Nat. Symp. Biochem. Sci. Hlth. Environ. Asp.*, 477–479.

Rana, M.A. Yeragi, G.S. and Koli, A.V., 2003. Effect of pesticides on acid phosphatase activity of *Mudskipper Boleophalmus dussumieri. J. Ecotoxicol. Environ. Monit.*, 13 (2): 155-158

Saxena, K.K. and seth, N. 2002. Toxic effects of cypermethrin on certain haematological aspects of freshwater fish *Channa punctatus, Bull. Environ. Contam. Toxicol.*, 69: 364–369.

Saxena, K.K. and Gupta, P., 2005. Impact of carbamates on glycogen contents in the muscles of freshwater fish *Channa punctatus. Poll. Res.*, 24(3): 669-670.

Sirohi, V. and Saxena, K.K., 2006. Toxic effect of l-cyhalothrin on biochemical contents of freshwater fish *Channa punctatus. Journal of Fisheries and Aquatic Science*, 1(12): 112-116.

Singh, K. and Kaushal, P., 1995. Effect of fenbendazole and ivermectin on carbohydrate metabolism of swine kidneyworm *Stephanurus dentatus. J. Parasitic Diseases*, 19: 191-192.

Saxena, K.K. and Sirohi, V., 2007. Effect of 1–cyhalothrin on the activities of Trypsin and Lypase in freshwater fish *Channa punctatus. Journal of fisheries and Aquatic Science*, 2(2): 168-172.

Chapter 18

Traits Related to Social Hierarchy in Ruminant Ungulates: Dairy Goats

*R. Ungerfeld**

Departamento de Fisiología, Facultad de Veterinaria, Lasplaces 1550, Montevideo 11600, Uruguay

ABSTRACT

In ruminants, social hierarchies determine an unequal access to different resources. Social rank has been related to body weight, age and horn presence in wild small ruminant populations; thus to determine: 1) if social rank is related to body weight, horn presence and length, and primiparity, and 2) the role of visual signals in maintenance of hierarchy and in agonistic behaviours used to maintain hierarchical relationships in dairy goats under intensive management, two trials were performed. In Trial 1, the relation between the defined traits and success index was determined in three herds of dairy goats comprised by 45, 43 and 51 goats. Body weight was the only characteristic positively associated with success index. In the second Trial, the success index of 11 goats was compared with data obtained after temporary blindness with a bandage. Although success index was modified during the period in which goats cannot see each other, modifications in their hierarchical positions were only temporarily affected. It was

* Corresponding author: E-mail: rungerfeld@gmail.com

also observed that the agonistic behaviours used by goats to displace each other were modified during the period in which artificial blindness was provoked. It was concluded that social rank is positively related with body weight in dairy goats, and this relation may be consequence of how each goat view other goats, as visual signals were necessary to maintain the hierarchical positions into the group. Displacement strategies were modified when view communication was not possible.

Keywords: *Dominance, Social rank, Hierarchy, Agonistic interactions, Caprine.*

Introduction

Most ungulates exhibit high levels of social organization, including in most species social hierarchies. In ruminants, social hierarchies determine an unequal access to different resources, like food (Loretz *et al.*, 2004; Jørgensen *et al.*, 2007), water (Andersson *et al.*, 1984), lying space (Andersen and Bøe, 2007), shade (Sherwin and Johnson, 1987), or individuals from the other gender (Alvarez *et al.*, 2003). Social dominance has been investigated in wild goats (Côte, 2000), in which it has been reported a direct effect of social rank on reproductive success (Côte and Festa-Bianchet, 2001). In dairy goats a positive relationship has been observed between milking order and milk yield (Patón *et al.*, 1995), and social rank and response to superovulatory treatments (Ungerfeld *et al.*, 2007) or control of gastrointestinal parasites (Ungerfeld and Correa, 2007).

In females from wild small ruminants, social rank has been related with different traits. For example, in chamois age, body weight, and horn size are related to social rank (Locati and Lovari, 1991). In other small ruminants social rank has been linked to age (bighorn sheep: Festa-Bianchet, 1991; red deer: Thouless and Guinness, 1986; Nubian ibex: Greenberg-Cohen *et al.*, 1994; Ammotragus: Cassinello, 1995). Also the presence of weapons, as horns or antlers, has been related to social rank (Kojola, 1977; Greenberg-Cohen *et al.*, 1994). However, other authors did not find a relation between social rank and the presence of antlers (Jennings *et al.*, 2004).

On the other hand, social behaviour of farmed goats, which is highly influenced by domestication processes, has been scarcely studied (Miranda de la Lama and Mattiello, 2010). The domestication process is accompanied by important changes in the biological and physical environment of the populations, which interact with selective pressures. In semi-extensive managed goats it was observed that age, horn presence and horn length seem to be positively related to social rank (Barroso *et al.*, 2000). However, information characterizing traits related to social rank in intensively managed dairy goats is lacking. Therefore, the objectives were to determine: 1) if social rank is related to body weight, horn presence and length, and primiparity (multiparous/primiparous), and 2) the role of visual signals in maintenance of hierarchy and in agonistic behaviours used to maintain hierarchical relations in diary goats under intensive management.

Materials and Methods

Trial 1

The study was performed in 3 herds from commercial dairy farms located in the southern region of Uruguay (34.5 to 35° S). Goats were intensively managed, milked twice daily, and grazed on native or improved pastures during part of the day, receiving ration during milking and at night enclosure. Commercial ration for dairy goats (100-200 g/animal/d) was daily provided on a group basis. For individual identification, all does were tagged, and the tag number was painted on both sides of the body.

Goats were from the Anglo Nubian, Alpine and Saanen breeds. Herds I, II and III were comprised by 45, 43 and 51 does, respectively. At the beginning of the study goats were weighted, and horn length was individually measured. As age data from all animals was not available, but farmers have recorded parturitions, primiparity was considered for the analysis. The body weight, the number of multiparous goats, and the number of horned goats and horn length (mean of both horns) for each herd are presented in Table 18.1. In herds I and III farmers began to cut off offspring horns shortly before trial recording started. Therefore, all younger goats were dehorned.

Trial 2

Eleven Saanen and Anglo Nubian crossbreed goats (seven multiparous and four primiparous), weighting 29.5 ± 3.7 kg were moved to the Departamento de Fisiología, Facultad de Veterinaria, Montevideo, Uruguay (35° S) more than one month before the trial onset. Nine of these goats were horned, with a horn length of

17.5 ± 1.7 cm (range: 7.8 to 22.5 cm). Goats were maintained in a 10 X 10 m pen during all the trial, and received alfalfa hay, 900 g of ration per day and water was administered *ad libitum*.

Table 18.1: Characteristics of the Goats Used in Trial 1. Weight and horn length (from H goats) are presented as mean ± SEM.

Herd	*n*	*Weight (kg)*	*Multiparous (%)*	*Horned (%)*	*Horn Length (cm)*
I	45	46.8 ± 1.6	30 (66.7)	29 (64.4)	18.2 ± 0.9
II	43	43.2 ± 1.5	35 (81.4)	28 (65.1)	20.0 ± 1.0
III	51	47.8 ± 1.2	43 (84.3)	25 (49.0)	27.3 ± 1.5

Agonistic interactions were recorded during 3 periods: I) initial, II) temporary blinded with a bandage, III) without bandage. Goats were blinded all at the same time during one-two h per day during five-six days before beginning period II. In periods II and III the agonistic behavioural interactions used for displacement were also recorded.

Behavioural Recordings

In Trial 1 agonistic behaviours were recorded during the days following characterization of the herd. In both trials, two observers recorded all agonistic interactions that determined the physical displacement of one individual by other. The behaviours recorded included threat, physical contact (*e.g.* butting) or the mere approach of a goat accompanied by retreat or avoidance of another goat, in all cases determining the displacement of the later.

In Trial 1, goats from each herd were observed during one h, three-four times/day, in five-six non-consecutive days. The observation periods were separated each one by at least one hour, during which the animals were grazing on pasture. During observational periods, the animals were allocated in the same pen which was normally used twice daily for food supply before milking and for overnight enclosure, and all agonistic interactions were recorded. Most times behaviour was spontaneous, but in some opportunities it was stimulated by food administration. In herds I, II and III, 1840, 1730 and 1925 agonistic interactions were recorded.

In Trial II, animals were observed during one-1.5 h periods. In period II, most interactions were consequence of the competition for food. Overall, 196, 139 and 102 agonistic interactions were recorded during periods I, II and III.

The following agonistic interactions were recorded in periods II and III: front butting with horns (FB), butting with the horns from the side (SB), upper butt with horns (UB), butt from the back (BB), bite (Bi), push with the horns (PH), butt with the head (BH), displacement by the mere presence (P) and attempt of bite (AB).

Determination of Success Index

The success index (SI) for each goat was calculated according to Mendl *et al.* (1992) and Alvarez *et al.* (2003) according to the results of dyad relationships:

$$SI = \frac{\text{Number of individuals displaced}}{\text{Number of individuals displaced} + \text{Number of individuals that displace it}}$$

Statistical Analysis

In Trial 1, variance components (the influence of body weight, reproductive status, presence of horns and horn length, breed and herd) on success index was determined by an ANOVA. In Trial 2, regressions between body weight and social ranks calculated from interactions recorded during periods I and II were performed. Also regressions between the success indexes calculated during periods I and II, and I and III were performed. Individual rank positions in periods I and II and I and III were compared with Kendall's test. Results were considered significant at $P \leq 0.05$.

Results

Trial 1

In Trial 1, the ANOVA values were $r^2=0.76$ and $P<0.0001$. Only the body weight ($P<0.0001$) and the herd ($P=0.003$) influenced significantly the success index. Horn presence ($P=0.26$), length ($P=0.13$), parity ($P=0.47$) and breed ($P=0.58$) were not significantly related with success index. The body weight in relation to success index is presented in Figure 18.1.

Trial 2

Initially, body weight was related with social index ($r^2 = 0.94$, $P < 0.001$). However, that relation was not maintained during period

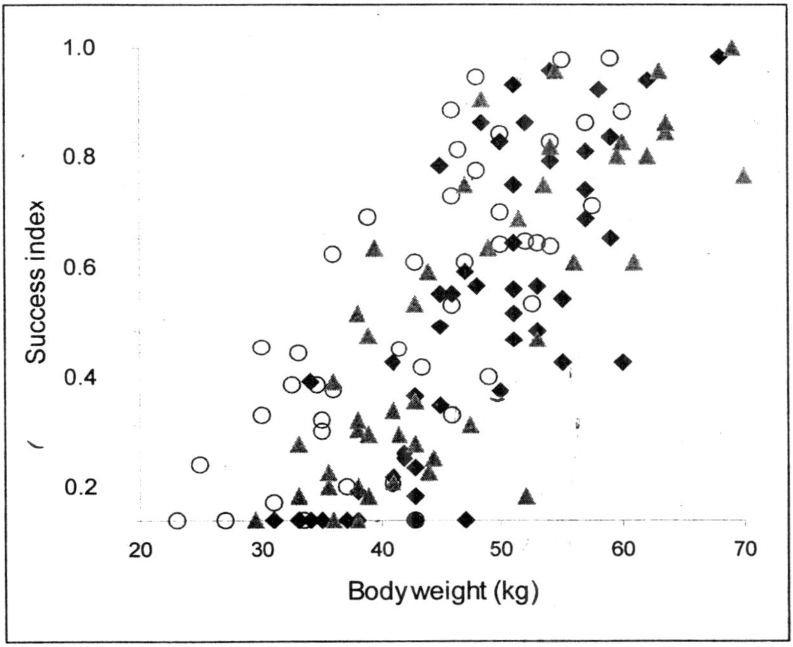

Figure 18.1: Relationship Between Body Weight and Success Index in 3 Herds of Dairy Goats

II ($P>0.05$). Success indexes for periods I and II were not related ($r^2=0.17$; $P=0.13$). However, those from periods I and III were similar ($r^2=0.95$; $P<0.001$). The individual position according to success index varied between periods I and II ($P<0.05$), but not between periods I and III ($P>0.05$).

The frequency of FB, SB and PH was greater in period II than III (Table 18.2). However, the frequency of BH, P and AB was lower in period II than III (Table 18.2). There were no differences on the frequency of UB, BB and Bi.

Discussion

We observed that there is a direct relationship between the individual body weights with social rank. This partially agrees with Barroso *et al.* (2000) who observed a direct relationship between social rank and body size in semi-extensively managed goats. However, Barroso *et al.* (2000) also reported a positive relation

Table 18.2: Frequency of Agonistic Interactions in Goats with (Period II) or without their Eyes Temporarily Blinded (Period III)

Agonistic Unit	*Period II (%)*	*Period III (%)*	*P*
FB	44 (31.7)	8 (7.8)	*
SB	36 (25.9)	12 (11.8)	**
UB	9 (6.5)	5 (4.9)	
BB	1 (0.7)	0 (0)	
Bi	5 (3.6)	2 (2)	
PH	18 (12.9)	3 (2.9)	**
BH	26 (18.7)	33 (32.4)	*
P	0 (0)	31 (30.4)	**
AB	0 (0)	8 (7.8)	**
Total	139	102	

For the same row: *: P<0.05; **: P<0.01

FB: Front butting with horns; SB: Butting with the horns from the side; UB: Upper butt with horns; BB: Butt with the horns from the back; Bi: Bite; PH: Push with the horns; BH: Butt with the head; P: Displacement by the mere presence; AB: Attempt of bite.

between social rank and age, characteristics that have also been related in non-domestic small ruminants. Though, as in our work exact age data was not available we cannot discard a possible relation between social rank and age. Moreover, age may possibly explain variations in body mass. It should be considered that domestication implies lighter pressures for animals compared with wild life (Price, 2002) and influences of farming decisions as dehorning, might result in differences in the relative importance of social hierarchical traits (Miranda de la Lama and Mattiello, 2010).

As body weight is related to visual size, probably subordinated animals visualize dominant individuals as having bigger size. This agrees with the observations from the second trial, in which it was determined that visual signals affect hierarchical positions. In effect, some high-ranked goats lost their rank, and lower ranked individuals gained rank. Therefore, this results supports the concept that visual signals are necessary for goats to determine how individuals perceive each other in relation to dominance relationships. In cows Klemm *et al.* (1984) have reported that the vomeronasal and thus pheromones are main determinants of hierarchical positions. As animals in Trial 2 maintain their olfactory systems intact, although we cannot discard an influence of chemical signals, we suggest that in these goats chemical signals alone cannot replace visual signals in goats' dominance relationships.

When animals recover their view, the hierarchical positions were similar to those recorded before the period in which animals cannot view each other. This supports the concept that the privation of visual signals determines hierarchical changes only during that period, and not permanent changes. As a consequence of the temporal blindness, goats also changed their strategies on how to displace other individuals. As it may be expected, the frequency of interactions that depend completely on view disappeared (attempt of bite, mere presence), and the frequency of the interactions that may be consequence of physical contact (butts, push) increased. The differences in displacement strategies may also be related with the changes in hierarchical positions. In effect, Tölü and Sava (2007) reported that the displacement strategies of goats vary in relation to hierarchical positions. However, our data indicated that temporal changes in displacing strategies provoked by temporal blindness do not have permanent effects.

Overall, it was concluded that social rank is positively related with body weight in dairy goats. This relation may be consequence of how each goat views other goats, as visual signals were necessary to maintain the hierarchical positions into the group. Displacement strategies were modified when view was not active, but those modifications were observed only while artificial blindness was maintained.

Acknowledgments

Author acknowledges Luis Puig, Fernando Aldama, and Daniel Manzione, owners of the farms used for Trial 1. Laura Dago, Alejo Menchaca and Vanina Panossián helped with data recording. Milton Pintos took care of the animals in Trial 2. Financial support: CIDEC (Facultad de Veterinaria, Uruguay).

References

Alvarez, L., Martin, G.B., Galindo, F. and Zarco, L.A., 2003. Social dominance of female goats affects their response to the male effect. *Appl. Anim. Behav. Sci.*, 84: 119-126.

Andersen, I.L. and Bøe, K.E., 2007. Resting pattern and social interactions in goats: The impact of size and organisation of lying space. *Appl. Anim. Behav. Sci.,* 108: 89-103.

Andersson, M., Schaar, J. and Wiktorsson, H., 1984. Effects of drinking water flow rates and social rank on performance and drinking behaviour of tied-up dairy cows. *Livest. Prod. Sci.,* 11: 599-610.

Barroso, F.G., Alados, C.L. and Boza, J., 2000. Social hierarchy in the domestic goat: effect on food habits and production. *Appl. Anim. Behav. Sci.,* 69: 35-53.

Cassinello, J., 1995. Factors modifying female social ranks in *Ammotrangus. Appl. Anim. Behav. Sci.,* 45: 175-180.

Côte, S.D., 2000. Dominance hierarchies in female mountain goats: stability, aggressiveness and determinants of rank. *Behav.,* 137: 1541-1566.

Côte, S.D., and Festa-Bianchet, M., 2000. Reproductive success in female mountain goats: the influence of age and social rank. *Anim Behav.,* 62: 173-181.

Festa-Bianchet, M., 1991. The social system of bighorn sheep: grouping patterns, kinship and female dominance rank. *Anim. Behav.,* 42: 71-82.

Greenberg-Cohen, D., Alkon, P.U. and Yom-Tov, Y., 1994. A linear dominance hierarchy in female Nubian ibex. *Ethol.,* 98: 210-220.

Jennings, D.J., Gammell, M.P., Carlin, C.M. and Hayden, T.J., 2004. Effect of body weight, antler length, resource value and experience on fight duration and intensity in fallow deer. *Anim. Behav.,* 68: 213-221.

Jewell, P.A., 1976. Selection for reproductive success. In: Austin CR, Short RV (eds) Reproduction in mammals. Book 6: The evolution of reproduction, Cambridge University Press, UK, pp. 71-109.

Jørgensen, G.H.M., Andersen, I.L. and Bøe, K.E., 2007. Feed intake and social interactions in dairy goats: The effects of feeding space and type of roughage. *Appl. Anim. Behav. Sci.,* 3-4: 239-251.

Klemm, W.R., Sherry, C.J., Sis, R.F., Schake, L.M. and Waxman, A.B., 1984. Evidence of a role for the vomeronasal organ in social hierarchy in feedlot cattle. *Appl. Anim. Behav. Sci.*, 12: 53-62.

Kojola, I., 1977. Behavioural correlates of female social status and birth mass of male and female calves in reindeer. *Ethol.*, 103: 809-814.

Locati, M. and Lovari, S., 1991. Clues for dominance in female chamois: age, weight, or horn size? *Aggress. Behav.*, 17: 11-15.

Loretz, C., Wechsler, B., Hauser, R. and Rüsch, P., 2004. A comparison of space requirements of horned and hornless goats at the feed barrier and in the lying area. *Appl. Anim. Behav. Sci.*, 87: 275-283.

Mendl, M., Zanella, A.J. and Broom, D.M., 1992. Physiological and reproductive correlates of behavioural strategies in female domestic pigs. *Anim. Behav.*, 44: 1107–1121.

Miranda-de la Lama, G.C. and Mattiello, S., 2010. The importance of social behaviour for goat welfare in livestock farming. *Small Rum. Res.*, 90: 1-10.

Patón, D., Martin, L., Cereijo, M., Rota, A., Rojas, A. and Tovar, J., 1995. Relationship between rank order and productive parameters in Verata goats during milking. *Anim. Sci.*, 61: 545-551.

Price, E.O., 2002. *Animal domestication and behavior*. CAB Publishing, Wallingford, UK.

Sherwin, C.M. and Johnson, K.G., 1987. The influence of social factors on the use of shade by sheep. *Appl. Anim. Behav. Sci.*, 18: 143-155.

Thouless, C.R. and Guinness, F.E., 1986. Conflict between red deer hinds: the winner always wins. *Anim. Behav.*, 34: 1166-1171.

Tölü, C. and Sava, T., 2007. A brief report on intra-species aggressive biting in a goat herd. *Appl. Anim. Behav. Sci.*, 102: 124-129.

Ungerfeld, R. and Correa, O., 2007. Social dominance of female dairy goats influences the dynamics of the number gastrointestinal parasite eggs. *Appl. Anim. Behav. Sci.*, 105: 249-253.

Ungerfeld, R., González-Pensado, S.P., Dago, A.L., Vilariño, M. and Menchaca, A., 2007. Social dominance of female dairy goats and response to oestrous synchronisation and superovulatory treatments. *Appl. Anim. Behav. Sci.*, 105: 115-121.

Section II

Animal Reproduction

Chapter 19

Feeding and Reproduction of *Artemia franciscana* from Basrah

Dawood S.M. Abdullah, Salman D. Salman and Malik H. Ali*
Dept. Marine Biology, Marine Science Centre, Basrah University, IRAQ

ABSTRACT

The effects of varied types of live and inert food items on the survivals, growth and biomass of *Artemia franciscana* a 30°C and 55 ‰ were investigated for 14 days. The first food item used was a living alga (*Dunaliella* sp.) at 5 densities (25×10^4, 50×10^4, 75×10^4, 100×10^4 and 125×10^4 cells/ml), the second, third and forth food items used were powdered *Chlorella*, bakers yeast, rice bran at 5 concentrations (0.1, 0.15, 0.2, 0.25 and 0.3 mg/ml). The rest four food items were combinations of two or three foods given at a concentration of 0.3 mg/ml.

The results of live food revealed that growth rates, survivals and biomass increased positively with the increase of food concentration and the highest values were 5.67 mm, 90 per cent and 178.31 mg DW/L at 125×10^4 cells/ml, respectively, whereas the lowest values were 3.91 mm, 50 per cent and 48.6 mg DW/L at 25×10^4 cells/ml, respectively. While in the three

* Corresponding author: E-mail: dr_salmands@yahoo.com

dried food items the highest average growth, survivals and biomass achieved at the use of powdered *Chlorella* (6.2 mm, 93.3 per cent and 204.5 mg DW/L, respectively), and the lowest growth (3.3 mm) at the use of rice bran, while the lowest survivals and biomass were occurred at 0.1 mg/ml powdered *Chlorella* (40 per cent and 30.11 mg DW/L, respectively). Whereas, the results of the 4 mixtures of the dried food items were far more better than the single food item, the highest average growth was 6.7 mm with the use of a mixture of *Chlorella* and rice bran, and the highest survivals and biomass were 93.3 per cent and 250.46 mg DW/L, respectively with the use of a mixture of baker's yeast and rice bran.

The effect of 3 salt concentrations (60, 90 and 120 ‰) on 10 reproductive traits and the life span of *A. franciscana* were also investigated. A sufficient food of the live alga, *Dunalielle* sp. at 30 °C was used. It was noticed that *Artemia* undergoes two different reproductive patterns. In some times it was ovoviviparous (gives birth to nauplii) and in other times it was oviparous (produce cysts). At 60 ‰ the female tends to produce nauplii (73.25 nauplius/brood) more than cysts (10.49 per cent per brood), whereas the female produce cysts (51.5 per cent per brood) more than nauplii (39.50 nauplius/brood) at 90 ‰ and not at 120 ‰ as was believed before as the *Artemia* produces cysts only at higher salt concentrations. The statistical analyses indicated significant differences ($p<0.05$) between 60 and 120 ‰ at all reproductive traits studied. The differences were also significant ($p<0.05$) between 60 and 90 ‰ at most reproductive traits, except the average offspring per brood, the average number of broods per female and the incubation period. In comparing the reproductive traits at 90 and 120 ‰ they were significantly different ($p<0.05$) in the average number of offspring per brood, the average number of broods per female and the incubation period only.

Keywords: *Brine shrimps, Artemia franciscana, feeding, reproduction, Basrah, Iraq.*

Introduction

The brine shrimp *Artemia* is widely distributed on every continent except the Antarctica (Browne and MacDonald, 1982). It is found abundantly in both saline and hypersaline (coastal and inland) environments at salinity levels between 80 and 220 g/L, depending on the strain/or species, but it cannot survive for more than 1 day in the extreme salinity of 340 g/L (Camargo *et al.*, 2003). The genus *Artemia* consist of a complex of sibling species and superspecies (group of species and semispecies distinctive physiologically and isolated ecologically) (Bowen *et al.*, 1985). The largest population of *Artemia* is found in the Great Salt Lake, Utah, USA, where *A. franciscana* is the only dominant planktonic invertebrate species in the lake (Stephens and Gillespire, 1976). *A. franciscana* is probably the most popular live food in aquaculture. It's nauplii and metanauplii are used as food for larvae and juveniles of crustaceans and fish, and the use of mass-produced juveniles and adult of *Artemia* has become increasingly popular in ornamental fish culture and in hatcheries, where they are used to enrich the maturation diets for shrimps and fish broodstock (Dhont and Lavens, 1996). *Artemia* exhibit remarkable reproductive flexibility, as reproductive mode can vary between parthenogenesis and obligate cross-fertilization and all strain were found to utilize a combination of viviparous and encysted zygotic production, with the ratio of the two varying widely with the strain (Browne, 1980). Vos and La Rosa (1980) discussed the reproductive strategy of *Artemia* in some detail and pointed out that there are more than 50 known strains of *Artemia* in the world differing from each other in some important features like hatching rate, egg size, optimum temperature and salinity for survival, and some of these strains are parthenogenetic (females only) but most of them are bisexual (male and female).

The ovoviviparity may be considered as an adaptation to the hypersaline condition experienced by *Artemia* since this reproduction rate has not been reported for freshwater Anostracans (see Barata *et al.*, 1996). Nauplius reproduction allows for rapid population growth, whereas, the production of diapause cysts ensures the survival of a population through unfavourable conditions (Versichele and Sorgeloos, 1980; Lenz and Dana, 1987). Moreover, it has been suggested that dormant cysts tend to be produced under deleterious environmental conditions, such as high salinity and low oxygen levels, whereas nauplii tend to be released under more favourable conditions (Clegg and Trotman, 2002).

The factors inducing females to produce cysts have up to now, been believed to be environmental in nature (Browne *et al.*, 1984). Natural selection compares heritable variants for their effect on reproduction, and so different localities are interpreted as the result of adaptation to different environmental conditions (Browne, 1982), for individuals with reproductive patterns most suited to maximize fitness in their environment will be favoured by natural selection (Streans, 1993; Rose and Lauder, 1996). This view, however, contradicted by Gajardo and Beardmore (1989), who suggested that the ability of *Artemia* females to produce dormant cysts instead of free swimming nauplii is a reproductive device that gives survival advantage to the species. But Versichele and Sorgeloos (1980) managed to increase the production of encysted eggs of *A. franciscana* through exposure to oxygen deficiency or addition of iron to the culture medium in the form of chelating compounds.

Most of the physical factors affecting the age and reproduction of the animals in the saline habitat are temperature and salinity as well as the genetic traits of the animals, therefore, most of the studies carried out stressed upon the role of these factors and their effects on the age and reproduction of *Artemia* (Browne *et al.*, 1984; Wear *et al.*, 1986; Abatzopoulos *et al.*, 1993; Triantaphyllidis *et al.*, 1995; Browne and Wanigasekera, 2000). Moreover, Browne *et al.* (2002) studied the effects of various environmental and genetic factors on the age and the reproductive traits of *Artemia*. Abatzopoulos *et al.* (2003) investigated the effect of salinity and temperature on a parthenogenetic strain of *Artemia* in Cyprus, whereas Baxevanis *et al.* (2004) tested the effects of salinity upon the maturation, reproduction and lifespan of 4 populations of *Artemia* from Egypt.

A. franciscana has been found in the temporary pools in various places of Basrah district, south of Iraq. Since no previous work was done on the feeding requirements of the present species in Basrah. It was then decided to undertake a comprehensive work trying to mass–culture it. One of the experiments to be done in this respect is the evaluation of different feeds on the growth, survival and biomass of *A. franciscana.*

Therefore, several attempts were tried with various live and inert food items. Live *Dunaliella* sp., powdered *Chlorella vulgaris*, backer's yeast, rice bran and various combinations of these inert feeds were tried in order to gather quantitative data on the changes in the growth and survival rates when *A. franciscana* was grown from nauplii to adult stage under different concentrations of live and inert food. Based on growth and survival experiments, the present paper describes changes in the growth and survival rates and determines the food concentrations to when the growth and survival rates are maximal and when the growth and survival rates are limited by the food intake.

The reproductive traits and lifespan have not been in focus in Iraq, the present study, therefore, is concerned also with the effects of salinity on various reproductive characters and on the lifespan of a local strain of *A. franciscana*.

Material and Methods

Food and Feeding

Rearing *A. franciscana* Nauplii on Live *Dunaliella* sp.

Previous observations on the effect of temperature and salinity on the survival and growth of *A. franciscana* nauplii revealed that the optimal temperature and salinity were 30 °C and 55 ‰(Abdullah, 2007). Therefore, these two parameters were used in the following experiments.

Live *Dunaliella* sp. were brought from a temporary pool inside the University Campus of Garmat-Ali. Five concentrations of *Dunaliella* sp. were prepared, these are: 25 x 10^4, 50 x 10^4, 75 x 10^4, 100 x 10^4, and 125 x 10^4 cell/20 nauplii/day. Counting of the cells was carried out with the aid of a Haemocytometer. Three replicates of each concentration were prepared.

Nauplii were reared in 100 ml beakers for 14 days. Filtered biotope water was used for rearing the nauplii. The animals were then fixed with Lugal´s solution. Total length of each larva remained alive at the end of the experiment was measured, to the nearest 0.01 mm, from the anterior end of the body to the tip of telson excluding the setae.

Survivals and dead larvae were recorded every two days throughout the period of the experiment.

Rearing *A. franciscana* Larvae on Dried Extract of *Chlorella vulgaris*

C. vulgaris was isolated from Shatt Al-Arab River waters, cultured and purified on F/2 media (Guillard and Ryther, 1962). Graduated cylinders of a capacity of 2000 ml were used in culturing *C. vulgaris*, maintained at 28 ± 1 °C and at illumination of 2000 lux, for 12 light: 12 dark hour periods.

When the culture attained the exponential growth phase, the algae were collected, oven dried at 60 °C for 24 h (Vollenweider, 1974), ground to powder and sieved through 35 μ sieve.

Five concentrations of *C. vulgaris* powder were prepared in the following sequences: 0.1, 0.15, 0.2, 0.25 and 0.30 mg DW/ml. Three replicates of each concentration were used. Twenty nauplii were placed into a 100 ml glass beaker containing 20 ml filtered biotope water at temperature of 30 °C and a salinity of 55 ‰Survivals and dead larvae were recorded every other day, throughout the 14 days of the experiment. Water of the culture media was changed every two days. Total length of each living *Artemia* were measured after fixation in Lugal´s solution.

Rearing *A. franciscana* Nauplii on Different Dried Food

Among other types of food used for rearing *A. franciscana* nauplii are baker's yeast and rice bran. Bakers yeast was used as a unicellar protein source for feeding the nauplii after being dissolved in warm water. Rice bran was used, ground and sieved through a 35 μm mesh-size sieve.

The same sets of concentrations and experimental conditions used in *C. vulgaris* are used in these two dried food experiments. Another series of experiments were done using mixed food of the above food items, these are:-

1. Rice bran and baker's yeast.
2. Rice bran and powdered *C. vulgaris*
3. Baker's yeast and powdered *C. vulgaris*
4. Rice bran and baker's yeast and powdered *C. vulgaris.*

Equal weights of these foods were made and given as 0.3 mg DW/ml/larva. The following steps and experimental conditions used here are similar to those used in *C. vulgaris*.

Biomass of *A. franciscana* was calculated by measuring the mean total length of the surviving animals at the end of the experiments, converted into weight, according to the length-weight expression (Ahmed, 2002):

$$\text{Log W} = -2.135 + 1.9 \text{ Log L}$$

Then the biomass was obtained from the following equation (Marques *et al.*, 2004):

Total length (mm) =

Number of survivors × mean individual length.

Reproduction

Pre-adult *Artemia* were selected from a rearing tank in which the animals were cultured on conditioned sheep manure through visual inspection of *Artemia* and classified into males and females after the modification, in the males, of the second antennae into claspers, and the growth, in the females, of the ovisac, and every pair were isolated alone (Browne *et al.*, 1988), in a vial containing 50 ml of filtered biotope waters. Three salinity treatments were chosen: 60, 90 and 120 ‰and the temperature was fixed at 30 °C. Five replicates of each treatment were done.

The animals were daily fed on a sufficient quantity of *Dunaliella* sp. The vials were examined every day for records of the number of nauplii or encysted eggs released by each female. Water was changed after each counting occasion. Records of the data on the reproductive characters and the lifespan were made, these include: offspring per brood, brood per female, offspring per day per female, time between broods(d), female pre-reproductive period (d); estimated from the day of hatching to the release of the first batch of eggs by the female, female reproductive period (d); represents the period from the time of the first brood till the last one, female post-reproductive period (d); recorded from the time of the last brood till the death of the female, total female lifespan (d), total male lifespan (d), live offspring per brood and the gestation period (d); starting from the time of copulation till the release of the fist brood.

Chemical Constituents of the Various Food Items

Nitrogen contents of the different food items were estimated by the semi-microkjeldahl (Pearson, 1970), and multiplied by 6.25 to get the protein contents. Lipids were determined according to the A.O.A.C. (1984) method.

A one way analysis of variance was used to test statistical significance between the various treatments using the SPSS program.

Results

The growth of *A. franciscana* nauplii fed on live *Dunaliella* sp. at 30°C and 55‰increased gradually from 3.91 mm at 25 × 10^4 cells/

ml to 5.67 mm at 125 × 10^4 cells/ml (Figure 19.1). The survival rate fluctuated between 50 per cent at the lowest cell concentration and 90 per cent at the highest.

The growth, in terms of total length, at different concentrations of dried food at 30 °C and 55 %in case of *C. vulgaris,* increased from 3.4 ± 0.09 mm at 0.10 mg/ml to 6.2± 0.03 mm at 0.30 mg/ml (Table 19.1).

However, at the baker's yeasts, the growth ranged from 4.4 ± 0.14 to 6.0 ± 0.12 mm at 0.10 and 0.30 mg/ml, respectively. Whereas at the rice bran the growth changed from 3.3 ± 0.16 to 4.5 ± 0.05 mm at concentrations of 0.1 and 0.3 mg/ml, respectively (Table 19.1).

The one way ANOVA test indicates that there were significant differences ($p<0.05$) in total length of the larvae fed baker's yeast and both *C. vulgaris* and the rice bran at the concentration 0.1, while in the following three concentrations of rice bran the total lengths of the larvae were significantly different from both *C. vulgaris* and baker's yeast (Table 19.1). Whereas the results of the three kinds of food were significantly different ($p<0.05$) from each other at 0.30 mg/ml.

It is apparent that the algal food (*C. vulgaris*) is the most efficient food for *A. franciscana* larvae.

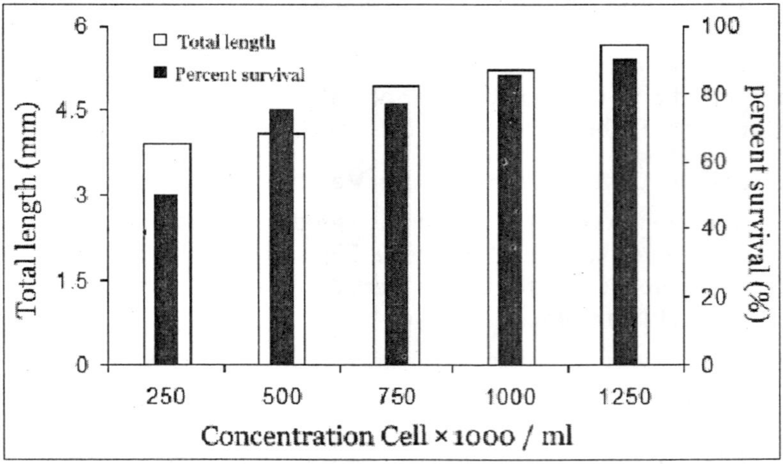

Figure 19.1: Total Length (mm) and percent Survivors of Larvae of Local *Artemia franciscana* Reared on Different Concentrations of *Dunaliella* sp. at 30 °C and 55‰ for 14 Days

Table 19.1: Mean Values of Total Length (mm) and Standard Deviation (±) of Larvae of Local *Artemia franciscana* at Different Concentrations of Food Items at 30°C and 55 ‰ during 14 Days. Different letters at the same horizontal raw denote significant difference (P<0.05) between treatments made by the one way ANOVA, n = 20.

Conc. (mg/ml)	*Food Item*		
	Chlorella vulgaris Powder	*Baker's Yeast*	*Rice Bran*
	$\bar{x} + s.d.$	$\bar{x} + s.d.$	$\bar{x} + s.d.$
0.10	3.4±0.09 a	4.4±0.14 b	3.3±0.16 a
0.15	4.8±0.02 a	4.8±0.15 a	3.6±0.18 b
0.20	5.3±0.31 a	5.3±0.24 a	4.0±0.18 b
0.25	6.0±0.53 a	5.5±0.42 a	4.1±0.13 b
0.30	6.2±0.03 a	6.0±0.12 b	4.5±0.05 c

The results of mixed food on the growth of *A. franciscana* nauplii showed no significant differences (p>0.05) between the different trials (Table 19.2). However, the growth performed by mixed food was generally slightly higher but not significantly different (p>0.05) from that of a single food item at 0.30 mg/ml concentration.

It is also obvious that the length increased with increasing concentration of food.

Table 19.2: Mean Total Length (mm) and Standard Deviation (±) of Larvae of Local *Artemia franciscana* at Concentration of 0.3 mg/l of Different Mixtures of Food Items During 14 days. Various letters indicate significant difference between treatment (P<0.05) using one way analysis of variance (n=20).

Food Item	*Average Length (mm) ± sd.*
Chlorella vulgaris powder + Baker's yeast	6.3 ± 0.52 a
Chlorella vulgaris powder + Rice bran	6.7 ± 0.33 a
Baker's yeast + Rice bran	6.5 ± 0.43 a
Chlorella vulgaris powder + Baker's yeast + Rice bran	6.3 ± 0.41 a

Effect of Dried Food on the Survivals of *A. franciscana*

The highest percent survivals (80 per cent) at a concentration of 0.1 mg/ml was attained by the rice bran followed by the baker's yeast (70 per cent), whereas the dried algae showed the least survivals (40 per cent) (Table 19.3).

Table 19.3. Mean value of percentage survivors and sd (±) of larvae of local *Artemia franciscana* at different food items at 30 °C and 55 ‰ during 14 days. Similar letters indicate no significant (P>0.05) differences between treatments. A one way analysis of variance was used (20= n).

Conc. (mg/ml)	*Food Item*		
	Chlorella vulgaris Powder	*Baker's Yeast*	*Rice Bran*
	$\bar{x} + s.d.$	$\bar{x} + s.d.$	$\bar{x} + s.d.$
0.10	40.0 ± 10.0 a	70.0 ± 5.00 b	80.0 ± 8.60 b
0.15	93.3 ± 5.77 a	88.0 ± 7.60 a	81.0 ± 7.60 a
0.20	93.3 ± 7.63 a	71.0 ± 15.0 b	85.0 ± 5.00 ab
0.25	96.6 ± 2.88 a	88.0 ± 7.60 a	85.0 ± 8.60 a
0.30	83.3 ± 2.88 a	75.0 ± 17.3 a	80.0 ± 17.0 a

In the rest cf treatments, the results were on the contrary *i.e.* in the dried algae, the survivals were higher in the concentrations 0.15-0.30 mg/ml than in the baker's yeast and the rice bran (Table 19.3).

There were no significant differences (p>0.05) in the percent survivals in the three diets at the concentrations 0.15, 0.25 and 0.30 mg/ml. But at 0.10 mg/ml, the dried algae showed a rather significant result (p<0.05) from the other 2 treatments. At 0.20 mg/ml, the survival in the dried algae was significantly higher than the baker's yeast, but not at the rice bran. Whereas, no significant difference (p>0.05) was noticed between the baker's yeast and the rice bran at the 0.20 mg/ml treatment (Table 19.3).

The results of the mixed food at a concentration of 0.30 mg/ml at 30 °C and 55 %indicate that the mixture of baker's yeast and rice bran exhibited the highest survival (93 per cent) and the lowest was attended at the mixed dried algae and baker's yeast (61 per cent) (Table 19.4).

It is apparent that the survivals at the dried algae and baker's yeast were not significantly different from the dried algae and the rice bran, whereas the differences were significant between the mixture of dried algae and baker's yeast and that of the baker's yeast and rice bran and dried algae, baker's yeast and rice bran ($p<0.05$) (Table 19.4).

Table 19.4: Average per cent Survivors and the Standard Deviation (±) of Larvae of Local *Artemia franciscana* Fed at Various Mixtures of Food Items at a Concentration of 0.3 mg/ml at 30 °C and 55 ‰ during 14 Days. Different letters denote significant difference (P<0.05) among different treatments by the one way analysis of variance (n = 20).

Food Item	*Percent Survival (%)* $\bar{x} + s.d.$
C. vulgaris powder + Baker's yeast	61.0 ± 16.1 b
C. vulgaris powder + Rice bran	73.3 ± 10.4 ab
Baker's yeast + Rice bran	93.3 ± 2.88 a
C. vulgaris powder + Baker's yeast + Rice bran	86.0 ± 7.63 a

Effect of Type of Food on the Biomass

The biomass (mg DW/L) of *A. franciscana* larvae was estimated using live food (*Dunaliella* sp.), dried algae (*C. vulgaris*), baker's yeast and rice bran, at 30 °C and 55 ‰

The biomass, when *Dunaliella* sp. was supplied, increased linearly from 48.6 mg/l at 25×10^4 cells/ml to 178.31 mg/l at 125×10^4 cell/ml (Figure 19.2).

In case of the dried food, the dried algae showed the best results. The biomass increased from 30 mg DW/L at 0.10 mg/ml to 204.5 mg DW/L at 0.25 mg/ml and decreased at 0.30 mg/ml. The lowest values were 57.64 and 104.75 mg DW/ml obtained when the rice bran was given as food at 0.1 and 0.25 mg/ml. The biomass gained from rearing *A. franciscana* nauplii on the three mixed food indicates that the lowest value was 151.2 mg DW/L when fed on mixture of dried algae and baker's yeast. The highest value was 250.5 mg DW/L when fed on a mixture of baker's yeast and rice bran (Figure 19.3).

Chemical Composition of Different Food Items

Table 19.5, showed the chemical composition of the three types of food used in rearing *A. franciscana* larvae. The lowest value of lipids was 1.70 per cent DW supplied by the rice bran and the highest was 6.26 per cent DW obtained from the dried algae.

Figure 19.2: Biomass (mg DW/ml) of Larvae of Local *Artemia franciscana* Reared on *Dunaliella* sp., Powdered *Chlorella vulgaris*, Baker's Yeast and Rice Bran at 30°C and 55 ‰ during 14 Days

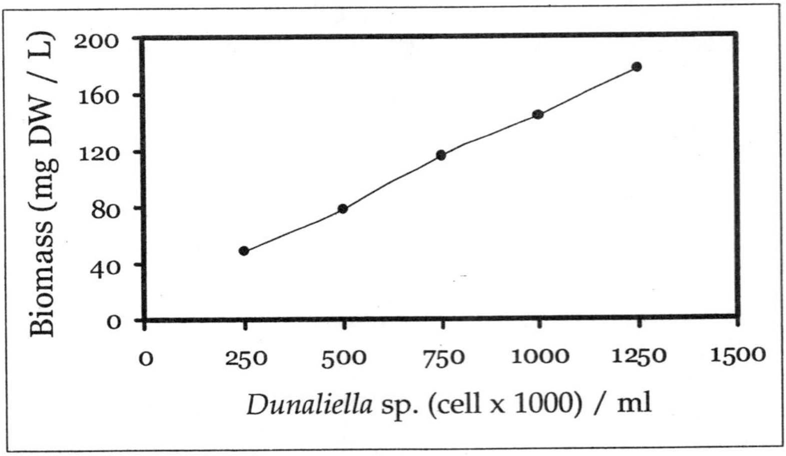

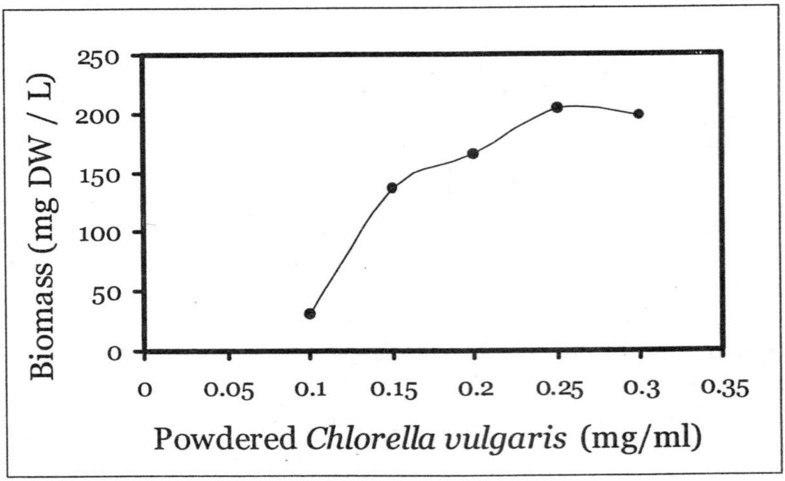

Contd...

Figure 19.2–Contd...

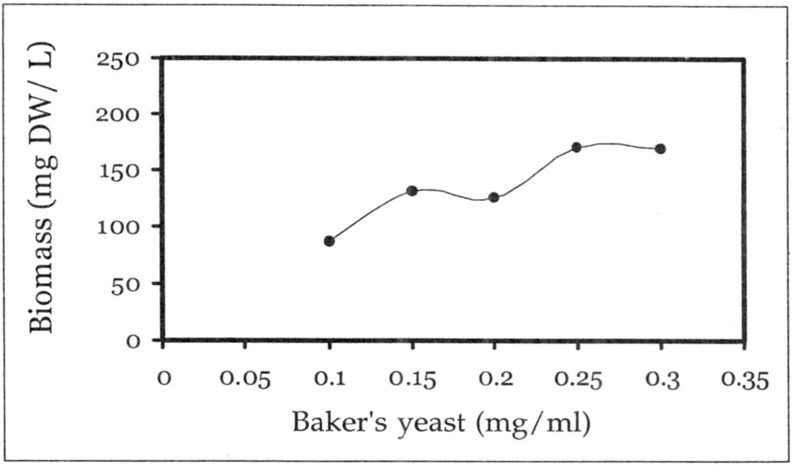

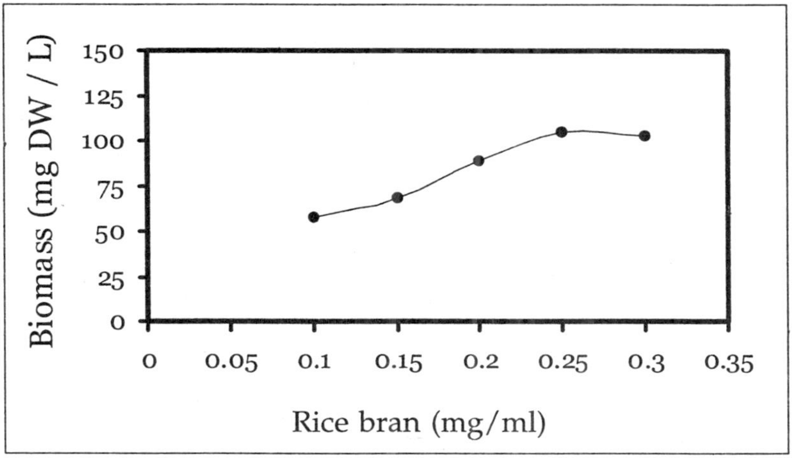

The lowest protein value was 14.09 per cent DW given by the rice bran and the highest was 58.14 per cent DW supplied by the dried algae. The dried alga was containing the least levels of carbohydrates (15.05 per cent DW), whereas the rice bran showed the highest value (79.84 per cent DW).

Reproduction

Table 19.6 summarizes the effect of 3 salinity levels on the reproductive characters of a local strain of the bisexual *Artemia*

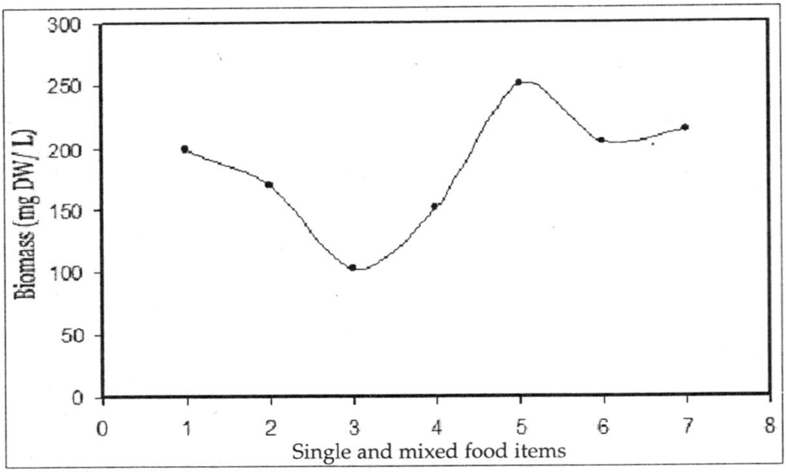

Figure 19.3: Biomass (mg DW/L) of Larvae of a Local Strain of *Artemia franciscana* Reared on Various Food Items (Single and mixed) at a Concentration of 0.3 mg DW/L, during 14 Days at 30°C and 55 ‰. 1, powdered *Chlorella*; 2, baker's yeast; 3, rice bran; 4, mixed *Chlorella* and baker's yeast; 5, mixed baker's yeast and rice bran; 6, mixed powdered *Chlorella* and rice bran; 7, mixed powdered *Chlorella*, baker's yeast and rice bran.

franciscana in Basrah. Obviously at salinities of 60 and 120 ‰, the one-way analysis of variance indicated significant differences ($p<0.05$) among all the studied reproductive traits. However, the differences between the 90 and 120 ‰ treatments, were significant ($p<0.05$) in the offspring per brood, brood per female, and gestation period, and at 60 and 90 ‰ also the rest of traits exhibited significant differences ($p<0.05$). There were inverse relations between salinity and the offspring per brood (81.77 at 60 ‰ versus 54.73 at 120 ‰), offspring per day per female (31.07 at 60 ‰ versus 16.09 at 120 ‰), female post-reproductive period (3.50 d at 60 ‰ as opposed to 3.00 d at 120 ‰), live offspring per brood (73.25 at 60 ‰ against 36.33 at 120 ‰) and the gestation period (6.25 d at 60 ‰ versus 4.00 d at 120 ‰). Meanwhile, there were positive relations between salinity and brood per female (4.37 at 60 ‰ and 7.50 at 120 ‰), time between broods (2.26 d at 60 ‰ against 3.85 d at 90 ‰), female pre-reproductive period (18.40 d at 60 ‰ versus 21.00 d at 90 ‰), female reproductive period (11.50 d at 60 ‰ as opposed to 25.50 d at 120 ‰), total female lifespan (33.40 d at 60 ‰ against 49.00 at

120 ‰), total male lifespan (23.63 d at 60 ‰ and 43.50 d at 120 ‰) and percent offspring encysted (10.40 at 60 ‰ versus 51.50 at 90 ‰).

Discussion

As long as the *Artemia* is non-selective filter feeders, the diet of it includes a wide range of living and inert food. Therefore, the important limiting factors in culturing it are the particle size and food value. The present study was performed to test the growth and survival of *A. franciscana* on living food for instance *Dunaliella* sp. and inert food, like powdered *C. vulgaris*, baker's yeasts and rice bran which were filtered through a sieve of 35 µm mesh- size, supposedly small enough to be ingested by the *Artemia*.

The present results indicated that the best concentration of live food for optimum growth in terms of biomass was 125×10^4 cells/ml, as the biomass increased linearly with increasing food concentration. This is in agreement with the results of Evjemo and Olsen (1999) who concluded that the food concentration (*Isochrysis galbana*) had a more pronounced effect on the production rate of *A. franciscana*. Similarly Johnson (1980) found that the larvae of *Artemia* increased in mean length with increasing algal concentration (*Dunaliella tertiolecta*).

The use of live algae in feeding *Artemia* nauplii is encouraging, as live food is highly nutritive, thereby increasing the growth rates of the animal and improving its quality, which is a very important step in fish culture. Wickins (1972) noticed that the newly hatched *Artemia* from a Utah strain, when fed to the larvae of the shrimp *Palaemon serratus*, the results were not promising, but when the larvae of *Artemia*, fed on live algae for 24h, it's nutritional value increased. Moreover, Tachibana *et al.* (1997) indicated that the use of live *Dunaliella* to feed the larvae of *Artemia*, gave the animal a bright colour and increased their nutritional value, as it contain carotenes which is transferred to the animals feeding upon the *Artemia*. Sick (1976) had a good result in rearing *Artemia* on live *Dunaliella* and suggested that this result is due to it's high degree of digestibility and acceptance by the larvae, for it lacks cell wall and had a high contents of proteins and energy. Furthermore, it can live in highly saline habitats, far more saline than that where *Artemia* lives (Borowitzka *et al.*, 1984). However, it was noticed here that the algal cells out of the needs of *Artemia* remain swimming in the water column and didn't spoil the medium as the dried food may do. Neagel (1999) indicated that live food (*Chaetocerus* sp.) has a preference upon commercial diets in culturing *Artemia* larvae, as there is no need for cleaning the culture units or using filtration units for removing excess food remains. In the present experiments, the water was changed every two days particularly in those using inert food.

The lowest rates of growth and survivals of the larvae of *A. franciscana* fed live *Dunaliella* sp. during the 14 days of the present experiment was at the lowest concentration of the alga (25 x 10^4 cells/ml). This is seemingly due to the insufficient food supply, which is in accordance with the conclusion of Dhont and Lavens (1996), that when *Artemia* is fed on unicellular proteins like algae, the concentration of algae must be above the critical lower quantity for *Artemia*, otherwise the contrary would occur.

The present results revealed that the lowest growth rates of *A. franciscana* were achieved when fed on rice bran. However, Sorgeloos *et al*. (1980) concluded that the rice bran is the most suitable diet for brine shrimp, which is perhaps the cheapest food ever used for feeding brine shrimp. They suggested that, although the nutritional value of rice bran as a sole food for fishes and crustaceans is known to be very low, it appears to be an excellent diet for brine shrimp, provided that it is free of contamination by pesticides, which are often applied during long-term transport or storage of the crude product. Furthermore, preliminary tests, in flow-through experiments of culturing brine shrimp, indicate that when rice bran is used, up to five times more biomass can be produced per unit of tank volume as compared with results obtained with batch culturing (Sorgeloos *et al.*, 1980).

Some other results have further indicated the efficiency of rice bran in achieving optimum growth and survivals when used to feed *Artemia* larvae. Dobbeleir *et al.* (1980) had achieved 80 per cent survivals and 4.26 mm total length when rice bran was used. Johnson (1980) had shown that the rice bran fed to the larvae of *Artemia* fulfilled the nutritional needs at least during the first three days. He achieved a growth of 2 mm for larvae of *Artemia* when reared in out door tanks, by doubling the concentration of diet (0.11 mg/ml) daily until the seventh day of the experiment. Despite that, Vijayaraghavan *et al*. (1987) reported a growth of *Artemia* larvae of 2.25 mm during the eight day of the experiment using 0.5 mg/ml of rice bran daily.

The apparent contradiction of the present results of the rice bran experiment with the previous results may entirely be due to a direct comparison made here between the rice bran and some other good food items like dried algae and baker's yeasts which have high protein and lipid contents compared with those of the rice bran (Table 19.5), and have the essential nutrients required during the early developmental stages of *Artemia,* whereas carbohydrates, together with proteins, are more important for juveniles and adults (D´Agostino and Provasoli, 1980).

Table 19.5: Approximate Chemical Composition of (per cent DW) of Various Food Items Used in Feeding the Local Strain of *Artemia franciscana*

Composition	*Chlorella vulgaris* Powder*	*Baker's Yeast*	*Rice Bran*
Fat	6.26	2.80	1.70
Protein	58.14	49.37	14.09
Carbohydrate	15.08	34.70	79.84

* Abdullah and Rajab, 1998.

The present results revealed that the mixture of food used gave rather positive results in terms of growth than the single food items. This is in agreement with the conclusion of Lavens and Sorgeloos (1991) and Al-Obaidy (2005). It is apparent here that the growth of *A. franciscana* larvae increased with increasing food concentration in the three different feeds, *viz*, dried algae, baker's yeasts and rice bran. Similarly, Evjemo and Olsen (1999) illustrated an increase in the growth rate of *A. franciscana* with increasing food concentration up to 10 mgC/L, and the growth rate appeared to be constant above this concentration.

It is inferred here that the growth rates of *A. franciscana* larvae increased with increasing dried algae concentration up to 0.25 and 0.3 mg/ml and this was higher than the results obtained from using baker's yeast and rice bran. This is seemingly due to the high nutritive value of this food, which is in concurrence with the conclusion of Lora-Vilchis *et al.* (2004), that the nutritional value of the alga depends on the chemical constituents of the alga and the nutritional requirements of the animals feeding upon it.

Moreover, *Artemia* can take up and digest exogenous microflora as part of the diet. Bacteria and protozoans which develop easily in the *Artemia* cultures are able to biosynthesize essential nutrients using the supplied brine shrimp food as a substrate, in this way they compensate for any possible deficiencies in the diet's composition (Dhont and Lavens, 1996).

Table 19.6: Average Values of the Reproductive Traits of a Local strain of *Artemia franciscana* at 60, 90 and 120 ‰ and 30°C. (±) standard deviation. Different letters indicate significant differences between treatments ($p<0.05$) tested by the one way analysis of variance, n= 10 for each treatment

Reproductive Traits	*Salinity (‰)*		
	60	*90*	*120*
Offspring per brood	81.77a (21.03)	82.40a (17.7)	54.73b (15.18)
Brood per female	4.370a (1.03)	5.60a (1.5)	7.500b (0.5)
Offspring per day per female	31.07a (3.89)	21.36b (10.6)	16.09b (1.34)
Time between brood (d)	2.600a (0.04)	3.850b (1.15)	3.400b (0.3)
Female pre reproductive period (d)	18.40a (1.64)	21.00b (3.0)	20.50b (0.5)
Female reproductive period (d)	11.50a (1.26)	21.00b (5.8)	25.50b (1.5)
Female post reproductive period (d)	3.500a (0.54)	3.000b (0.2)	3.000b (0.0)
Total female lifespan (d)	33.40a (1.87)	45.00b (3.36)	49.00b (1.5)
Total male lifespan (d)	23.62a (2.86)	42.00b (2.4)	43.50b (3.0)
Per cent offspring encysted	10.40a (4.1)	51.50b (8.05)	33.60b (2.85)
Live offspring per brood	73.25a (20.2)	39.90b (15.2)	36.33b (5.8)
Gestation period (d)	6.250a (1.94)	6.800a (2.1)	4.000b (0.5)

Optimum survivals of *A. franciscana* larvae were obtained in the present study at the concentration of 0.15-0.30 mg/ml of dried algae. Again, this suggests that algal food is efficiently meeting the requirements of the larvae than the other two types of food.

The survival rates of the larvae of *A. franciscana* become higher with increasing food concentration to 0.25 mg/ml at the three kinds of food. The concentration of 0.3 mg/ml obviously did not achieve a better result than the previous concentrations, especially when using dried food. This is perhaps, due to an increase of food particles in the culture medium, which ultimately affects the filtration mechanism of the feeding appendages, hence negatively affecting the growth of *Artemia* (Prema and Palavesam, 2004). It has also been previously found that, at certain higher concentrations of the four types of food used for rearing *Artemia* larvae, no significant increase in survivals or in growth had been achieved (Vijayaraghavan *et al.*, 1987).

It has been stated that when natural or commercial food were given to the *Artemia* they should fulfill the requirements for survivals and for further production so that *Artemia* can express maximum commercial growth during the time of the experiment. The excess of food beyond the required level tends to be economically invalid, due to it's loss in the culture medium and also becomes a source of increase pollution (Branie *et al.*, 1997).

The present results show that survivals in the four mixed food at a concentration of 0.3 mg/ml, was minimal at the dried algae and baker's yeast, although each of these food items alone had a higher protein contents (Table 19.5), but the level of protein achieving optimum survivals and growth is limited. In this respect, Hanaoka (1973) stated that the rates of growth and survivals of *Artemia* larvae are positively correlated with the protein contents of the food to a certain level, and when the level exceeds 28 per cent, the rate of growth and survivals decreased by increasing ammonia resulted from the decomposition of protein. This fact is in accordance with the results obtained in the present study in regards to the survivals, growth and biomass of the larvae of *A. franciscana* when a mixture of dried algae and baker's yeast were given. On the other hand, it is well documented that the nutritive value of the baker's yeast did not support optimal growth and survivals and in some cases it is not suitable for feeding larvae of *Artemia*, even for one week (Hirayama,

1987; Al-Obaidy, 2005). In order to improve the nutritional value of the food it should be mixed with some others to give better results of growth and survivals (Coutteau, 1992). It has been found here that the mean biomass achieved by the larvae of *A. franciscana* when fed on a single food were higher at the dried algae than at the live food or the other kinds of food.

When the food requirements are met for optimum growth and survivals, the time for sexual maturity is shortened. It is reported here that, larvae of *A. franciscana* attain sexual maturity within 11 days at all food items, except the rice bran in which the larvae failed to reach maturity during the period of the experiment which lasted for 14 days.

In comparing the present results of the time of attainment of sexual maturity of *A. franciscana* with those of other *Artemiidae* for which such data are available, it is obvious that the parthenogenetic *Artemia* took very much longer time than the bisexual *Artemia* (Table 19.7). Parthenogenetic *Artemia* from India may take 32, 23 and 27 days to reach maturity depending upon the experimental conditions and food items given. The shortest time was that of Urmia strain (18 days) (Manaffar *et al.*, 2004). Whereas the bisexual *Artemia* may take 9.5-19 days for attaining sexual maturity, depending upon the experimental conditions, food items given and the strain of *Artemia* (Table 19.7).

Two modes of reproduction were exhibited by *Artemia*, oviparity and ovoviviparity, and it has long been thought that *Artemia* produces cysts only at the extreme environmental condition, whereas it produces larvae when the conditions become favourable. However, in the present study, it was found that *Artemia* produces cysts and larvae simultaneously at different proportions at the three salinity levels used (60, 90 and 120 ‰). Similar results of the correlation between increases in salinity, pigmentation of the haemolymph and shell gland, and a switch to production of cysts in *A. franciscana* were obtained (Lochhead, 1961; Versichele and Sorgeloos, 1980), but the nature of this relationship remains unclear (Drinkwater and Clegg, 1991). Furthermore, Berthelemy-Okazaki (1986) suggested that the determination of the reproductive mode occurs at some unknown point in the reproductive cycle of the female, possibly even before the eggs have entered the ovisac.

Table 19.7: Comparison of Sexual Maturity of *Artemia* Fed on Different Food Items from Various Sources

Strain	*Food Items*	*Salinity (‰)*	*Temp. (°C)*	*Age at Maturation (d)*	*Reference*
Parthenogeneic, India	Baker's yeast	125	–	32	Baid (1963)
Parthenogeneic, India	*Spirulina* powder	35	27	23	Royan (1980)
Parthenogeneic, India	Rice bran	35	27	27	Royan (1980)
Parthenogeneic, Urmia Lake	–	–	–	18	Manaffar *et al.* (2004)
A. franciscana, New Zealand	*Dunaliella* sp.	140	26	9.5	Wear *et al.* (1986)
Sexual strain, Belgium	Chicken manure	60	–	12	Basil *et al.* (1995)
A. salina, Eygpt	*Dunaliella* sp. + Baker's yeast	120	25	17	Baxevanis *et al.* (2004)
A.? franciscana, Iraq	Rice bran + Soya bean	35	25	19	AL-Obaydi (2005)
A. franciscana, Iraq	Powdered *Chlorella vulgaris* + Baker's yeast	55	30	11	Present study

The influence of some environmental factors on a reproductive mode has been determined in a laboratory experiment. In mass cultures, Versichele and Sorgeloos (1980) were able to increase cysts production significantly in SFB population by subjecting female to cyclic low oxygen levels or a medium containing chelated iron. Based on these results and the observations of others, they concluded that the decreasing levels of oxygen that parallel increasing salinities in the natural habitat induced female to switch to oviparous reproduction. Moreover, Berthelemy-Okazaki and Hedgecock (1987) found that a number of factors, both intrinsic and extrinsic, affect reproductive mode in SFB *Artemia* including brood number, density, photoperiod, salinity, and temperature. Recently, Baxevanis *et al.* (2004) concluded that cysts production is enhanced by salinity and it seems that there is a salinity threshold that triggers ovoviviparity. Nambu *et al.* (2004) obtained high percentages of production of cysts in *A. franciscana* even in very low levels of salinity (2 per cent), and suggested that higher levels of salinity may not always be suitable for cysts production, as the mode of reproduction in the above study was affected greatly by the photoperiod and was less affected by temperature. It is evident from the present results that the percentage of cysts production decreased with increasing of salinity from 90 to 120 ‰, which suggests the presence of optimum salinity level achieving the highest percentage production of cysts. This is supported by the findings of Berthelemy-Okazaki (1986) who illustrated that the production of encysted eggs by *A. franciscana* is inhibited at 120 ‰ or more and the female at this salinity level produces the least number of cysts, and these cysts can not hatch at this salinity. This was formerly indicated by the results of Clegg (1974). However, cysts production seems to be more favoured in the salinities of 35, 140 and 180 ‰(Triantaphyllidis *et al.*, 1995). Despite that Drinkwater and Clegg (1991) stated that this surprising result disagreed with the prevailing views on *A. franciscana* life-cycle regulation and illustrates the need for much more research on the physiological ecology of that population. In addition, there is at least circumstantial evidence that cysts production is more costly than viviparous reproduction, as a typical cysts brood is only 74 per cent the size of viviparous broods (Browne, 1980). The present results showed significant differences in most of the reproductive traits and the lifespan specific characters in *A. franciscana* and the salinity level tested. More specifically, with the increase of salinity, an increase was observed in brood per female, time between broods (d), female pre reproductive period (d), female reproductive period (d), total female lifespan (d), total male lifespan (d), and percent offspring encysted, and a decrease was obvious in offspring per brood, offspring per day per female, female post reproductive period (d), live offspring per brood, and gestation period (d). These changes caused by the elevation of salinity may be attributed to the fact that part of the available energy of the individuals was consumed for homeostasis and not for reproduction (Baxevanis *et al.*, 2004). Vanhaeke *et al.* (1984) and Triantaphyllidis *et al.* (1995) obtained similar results, as at salinities above 140 ‰ the mortality of *A. franciscana* is high and the survival display retarded growth and poor reproductive characteristics.

On the contrary *A. urmiana* shows better reproductive performance in a high salinity and it can be considered a very efficient osmoregulator within the *Artemia* genus (Abatzopoulos *et al.*, 2006).

The physiological activities of the organism are usually correlated with the limiting environmental conditions, and temperature and salinity are the effective of these conditions. Sexual maturity in *Artemia,* as in other organisms is a physiological activity and is affected by these conditions.

The determination of the optimum period for the production of larvae and cysts in *Artemia* is depending on the sexual maturity. The previous results showed that nauplii of local *A. franciscana* reached sexual maturity within 11 days at 30 °C and 55 ‰(Abdullah, 2007). Results of the ANOVA test showed that in *A. franciscana*, the time to 50 per cent maturity, egg development time, and generation time were more heavily influenced by temperature than by salinity as they decreased with increasing temperature up to 26 °C but is extended at 30 °C, and Wear *et al.*(1986) found the most favourable temperature *i.e.* 24°C prevaling in the region. However, Von Hentig (1971) showed that the Great Lake strain of *A. franciscana* matured rapidly at 30 °C than at 20 °C in all the three salinities of: 15, 32, and 70 ‰but this dose not obviate the possibility that fastest maturation may have occurred at < 30 °C. Furthermore, increase in salinity enhanced the fecundity and maturity of *Artemia* in several cases (Dwivedi *et al.*, 1980, Basil *et al.*, 1987, Basil and Pandian, 1991), but the prereproductive period in *A. franciscana* and parthenogenetic *Artemia* decreased with increasing of salinity up to 100 ‰and increased again from 140-180 ‰(Triantaphyllidis *et al.*, 1995). Similarly the case of 3 parthenogenetic and one bisexual strain (*A. salina*) from Eygpt have revealed a decrease in the prereproductive period at 35 – 80 g/L and then an increase at 120 -200 g/L (Baxevanis *et al.*, 2004). These results are in agreement with the present results on *A. franciscana*. In addition, the female reproductive period and the number of brood per female increased with increasing salinity. Obviously, the elevation of salinity induced oviparity, which is expressed as significant increase of the produced encysted embryos; a similar results have been reported by Abatzopoulos *et al.* (2003) from elsewhere.

It is generally a known fact that the age of male is shorter than that of female in all populations of *Artemia* whether they are Old World species or New World species, this result is observed here also, Browne and Halanych (1989) suggested that males may be less efficient than female at food gathering and storage or may swim at a faster rate than females and hence expend energy more rapidly.

Female reproductive period of *A. franciscana* was significantly longer at higher salinity, so as female lifespan, this is in agreement with the findings of Abatzopoulos *et al.* (2006) on *A. urmiana*. Female's post- reproductive period observed in the present study significantly decreased with increasing salinity, this is opposite to the results of Abatzopoulos *et al.* (2006) but in accordance with those of Abatzopoulos *et al.* (2003) and Baxevanis *et al.* (2004). This is apparently due to the fact that the female consumed most of the reserved energy in the production of broods (Abatzopoulos *et al.*, 2003; 2006; Baxevanis *et al.*, 2004).

The life history traits of *Artemia* strains are considered as an important factor when the animal is being introduced into a new habitat, especially when competition is expected with a strain resident in that habitat. The possibility of competition depends on several factors *viz.* the length of the reproductive period, the pre and post- reproductive period, lifespan, brood size, and the time between broods. Generally, the bisexual New World populations produced more offspring per brood, more broods per female per day and with rapid sexual maturity, therefore, they are better than the Old World bisexual populations and even better than the parthenogenetic populations (Branie *et al.*, 1997).

Acknowledgements

We are grateful to the Director General of the Marine Science Centre for supports in various ways.

References

Abatzopoulos, T.J., Triantaphyllidis, G.C. and Kastritsis, C.D., 1993. Genetic polymorphism in two parthenogenetic *Artemia* populations from Northern Greece. *Hydrobiologia*, 250: 73-80.

Abatzopoulos, T.J., El-Bermawi, N., Vasdekis, C., Baxevanis, A.D. and Sorgeloos, P., 2003. International Study on *Artemia*. LXVI. Effects of salinity and temperature on reproductive and life span characteristics of clonal *Artemia*. *Hydrobiologia*, 492: 191-199.

Abatzopoulos, T.J., Baxevanis, A.D., Triantaphyllidis, G.C., Pador. E.L., Van Stappen. G. and Sorgeloos, P., 2006. Quality evaluation of *Artemia urmiana* Günther (Urmia Lake, Iran) with special emphasis on its particular cyst characteristics. *Aquaculture*, 254: 442-454.

Abdullah, D.S.M., 2007. Selection of the optimum conditions for the best production of the brine shrimp *Artemia franciscana* (Kellogg) in Basrah. Ph.D. Thesis, Basrah University. 166 pp. (in Arabic).

Abdullah, D.S. and Rajab, T.M.A., 1998. Chemical composition of *Chlorella vulgaris* Beijernick isolated from Shatt Al-Arab River. *Marina Mesopotamica*, 13(1): 121-127.

Ahmed, H.K., 2002. Effect of temperature on growth and survival of the brine shrimp *Artemia salina* (L.): a local strain from Basrah region. *Marina Mesopotamica*, 17(2): 329-339.

A.O. A. C., 1984. Association of official analytical chemists, 14th ed. Official methods of analysis. Inc. S. Willims, (ed). SA. 1141.p.

Al-Obaydi, T.S.M., 2005. Study of some biological aspects of the brine shrimp and it's use in feeding the larvae of common carp *Cyprinus carpio* and grass carp *Ctenopharyngodon idella*. Ph.D. Thesis, Univ. Baghdad: 185 pp.

Baid, I.C., 1963. The effect of salinity on growth and form of *Artemia salina* L. *J. Exp. Zool.*, 153: 279-284.

Basil, J. A. and Pandian, G. T., 1991. Culturing *Artemia* (Tuticorin strain) in organic and agricultural wastes at different salinities. *Hydrobiologia*, 212: 11-17.

Basil, J.A., Nair, V.K.S. and Thatheyus, A.J., 1995. Laboratory studies on the culture of the brine shrimp *Artemia* using organic wastes. *Bioresource Technology*, 15: 265-267.

Basil, J. A., Premkumar, D. R. D., Lipton, A. P. and Marian, M.P.(1987. Peliminary studies on the culture of *Artemia* using newable organic wastes. p. 275-278. In: Sorgeloos, P., Bengtson, D. A., Decleir, W. and Jaspers, E. (Eds.) *Artemia* Research and It's Applications, Vol. 3: Ecology, Culturing, Use in Aquaculture, Universa Press, Wetteren, Belgium.

Baxevanis, A.D., El-Bermawi, N., Abatzopoulos, T.J. and Sorgeloos, P., 2004. International Study on *Artemia*. LXVIII. Salinity effects on maturation, reproductive and life span characteristics of four Egyptian populations *Hydrobiologia*, 513: 87-100.

Berthelemy-Okazaki, N.J., 1986. Environmental, biochemical and genetic factors regulating oviparity in *Artemia*. Ph.D. dissertation, University of California, Davis.

Berthelemy-Okazaki N.J. and Hedgecock D., 1987. Effect of environmental factors on cyst formation in the brine shrimp *Artemia*. *In*: P. Sorgeloos, D.A. Bengtson, W. Decleir and E. Jaspers (eds.) P: 167-182. *Artemia* Research and it's Applications, vol 3. Ecology, Culturing, Use in aquaculture. Universa Press, Wetteren, Belgium.

Branie, A.A., Eesa, M.A., Abdul-Rahman, A., Othman, M.F. and Sadik, S.S., 1997. Scientific and practical principles of spawning of fish and crustacean caring in the Arab World. Vol. 1 Al-Daar Al-Arabiah for publication and distribution press, Cairo, U.E.R.872 pp. (in Arabic).

Borowitzka, L.J., Borowitzka, M.A. and Moulton, T.P., 1984. The mass culture of *Dunaliella salina* for fine chemicals: From laboratory to pilot plant. *Hydrobiologyia*, 116/117: 115-134.

Barata, C., Hontoria, F., Amat, F. and Browne, R.A., 1996.Competition between sexual and parthenogenetic *Artemia*: temperature and strain effects. *J. Exp. Mar. Biol. Ecol.*, 196: 313-328.

Bowen, S.T., Fogarion, A., Hitchner, K.N., Hitchner, G.L., Dana, G.L. Dana, H.S. Chow, Buonristiani, M.R. and Carl, J.R., 1985. Ecological isolation in: *Artemia* population differences in tolerance of anion concentrations. *J. Crust. Biol.*, 5: 106-129.

Browne, R.A., 1980. Reproductive pattern and mode in the brine shrimp. *Ecology*, 61: 466-470.

Browne, R.A., 1982. The costs of reproduction in brine shrimp. *Ecology*, 63: 43-47.

Browne, R.A., Davis, L.E. and Sallee, S.E., 1988. Temperature effects on life history traits and relative fitness of sexual and asexual *Artemia*. *J. Exp. Mar. Bio. Ecol.*, 124: 1-20.

Browne, R.A. and Halanych, K.M., 1989. Competition between sexual and parthenogenetic *Artemia*: A re-evaluation. *Crustaceana*, 57: 57-71.

Browne, R.A. and MacDonald, G.H., 1982. Biogeography of the brine shrimp, *Artemia*: distribution of parthenogenetic and sexual populations. *J. Biogeogr.*, 9: 331-338.

Browne, R.A., Moller, V., Forbes, V.E. and Depledye, M.H., 2002. Estimating genetic and environmental components of variance using sexual and clonal *Artemia*. *J. Exp. Mar. Biol. Ecol.*, 267: 107-119.

Browne, R.A., Sallee, S.E., Grosch, D.S., Sagrati, V.O., and Purser, S.M., 1984. Partitioning genetic and environmental components of reproduction and lifespan in *Artemia*, *Ecology*, 65: 949 – 960.

Browne, R.A. and Wanigasekera, G., 2000. Combined effects of salinity and temperature on survival and reproduction of five species of *Artemia*. *J. Exp. Mar. Biol. Ecol.*, 244: 29-44.

Camargo, W.N., Ely, J.S. and Sorgeloos, P., 2003. Morphometric characterization of thalassohaline *Artemia* population from the Colombian Caribbean. *J. Biogeog.*, 30: 697-702.

Clegg, J.S., 1974. Biochemical adaptations associated with the embryonic dormancy of *Artemia salina*. *Trans. Amer. Micros. Soc.*, 93 (4): 481-490.

Clegg, J.S. and Trotman, C.N.A., 2002. Physiological and biochemical aspects of *Artemia* ecology. P: 129-170. *In*: T.J. Abatzopolous, J.A. Beardmore, J.S. Clegg and P. Sorgeloos (eds). *Artemia* Basic and Applied Biology. Kluwer Academic publishers. Dordecht.

Coutteau, P., 1992. Baker's yeast as a substitute for micro-algae in the culture of filter-feeding organisms. Ph.D. Thesis, University of Ghent, Belgium.

D'Agostion, A.S. and Provasoli, L., 1968. Effects of salinity and nutrients on mono and diaxenic cultures of two strains of *Artemia salina*. *Biol. Bull.*, 134: 1-14.

Dhont, J. and Lavens, P., 1996. Tank production and use of ongrown *Artemia*. vol. 361, P. 164-195. *In*: P. Lavens and P. Sorgeloos (Eds.), Manual on the production and Use of Live Food for Aquaculture. FAO fish. Tech. Pap., Rome.

Dobbeleir, J., Adam, N., Bossuyt, E., Bruggeman, E., and Sorgeloos, P., 1980. New aspects of the use of inert diets for high density culturing of brine shrimp. P: 165-174. *In*: G. Persoone, P. Sorgeloos, O. Roels and E. Jaspers (eds). The Brine Shrimp *Artemia*. Universa Press. Wettern, Belgium.

Drinkwater, L.F., Clegg, J.S., 1991. Experimental biology of cysts diapause. In: *Artemia Biology*, Boca Raton, Ann Arbor, Boston: CRC Press, P. 93-117.

Dwivedi, S. N., Ansari, S. K. R. and Ahmed, M. G., 1980. Mass culture of brine shrimp under controlled conditions in cement pools in Bombay, India. P. 175-183. *In*: G. Persoone, P. Sorgeloos, O. Roels and E. Jaspers (eds.) The brine shrimp *Artemia*. vol. 3. Ecology Culturing, Use in Aquaculture, Universa Press, Wetteren, Belgium.

Evjemo, J.O. and Olsen, Y., 1999. Effect of food concentration on the growth and production rate of *Artemia franciscana* feeding on algae (T. Iso). *J. Exp. Mar. Bio. Ecol.*, 242: 273-296.

Gajardo, G.M. and Beardmore, J.A., 1989. Ability to switch reproductive mode in *Artemia* is related to maternal heterozygosity. *Mar. Ecol. Prog. Ser.*, 55: 191-195.

Guillard, R.L. and Ryther, J.H., 1962. Studies of marine plankton diatoms. I. *Cyclotella nana* Hustedt and *Detonula confervacea* (Cleve) Grun. *Can. J. Microbiol.*, 8: 229 – 239.

Hanaoka, H., 1973. Cultivation of three species of pelagic micro-crustacean plankton. *Bull. Plankton. Soc. Japan*, 20: 19-29.

Hirayama, K., 1987. A consideration of why mass culture of the rotifer *Brachionus plicatilis* with baker's yeast unstable. *Hydrobiologia*, 147: 269-270.

Johnson, D.A., 1980. Evaluation of various diets for optimal growth and survival of selected life stages of *Artemia*. P. 185-192. *In:* G.Persoone, P. Sorgeloos, O. Roels and E. Jaspers, editors. The brine shrimp *Artemia*. vol. 3. Universa Press, Wetteren, Belgium.

Lavens, P. and Sorgeloos, P., 1991. Production of *Artemia* in culture tanks. p. 317-350. *In*: A. Browne; P. Sorgeloos and C.N.A Trotman (eds.) *Artemia* Biology. CRC Press, Boca Raton, Florida, USA.

Lenz, P.H. and Dana, G.L., 1987. Life-cycle studies in *Artemia* a comparison between a subtropical and temperate population. Vol. 3: 89-100. In: P. Sorgeloos, D.A. Bengeston,W. Decleir and E. Jaspers (eds.). *Artemia* Research and its Applications. Universa Press, Wetteren, Belgium.

Lora-Vilchis, M.C., Cordero-Esquivel, B. and Voltolina, D., 2004. Growth of *Artemia franciscana* fed *Isochrysis* sp. and *Chaetoceros muelleri* during its early life stages. *Aquaculture Research*, 35: 1080-1091.

Lochhead, J.H., 1961. Oviparity versus ovoviviparity in the brine shrimp, *Artemia*. *Bio. Bull.*, 121, 396.

Manaffar, R., Atashbar, B., Mahimi, S., Hasani, A. and Agh. N., 2004. Successful pond culture of *Artemia parthenogenetica* and *Artemia urmiana* at the vicity of the Lake Urmia. Inco. –DEV. Project on *Artemia* biodiversity. Iran international workshop – Sep 21-25, 2004. Urmia, Iran. 110.

Marques, A., Dhout, J., Sorgeloos, P. and Bossier, P., 2004. Evaluation of different yeast cell wall mutants and micro algae strains as feed for gnotobiotically growth brine shrimp *Artemia franciscana*. *Exp. Mar. Biol. Ecol.*, 312: 115-136.

Nambu, Z., Tanaka, S. and Nambu, F.(2004. Influence of photoperiod and temperature on reproductive mode in the brine shrimp, *Artemia franciscana*. *Exp. Zool.*, 301 (A). 542-546.

Neagel, L.C.A., 1999. Controlled Production of *Artemia* biomass using an inert commercial diet compared with the micro algae *Cheatoceros* sp. *Aqua- cultural Engineering*, 21: 49-59.

Pearson, D., 1970. The chemical analysis of foods. 6thed. J. and A. Churchill, London, p. 7-11.

Prema, P. and Palavesam, A., 2004. Effect of Ayurvedic products on the growth, survival and reproduction of *Artemia parthenogenetica* (Abreu-Grobois and Beardmore). *Asian Fisheries Science*, 17: 61-69.

Rose, M.R. and Lauder, G.V., 1996. Post-spandred adaptionism. P. 1-8. *In:* Adaptation, Rose, M.R. and Luder, G.V. (Eds.). New york, Academic Press.

Royan, J.P., 1980. Laboratory and field studies on an Indian strain of the brine shrimp *Artemia*. P. 223-230. *In*: G. Persoone, P. Sorgeloos, O. Roels and E. Jaspers (eds.). The brine shrimp *Artemia*. vol. 3. Universa Press. Wetteren, Belgium.

Sick, L.V., 1976. Nutritional effect of five species of marine algae on the growth, development, and survival of the brine shrimp *Artemia salina*. *Mar. Biol.*, 35: 69-78.

Sorgeloos, P., Mesa, M., Bossuyt, E., Bruggeman, E., Dobbeleir, J., Versichele, D., Lavina, E. and Bernardino, A., 1980. Culture of *Artemia* on rice bran: The conversion of a waste-product into highly nutritive animal protein. *Aquacult.*, 21: 393-396.

Stephens, D.W. and Gillespire, D.M., 1976. Community structure and ecosystem analysis of the Great Salt Lake. In Riley, J.P. (ed.), the Great Salt Lake and Utah's water Resources. Utah water Research Lab., Utah State University. USA: 66-72.

Streans, S.C., 1993. The evolution of life histories. Oxford, Oxford Press.

Tachibana, K., Yagi, M., Haram K., and Mishima, T., 1997. Effects of feeding of B-carotene supplemented rotifers on survival and lymphocyte proliferation reaction of fish larvae Japanese parrot fish (*Oplegnathus fasciatus*) and spotted parrot fish (*Oplegnathus punctatus*): Preliminary trials. *Hydrobiologia*, 358: 313-316.

Triantaphyllidis, G.C., Poulopoulou, K. Abatzopoulos, T.J., Pinto Perez., C.A. and Sorgeloos, P., 1995. International study on *Artemia*, XLIX. Salinity effects on survival, maturity, growth, biometrics, reproductive and lifespan characteristics of a bisexual and a parthenogenetic population of *Artemia*. *Hydrobiologia*, 302: 215-227.

Vanhaecke, P., Siddal, S.E. and Sorgeloos, P., 1984. International study on *Artemia*. XXXII, Combined effects of temperature and salinity on the survival of *Artemia* of various geographical origin. *J. Exp. Mar. Biol. Ecol.*, 80(3): 259-275.

Versichele, D. and Sorgeloos, P., 1980) Controlled production of *Artemia* cysts in batch culture. P: 231–246. *In*: G. Persoone; O. Roels and E. Jaspers, (eds.). The brine Shrimp *Artemia* vol.3. Ecology, Culturing, use in aquaculture. Universa Press, Wetteren Belgium.

Vijayaraghavan, S., Krishnakumari, L. and Royan, J.P., 1987. Evaluation of Different feeds for optimal growth and survival of parthenogenetic brine shrimp, *Artemia*. *Indian J. Mar. Sci.*, 16: 253-255.

Vollenweider, R.A., 1974. A manual on methods for measuring primary production in Aquatic environments. Int. Biol. Program Hand book 12. Blackwell Scientific Publications Ltd. Oxford, 225p.

Von Hentig, R., 1971. Einfluss von Salzgehalt and temperature auf Entwieklug, Wachstum, fortplnzung and energiebilanz von *Artemia salina*. Biol., 9: 145-182. (cited by: Sorgeloos, P. and Persoon, G. 1975).

Vos, J. and De La Rosa, N.L., 1980. Manual on *Artemia* production in salt ponds in the Philippines. FAO/UNDP. BFAR, P. 1-48.

Wear, R.W., Haslett, S.T. and Alexander, N.L., 1986. Effects of temperature and salinity on the biology of *Artemia franciscana* (Kellogg) from Lake Grassmere, New Zealand. 2. Fecundity, and generation times. *J. Exp. Mar. Biol. Ecol.*, 98: 167-183.

Wickins, J.F., 1972. The food value of brine shrimp, *Artemia salina* L., to larvae of the prawn, *Palaemon serratus* Pennant. *J. Exp. Mar. Biol.*, 10: 151-170.

Chapter 20

Ultrastructural Studies of Vitellogenesis in *Asellus aquaticus* (Crustacea, Isopoda)

***Melike Erkan*[1*] *and Mário Sousa*[2]**

[1]*Istanbul University, Faculty of Sciences, Department of Biology, 34134 Vezneciler Istanbul, Turkey*

Laboratory of Cell Biology, Institute of Biomedical Sciences Abel Salazar, University of Porto, Portugal

ABSTRACT

During pre-vitellogenesis, nuage materials, free ribosomes, endoplasmic reticulum, dictyosomes and mitochondria accumulated at the nuclear periphery and then spread throughout the ooplasm. In early-vitellogenesis, lipid droplets were synthesized by the smooth endoplasmic reticulum and mitochondria, cortical vesicles appeared as expansions of the rough endoplasmic reticulum, and dense yolk vesicles were formed by endogenous and receptor-mediated endocytic pathways. At this stage, the oocyte surface developed small microvilli and a thin vitelline layer. During mid-vitellogenesis, a new type of yolk vesicles formed, and both dense and intermediate-dense yolk vesicles grew through

* Corresponding author: E-mail: merkan@istanbul.edu.tr

fusion and engulfment of smaller vesicles. At this stage, annulate lamellae were observed in the oocyte cortex in close association with nuage materials, and cortical vesicles, which increased in size through fusion, became concentrated at the oocyte periphery. At the oocyte periphery, microvilli increased in length, the vitelline layer increased in thickness, and smooth-pits appeared associated with the oolemma. During late-vitellogenesis, the ooplasm becomes filled with very large dense and intermediate-dense yolk vesicles as well as by very large lipid droplets. This was followed by the appearance of glycogen aggregates in follicular cells, which then disappeared as glycogen particles became concentrated in the ooplasm, simultaneously with the reappearance of receptor-mediated endocytosis, microvilli extension, and separation of the oocyte surface from the close contact with follicular cells.

Keywords: *Crustacea, Isopoda, Oogenesis, Reproductive biology.*

Introduction

In Arthropoda, the sites of yolk precursor synthesis varies considerably between different organisms, being either extra-oocytic or intra-oocytic (Kessel, 1968; Charniaux-Cotton, 1985; Norrevang, 1988; Suzuki *et al.*, 1989; Krol *et al.*, 1992; Wagele, 1992; Erkan *et al.*, 2001; Serrano-Pinto *et al.*, 2003; Kung *et al.*, 2004; Okumura *et al.*, 2007; Cabrera *et al.*, 2009).

In Xiphosura, the endogenous production of yolk vesicles by dictyosomes and the smooth endoplasmic reticulum was observed during vitellogenesis, simultaneously with endocytosis of exogenous components, giving rise to two different types of yolk vesicles (Dumont and Anderson, 1967). On the contrary, in Branchiopoda (Crustacea), yolk production in Notostraca was described to originate from endoplasmic reticulum activity (Scanabissi and Trentini, 1979), whereas in Conchostraca yolk vesicles formed from the Golgi and endoplasmic reticulum during pre-vitellogenesis, either without (Sabelli and Tommasini, 1990) or with participation of receptor-mediated endocytosis (Zeni and Zaffagnini, 1989) during vitellogenesis. A similar situation was observed in Balanidae (Crustacea, Cirrepedia), where endogenous yolk was first synthesized by the rough endoplasmic reticulum during pre-vitellogenesis, but then the same yolk vesicles received exogenous components during vitellogenesis (Lepore *et al.*, 1993; Ikuta and Makioka, 1997; Ikuta *et al.*, 1997).

The same phenomenon was observed in Isopoda (Crustacea, Malacostraca), where yolk vesicles are first formed by the rough endoplasmic reticulum, either at pre-vitellogenesis (*Saduria entomon*), early-vitellogenesis (*Porcellio scaber*), or mid-vitellogenesis (*Oniscus asellus*), and then later fused with endocytic vesicles at early-vitellogenesis (*S. entomon*; *P. scaber*) (Bilinski, 1979; Hryniewiecka-Szyfter and Babula, 1995), or during late-vitellogenesis (*O. asellus*) (Beams and Kessel, 1980). A very similar situation to Xiphosura was found in Amphipoda (Malacostraca), with yolk vesicles being simultaneously formed by the rough endoplasmic reticulum and endocytosis since early-vitellogenesis (Rateau and Zerbib, 1978; Zerbib, 1980).

On the contrary, species of Decapoda (Malacostraca) showed quite different aspects of yolk production (Quackenbush, 1989; Yano *et al.*, 1996). Thus, in the shrimps *Penaeus aztecus*, *P. setiferus* and *P. kerathurus*, the rough endoplasmic reticulum and the Golgi complex produced endogenous yolk vesicles since early-vitellogenesis, whereas during mid-vitellogenesis endocytosis gave origin to a distinct exogenous type of yolk vesicles (Duronslet *et al.*, 1975; Carvalho *et al.*, 1998a,b, 1999; Erkan *et al.*, 2001). In these latter species, another type of vesicles, the cortical vesicles, appeared formed by the rough endoplasmic reticulum during late-vitellogenesis, being later exocytated to be used in chorion formation after fertilization (Duronslet *et al.*, 1975; Carvalho *et al.*, 1998a,b, 1999; Erkan *et al.*, 2001). Other shrimps showed a similar process to Xiphosura, as in *Palaemon serratus*, where only one single type of yolk vesicles are formed since early-vitellogenesis by the cooperative action of the Golgi complex, rough endoplasmic reticulum and endocytosis (Papathanassiou and King, 1984). Finally, in *Penaeus vannamei*, no endocytosis was noticed, as in *Notostraca*, and the rough endoplasmic reticulum gave rise to two different types of yolk vesicles and to cortical vesicles during early-vitellogenesis (Rankin and Davis, 1990).

On the contrary, and as observed in Conchostraca, Branchiura and Isopoda, in crayfish (Beams and Kessel, 1962, 1963; Ganion and Kessel, 1972; Zerbib, 1979) and lobsters (Schade and Shivers, 1980), the rough endoplasmic reticulum gave rise to yolk vesicles during early-vitellogenesis to which exogenous material was later added. A more diverse situation occurs in crabs. As observed in Xiphosura and Amphipoda, in *Libinia emarginata* and *Coenobita clypeatus* only one type of yolk vesicles is synthesized during early-vitellogenesis through the combined action of the Golgi complex, rough endoplasmic reticulum and endocytosis (Hinsch and Cone, 1969; Goudeau and Lachaise, 1980a,b, 1983; Kom and Hinsch, 1985, 1987). However, as in some shrimps, in *Cancer pagurus* the rough endoplasmic reticulum gave origin to endogenous yolk vesicles, whereas the smooth endoplasmic reticulum, the Golgi complex and endocytosis formed a distinct mixed-type of yolk (Eurenius, 1973).

In the present study, the detailed fine structure of pre-vitellogenesis and vitellogenesis in *Asellus aquaticus* is presented, with particular emphasis to the pathways involved in the synthesis of lipids, yolk vesicles and cortical vesicles. The results of the study are compared to those obtained for other Crustacea.

Materials and Methods

Female specimens of *Asellus aquaticus* (Crustacea, Isopoda) were collected from rearing ponds of the Botany Department of the University of Istanbul in Turkey. *A. aquaticus* is a fresh-water species that is generally used for toxicological studies on the accumulation of heavy metals in environments contaminated with industrial waste products, being also an important element of the food-chain. Small pieces of ovary were prefixed in 2.5 per cent glutaraldehyde buffered with 0.1 M sodium-cacodylate, pH 7.2, for 2 h at 4°C, and rinsed in the same buffer for 2 h at 4°C. Specimens were then postfixed with 1 per cent osmium tetroxide in the same buffer, for 2 h at 4°C. After dehydration in a graded ethanol series, the material was transferred to propylene oxide and then embedded in Epon. Ultrathin sections were cut in a LKB ultramicrotome, and double stained with 3 per cent aqueous uranyl acetate for 20 min and lead citrate (Reynolds) for 10 min (Erkan, 1998). Grids were then examined in a JEOL 100CXII transmission electron microscope operated at 60 kV. All chemicals were of analytical grade and purchased from Merck.

Results

Pre-Vitellogenesis

At zygotene/pachytene, the nucleus presents evident synaptonemal complexes and occupies most of the cytoplasm. An intense synthesis and cytoplasmic exportation of ribonucleoproteins through nuclear pores was observed, with accumulation of nuage (granulo-fibrilar dense masses of ribonucleoproteins), free ribosomes, rough endoplasmic reticulum cisternae, and mitochondria in the perinuclear cytoplasm (Figure 20.1).

This phase is followed by transition to the diplotene stage, which appears characterized by the progressive loss of the synaptonemal complexes and appearance of a nucleolus. The nucleolus show seggregated components, being linked to the nuclear envelope through the granular component (Figures 20.2A,B). In the larger cytoplasm, all organelles and cytoplasmic structures increase in number and become distributed throughout the ooplasm, with fragmentation and dispersion of nuage materials, and appearance of dictyosomes (Figure 20.2C). Either during pachytene and early diplotene, the oocyte appears surrounded by a single layer of follicle cells, and at the oocyte surface, no microvilli, vitelline layer, or micropinocytosis were observed (Figures 20.1, 20.2C).

Early-Vitellogenesis

At this stage of oocyte growth, both the nucleus and ooplasm increased in size, with predominance of the latter. In the nucleoplasm, a very large and round nucleolus appears, being formed by a central dense fibrillar component, containing several small fibrillar centers, surrounded by a concentric external layer of the granular component. Several small round structures made of the nucleolar granular component were also observed near the nuclear envelope (Figure 20.3A). In the perinuclear cytoplasm, the number of small nuage aggregates increased, and these then dispersed throughout the entire ooplasm (Figure 20.3).

Besides mitochondria, endoplasmic reticulum, dictyosomes and microtubules, lipid droplets, dense yolk vesicles and cortical vesicles appeared in the ooplasm at this stage. Lipid droplets, closely surrounded by smooth endoplasmic reticulum cisternae and mitochondria, first appeared in the ooplasm and rapidly increased in size and number (Figure 20.3B). Simultaneously, small vesicles with floccular contents appeared, being formed through the combined activity of dictyosomes and the rough endoplasmic reticulum. These vesicles then aquire a dense core and fuse to originate large, irregularly-shaped, dense yolk vesicles, with cord-like

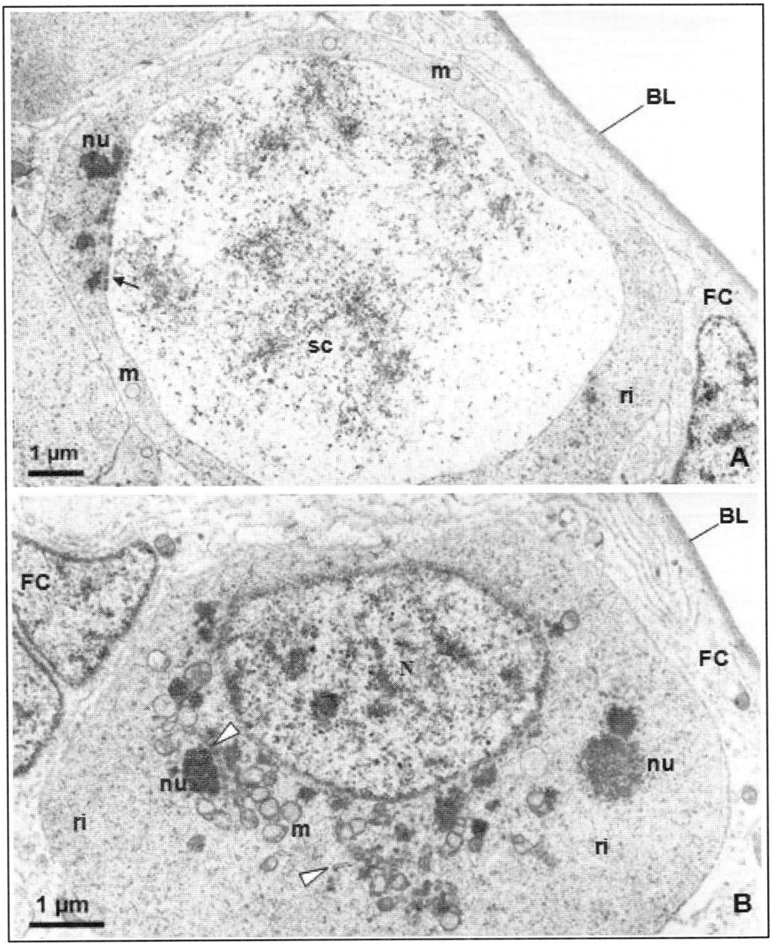

Figure 20.1: TEM Photographs of the Oocyte of *Asellus aquaticus* at the Zygotene/Pachytene Stage

(A), The oocyte is surrounded by a single layer of follicular cells (FC), and these are separated from the exterior by a thick basal lamina (BL). In the nucleus (N), there are synaptonemal complexes (sc), and ribonucleoproteins appear exported through the nuclear envelope (arrow) to accumulate in the perinuclear cytoplasm as small aggregates of nuage materials (nu). (B), At a late stage, the perinuclear cytoplasm also accumulates large nuage aggregates (nu), free ribosomes (ri), rough endoplasmic reticulum cisternae (arrowheads), and mitochondria (m).

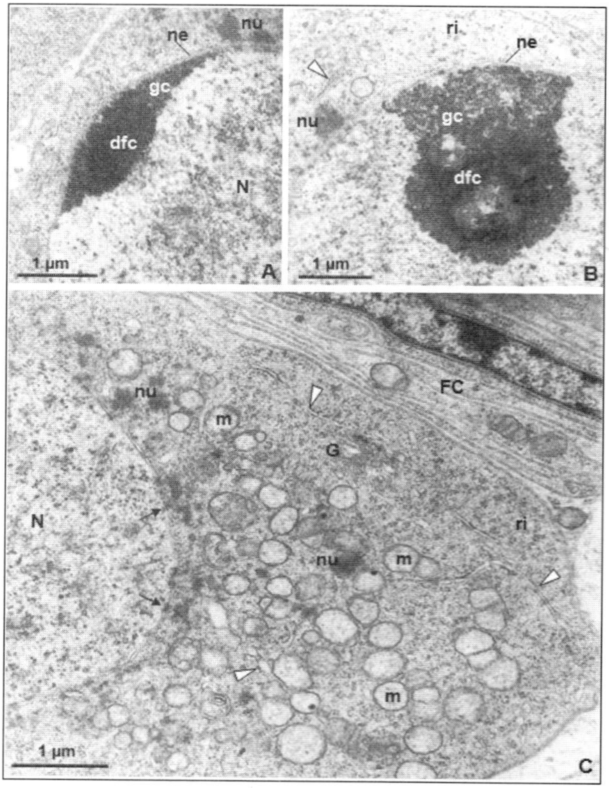

Figure 20.2: TEM Photographs of the Oocyte of *Asellus aquaticus* at the Early Diplotene Stage

(A), In the nucleus (N), the nucleolus first appears as an elongated structure attached to the nuclear envelope (ne). It is made of a central dense fibrillar component (dfc) and a lateral granular component (gc). (B), The nucleolus then increases in size, with a large and round dense fibrillar component (dfc) structure connected to the nuclear envelope (ne) by a stalk of the granular component (gc). (C), Ribonucleoproteins continue to be exported (arrows) from the nucleus (N), but nuage aggregates (nu) appear fragmented and dispersed. In the ooplasm there is a large increase in the number of free ribosomes (ri), mitochondria (m) and rough endoplasmic reticulum cisternae (arrowheads), and appearance of dictyosomes (G), which now fill the entire cytoplasm. At the oocyte periphery, there are no microvilli, vitelline layer, or endocytosis. FC, follicular cell.

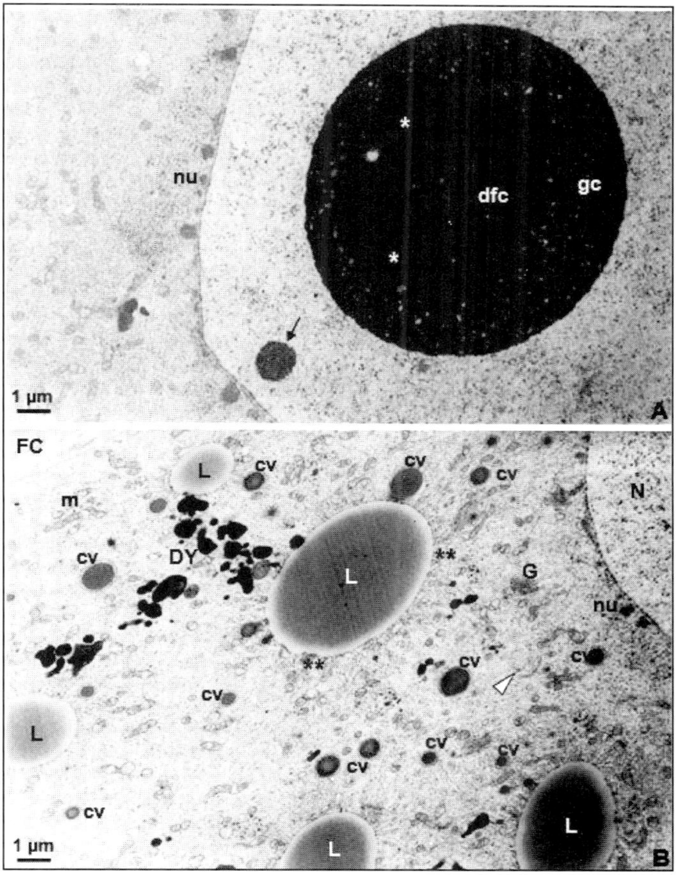

Figure 20.3: TEM Photographs of the Oocyte of *Asellus aquaticus* at Early-vitellogenesis

(A), The nucleus (N) contains a very large round nucleolus composed by a central dense fibrillar component (dfc), with numerous fibrillar centers (*), surrounded by an external layer of the granular component (gc). Small round structures (arrow) of the granular component are also observed near the nuclear envelope, and nuage aggregates (nu) accumulate in the perinuclear cytoplasm. (B), The ooplasm contains mitochondria (m), dictyosomes (G), rough endoplasmic reticulum cisternae (arrowhead), large lipid droplets (L), encircled by smooth endoplasmic reticulum cisternae and mitochondria (**), dense yolk vesicles (DY), and cortical vesicles (cv). FC, follicular cell.

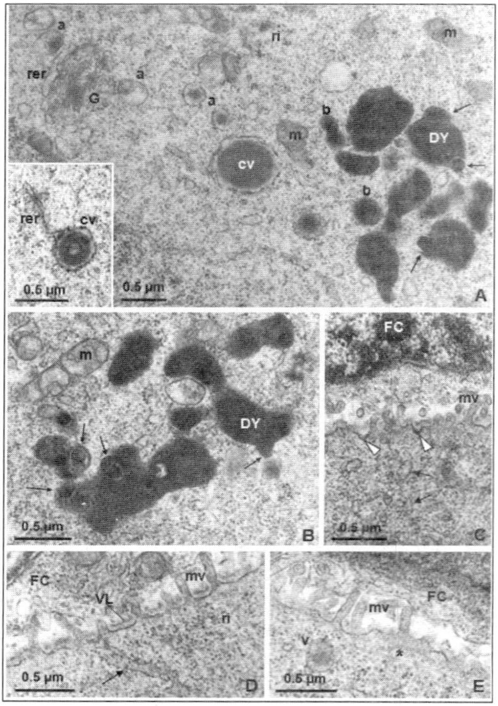

Figure 20.4: TEM Photographs of the Oocyte of *Asellus aquaticus* at Early-vitellogenesis

(A, inset), Cortical vesicles (cv) are formed from terminal expansions of the rough endoplasmic reticulum cisternae (rer), which accumulate concentric lamellae with a central pale region, and remain with associated ribosomes. (A, B), Dense yolk vesicles are formed from small vesicles, with flocular contents (a), which detach from dictyosomes (G) and rough endoplasmic reticulum cisternae (rer). These then acquire dense contents (b) and fuse to give origin to large and irregular, dense yolk vesicles (DY), with cord-like contents. Dense yolk vesicles then grow by engulfment (arrows) of the smaller vesicles, thus acquiring heterogeneous membranous contents. (C-E), The oocyte surface appears separated from the follicular cell (FC) by a narrow perivitelline space, and contains short microvilli (mv), a thin vitelline layer (VL), numerous coated-pits (arrowheads), endoplasmic reticulum vesicles and cisternae (arrows), small vesicles with flocular contents (v), free ribosomes (ri), and microtubules (*). m, mitochondria.

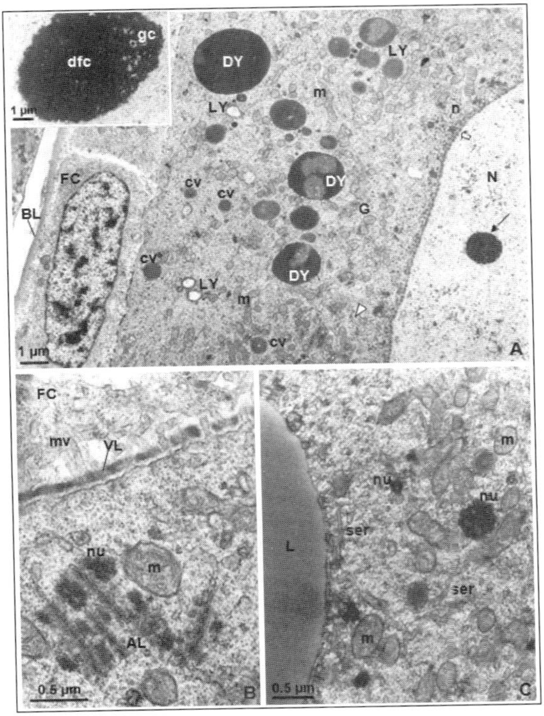

Figure 20.5: TEM Photographs of the Oocyte of *Asellus aquaticus* at Mid-vitellogenesis

(A, inset), The nucleolus is made of a large and round dense fibrillar component (dfc), with the granular component (gc) being decreased to a small patch at its periphery. (A), Small round granular masses (arrow) were observed near the nuclear envelope, through which ribonucleoproteins are exported (large arrow) from the nucleus (N) and then accumulate in the perinuclear region as nuage aggregates (nu). The ooplasm shows numerous mitochondria (m), endoplasmic reticulum cisternae (arrowhead), dictyosomes (G), cortical vesicles (cv), dense yolk vesicles (DY), and light vesicles of the forming intermediate-dense yolk vesicles (LY). (B), Annulate lamellae (AL) at the oocyte cortex, showing associated nuage materials (nu). At the oocyte surface, the length of microvilli (mv) and the thickness of the vitelline layer (VL) is increased. (C), Large lipid droplet (L), whose periphery is extensively associated with mitochondria (m) and smooth endoplasmic reticulum cisternae (ser). FC, monolayer of follicular cells; BL, basal lamina.

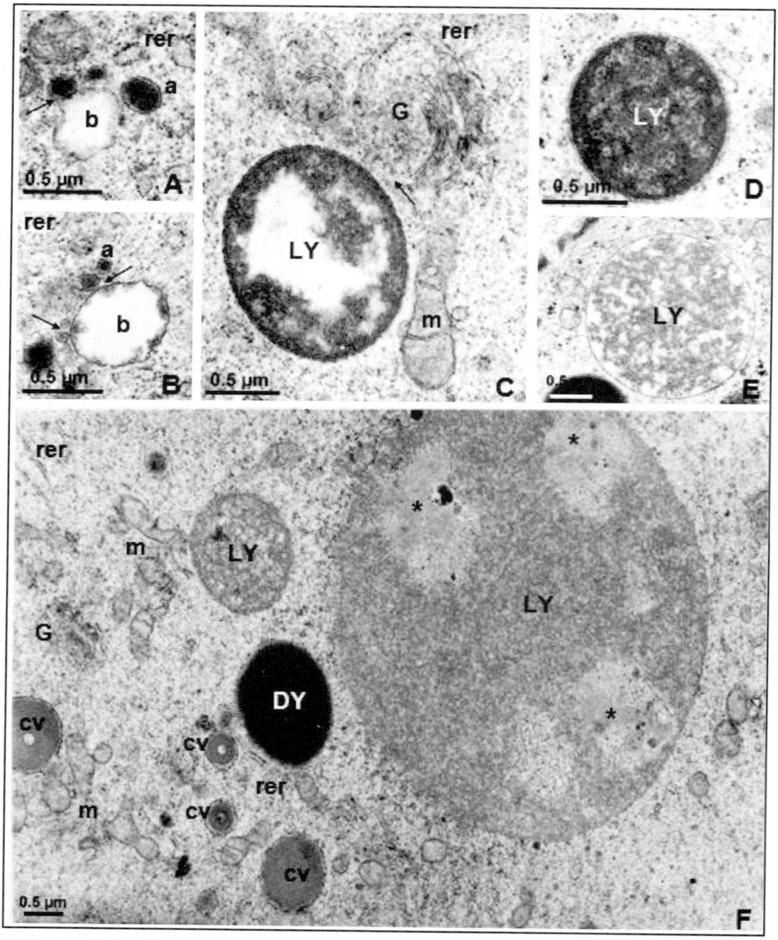

Figure 20.6: TEM Photographs of the Oocyte of *Asellus aquaticus* at Mid-vitellogenesis

(A-E), Continued fusion (arrows) of small dense vesicles (a) with a small light vesicle (b), derived from dictyosomes (G) and rough endoplasmic reticulum cisternae (rer), gives origin to large, intermediate-dense yolk vesicles (LY). (F), Engulfment of smaller vesicles, then originates very large, intermediate-dense yolk vesicles (LY), with heterogeneous membranous contents (*). DY, dense yolk vesicles; cv, cortical vesicles; m, mitochondria.

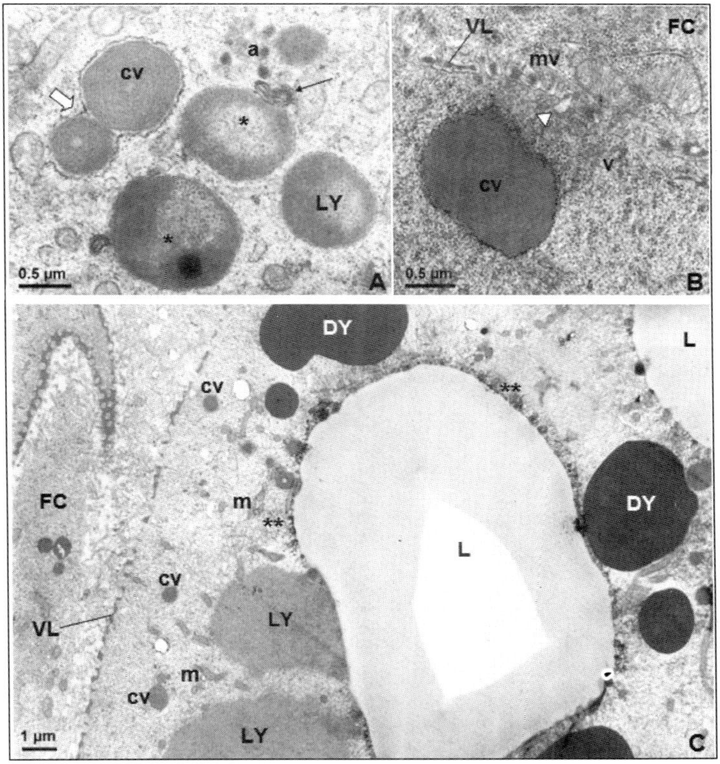

Figure 20.7: TEM Photographs of the Oocyte of *Asellus aquaticus* at Mid-vitellogenesis (A, B) and Late-vitellogenesis (C).

(A, B), During mid-vitellogenesis, cortical vesicles (cv) fuse between each other (large arrow) and transform into large cortical vesicles (cv) that become concentrated under the oolemma. Note engulfment (arrow) of small dense vesicles (a) by large intermediate-dense yolk vesicles (LY), with subsequent acquisition of heterogeneous membranous contents (*). At this stage, the oocyte surface shows numerous smooth-pits (arrowhead) and small vesicles (v). (C), At late-vitellogenesis, cortical vesicles (cv) appear concentrated at the oocyte cortex, and the ooplasm is filled with very large, dense (DY) and intermediate-dense (LY) yolk vesicles, as well as with very large lipid droplets (L), whose periphery remains extensively associated with mitochondria and smooth endoplasmic reticulum cisternae (**). FC, follicular cell; VL, vitelline layer; mv, microvilli; m, mitochondria.

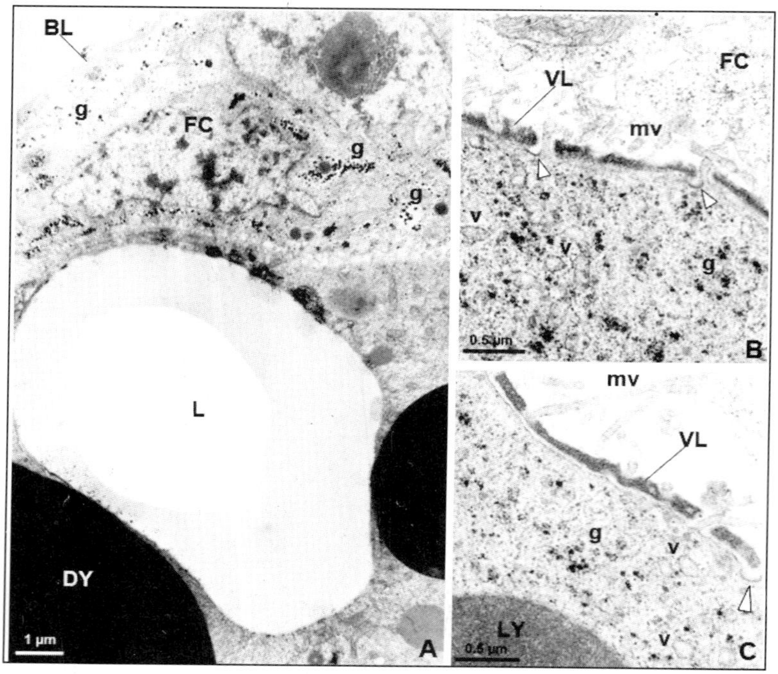

Figure 20.8: TEM Photographs of the Oocyte of *Asellus aquaticus* at Late-vitellogenesis

(A), Follicular cells (FC) accumulate glycogen aggregates (g). (B), At a later stage, follicular cells (FC) lose glycogen aggregates, whereas the oocyte develops numerous coated-pits (arrowheads) and accumulates glycogen (g). (C), Finally, microvilli (mv) expand, and the oocyte surface dettaches from follicular cells. L, lipid droplets; DY, dense yolk vesicles; LY, intermediate-dense yolk vesicles; VL, vitelline layer; v, small vesicles.

materials. Dense yolk vesicles then grew in size through engulfment of the smaller vesicles, showing heterogeneous membranous bodies (Figures 20.3B, 20.4A,B). Finally, the rough endoplasmic reticulum cisternae formed terminal vesicular expansions, with membrane-bound ribosomes, in which concentric lamellar contents appear accumulated. These then dettach to form cortical vesicles that remain with a few membrane-bound ribosomes (Figures 20.3B, 20.4A). During early-vitellogenesis, the oocyte surface develop small microvilli, a thin vitellin layer, and numerous coated-pits. The oocyte cortex appear rich in endoplasmic reticulum cisternae and vesicles, small vesicles containing floccular contents, and microtubules (Figures 20.4C-E).

Mid-Vitellogenesis

At this stage, the nucleolar granular component is reduced to a small patch that locates at one pole of the large and round structure made of the dense fibrillar component. Several small round structures made of the nucleolar granular component were also observed near the nuclear envelope, together with large amounts of small nuage aggregates in the perinuclear cytoplasm (Figure 20.5A).

Annulate lamellae were first observed at mid-vitellogenesis, being characteristically found at the oocyte cortex in close association with nuage materials (Figure 20.5B). Whereas lipid droplets (Figure 20.5C) and dense yolk vesicles (Figure 20.5A) continued to increase in size, a new type of yolk vesicles then formed through the combined activities of dictyosomes and the rough endoplasmic reticulum. Intermediate-dense yolk vesicles result from the fusion of small dense vesicles with a larger light vesicle (Figures 20.5A, 20.6). Their growth then proceed through engulfment of smaller vesicles, thus giving origin to large, intermediate-dense yolk vesicles, with heterogeneous membranous bodies inside a fibrillar matrix (Figure 20.6). During this stage, cortical vesicles also began to increase in size through fusion. They then progressively lose the associated ribosomes as they become concentrated at the oocyte periphery (Figures 20.5A, 20.6F, 20.7A,B). At the oocyte surface, only smooth-pits were observed, being associated with the presence of small vesicles in the oocyte cortex, and an increase in microvilli length and vitelline layer thickness (Figures 20.5A,B, 20.7B).

Late-Vitellogenesis

At this stage, most of the ooplasm appears filled with very large lipid droplets, dense yolk vesicles, and intermediate-dense yolk vesicles. Cortical vesicles, few in number, appear restricted to the oocyte periphery, and both annulate lamellae and nuage materials disappear (Figure 20.7C). During this phase of oocyte growth, the follicular cells appear highly activated and proliferate, and then begin to accumulate numerous glycogen aggregates (Figure 20.8A). This stage is followed by loss of glycogen deposits in follicular cells, simultaneously with glycogen deposition in the ooplasm, reappearance of oocyte receptor-mediated endocytosis, microvilli extension, and separation of the oocyte surface from the close contact with follicular cells (Figure 20.8B,C).

Discussion

The ultrastructural features of the primary oocyte during the zygotene/pachytene stage in *Asellus aquaticus*, demonstrate that cell division is halted and that a newly established metabolic activity is therein initiated to override meiosis and drive the oocyte to vitellogenesis. During that stage, an intense synthesis and exportation of ribonucleoproteins to the ooplasm was observed, together with concentration of mitochondria, ribosomes, and rough endoplasmic reticulum cisternae in the perinuclear region. These findings are similar to those previously described for other Isopoda, although we did not find annulate lamellae as in *Oniscus asellus* nor a large nucleolus as described in *O. asellus* and *Saduria entomon* (Beams and Kessel, 1980; Hryniewiecka-Szyfter and Babula, 1995). On the contrary, a nucleolus was developed only at early diplotene in *Asellus aquaticus*.

During pre-vitellogenesis, no evidence of yolk production was noticed in *Asellus aquaticus*. This situation is similar to findings in *Porcellio scaber* and *Oniscus asellus*, but contrary to *Saduria entomon* oocytes, which showed an endogenous synthesis of yolk during this stage (Bilinski, 1979; Beams and Kessel, 1980; Hryniewiecka-Szyfter and Babula, 1995). The absence of yolk synthesis during pre-vitellogenesis in *A. aquaticus* also contrasted with findings in some other species, in which the Golgi complex (Xiphosura: Dumont and Anderson, 1967), the Golgi apparatus and the rough endoplasmic reticulum (Branchiopoda-Conchostraca: Zeni and Zaffagnini, 1989; Sabelli and Tommasini, 1990; shrimps: Carvalho *et al.*, 1998a), or the rough endoplasmic reticulum (Branchiura-Balanidae: Lepore *et al.*, 1993; Amphipoda-Orchestia: Zerbib, 1980) begin an early synthesis of yolk vesicle precursors.

The endogenous synthesis of numerous and large lipid droplets in *Asellus aquaticus*, which is initiated during early-vitellogenesis, is a new description in the Isopoda. Synthesis of a few small lipid droplets during early-vitellogenesis was observed in the Isopoda *Oniscus asellus* (Beams and Kessel, 1980) and the shrimp *Palaemon serratus* (Papathanassiou and King, 1984), whereas in the

Branchiopoda-Conchostraca *Leptestheria dahalacencis* (Zeni and Zaffagnini, 1989) and in the shrimp *Penaeus kerathurus* (Carvalho *et al.*, 1998a,b, 1999) a limited amount of medium-size lipid droplets was observed during pre-vitellogenesis. Medium-size lipid droplets were also noticed in Xiphosura (Dumont and Anderson, 1967), in the Balanidae *Balanus amphitrite* and *B. perforatus* (Lepore *et al.*, 1993), the Amphipoda *Orchestia gammarellus* (Zerbib, 1980), and in crayfish (Beams and Kessel, 1963). On the contrary, very large lipid droplets were described in the crab *Coenobita clypeatus* (Komm and Hinsch, 1987).

Two different types of yolk vesicles were synthesized by *Asellus aquaticus* oocytes. Dense yolk vesicles first appear during early-vitellogenesis and their components derive from endogenous (Golgi and rough endoplasmic reticulum) and exogenous (receptor-mediated endocytosis) pathways. On the contrary, intermediate-dense yolk vesicles are formed during mid-vitellogenesis only by an endogenous mechanism, as at this stage no receptor-mediated endocytosis could be detected. This constitutes the first observation of two different types of yolk vesicles in the Isopoda. Pinocytosis has also been found to contribute to yolk production in other species of the Isopoda, during previtellogenesis in *Saduria entomon* (Hryniewiecka-Szyfter and Babula, 1995), during early-vitellogenesis in *Porcellio scaber* (Bilinski, 1979), and during mid-vitellogenesis in *Oniscus asellus* (Beams and Kessel, 1980). However, whereas endogenous and exogenous yolk production started at the same oogenetic stage in *Porcellio scaber* and *Asellus aquaticus*, in *Oniscus asellus* and *Saduria entomon* the endogenous mechanism preceeds the exogenous uptake of yolk precursors. Two different types of yolk vesicles were however observed in other arthropod species, one originated from an endogenous synthesis, and a second type derived from endocytosis, either alone or in conjunction with endogenous synthesis. This was found in Xiphosura (Dumont and Anderson, 1967), in the shrimps *Penaeus aztecus*, *P. setiferus* and *P. kerathurus* (Duronslet *et al.*, 1975; Carvalho *et al.*, 1998a,b, 1999), and in the crab *Cancer pagurus* (Eurenius, 1973).

Although yolk vesicles have been described to grow in size by vesicle fusion, in *Asellus aquaticus* yolk vesicles grew by two distinct mechanisms, one through fusion with precursor vesicles, and the other through direct engulfment of smaller vesicles. This latter aspect conferred the presence of membranous contents in large yolk vesicles, and is here signalled for the first time in Isopoda. Although being an infrequent finding, a very similar situation was also described in the shrimp *Penaeus kerathurus* (Carvalho *et al.*, 1998a,b, 1999). These observations thus confirm the existence of a novel behaviour of internal vesicles, as engulfment has only been ascribed to endogenous removal of organelles and phagocytosis.

In *Asellus aquaticus*, vesicles with concentric lamellar contents were directly formed by the rough endoplasmic reticulum during early-vitellogenesis. These vesicles remained connected to the rough endoplasmic reticulum cisternae and their membrane kept attached ribosomes. During mid-vitellogenesis, they fuse between each other and become concentrated under the oolemma. This type of vesicles is here first described in Isopoda, but similar vesicles have been observed in other crustacean species, including the Amphipoda *Orchestia gammarellus* (Zerbib, 1980) and the shrimps *Penaeus aztecus, P. setiferus, P. vannamei* and *P. kerathurus* (Duronslet *et al.*, 1975; Rankin and Davis, 1990; Carvalho *et al.*, 1999). These vesicles have been named cortical vesicles due to their final location under the oolemma, and have been shown to be exocytated before or during fertilization to create a protective external layer (Zerbib, 1980; Duronslet *et al.*, 1975; Rankin and Davis, 1990; Carvalho *et al.*, 1999).

Two waves of receptor-mediated endocytosis were observed in *Asellus aquaticus*. The first occurred during early-vitellogenesis and was coincident with dense yolk vesicle synthesis. The second took place during late-vitellogenesis, simultaneously with loss of glycogen by follicular cells and deposition of glycogen in the ooplasm. On the contrary, in the intermediate stage of mid-vitellogenesis no coated-pits were observed, but numerous smooth-pits appeared connected to the plasma membrane, at a time when the vitelline layer thickness increased. Although experimental approaches are needed to confirm these observations, results suggest that the oocyte may seggregate the materials which form the vitelline layer, and that glycogen may be imported from follicular cells. This contrasts with other species, such as in Xiphosura, where glycogen was not related to pinocytosis but synthesized during pre-vitellogenesis (Dumont and Anderson, 1967), or in the Branchiopoda *Leptestheria dahalacencis*, where glycogen was synthesized during vitellogenesis (Zeni and Zaffagnini, 1989).

Acknowledgements

We acknowledge Elsa Oliveira for technical assistance, and J.Carvalheiro for iconography.

References

Beams HW, RG Kessel. 1962. Intracisternal granules of the endoplasmic reticulum in the crayfish oocyte. *Journal of Cell Biology*, 13: 158-162.

Beams HW, RG Kessel. 1963. Electron microscope studies on developing crayfish oocytes with special reference to the origin of yolk. *Journal of Cell Biology*, 18: 621-649.

Beams HW, RG Kessel. 1980. Ultrastructure and vitellogenesis in the oocyte of the Crustacean, *Oniscus asellus*. *Journal of Submicroscopic Cytology*, 12: 17-27.

Bilinski S. 1979. Ultrastructural study of yolk formation in *Porcellio scaber* Latr. (Isopoda). *Cytobios*, 26: 123-130.

Cabrera A.R., K.V. Donohue, R.M. Roe. 2009. Regulation of female reproduction in mites: A unifying model for the Acari. *Journal of Insect Physiology*, 55: 1079-1090.

Carvalho F, M Sousa, E Oliveira, J Carvalheiro, L Baldaia. 1998a. Ultrastructure of oogenesis in *Penaeus kerathurus* (Crustacea, Decapoda). I. Previtellogenic oocytes. *Journal of Submicroscopic Cytology and Pathology*, 30: 409-416.

Carvalho F, M Sousa, E Oliveira, J Carvalheiro, L Baldaia. 1998b. Ultrastructure of oogenesis in *Penaeus kerathurus* (Crustacea, Decapoda). II. Vitellogenesis. *Journal of Submicroscopic Cytology and Pathology*, 30: 527-535.

Carvalho F, M Sousa, E Oliveira, J Carvalheiro, L Baldaia. 1999. Ultrastructure of oogenesis in *Penaeus kerathurus* (Crustacea, Decapoda). III. Cortical vesicle formation. *Journal of Submicroscopic Cytology and Pathology*, 31: 57-63.

Charniaux-Cotton, H. 1985. Vitellogenesis and its control in Malacostracan Crustacea. *American Zoology*, 25: 197-206.

Dumont JN, E Anderson. 1967. Vitellogenesis in the horseshoe crab, *Limulus polyphemus*. *Journal de Microscopie*, 6: 791-806.

Duronslet MJ, AI Yudin, RS Wheeler, WH Clark. 1975. Light and fine structural studies of natural and artificially induced egg growth of penaeid shrimp. *Proceedings of the Meeting of the World Mariculture Society*, 6: 105-122.

Erkan M. 1998. Ultrastructural study on the ovarian wall and the oviduct of *Asellus aquaticus* (Crustacea: Isopoda). *Tr J Zool.*, 22: 351-362.

Erkan M, M Sousa, F Carvalho, E Oliveira, L Baldaia. 2001. Fine silver staining analysis of the nucleolar organizer regions during oogenesis in *Penaeus kerathurus* (Crustacea, Decapoda). *Journal of Submicroscopic Cytology and Pathology*, 33: 47-57.

Eurenius L. 1973. An electron microscope study on the developing ooocytes of the crab *Cancer pagurus* L. with special reference to yolk formation (Crustacea). *Z. Morph. Tiere*, 75: 243-254.

Ganion LR, RG Kessel. 1972. Intracellular synthesis, transport, and packaging of proteinaceous yolk in oocytes of *Orconectes immunis*. *Journal of Cell Biology*, 52: 420-437.

Goudeau M, F Lachaise. 1980a. Fine structure and secretion of the capsule enclosing the embryo in a crab (*Carcinus maenas*)(L.). *Tissue and Cell*, 12: 287-308.

Goudeau M, F Lachaise. 1980b. Endogenous yolk as the precursor of a possible fertilization envelope in a crab (*Carcinus maenas*). *Tissue and Cell*, 12: 503-512.

Goudeau M, F Lachaise. 1983. Structure of the egg funiculus and deposition of embryonic envelopes in a crab. *Tissue and Cell*, 15: 47-52.

Hinsch GW, MV Cone. 1969. Ultrastructural observations of vitellogenesis in the spider crab, *Libinia emarginata* L. *Journal of Cell Biology*, 40: 336-342.

Hryniewiecka-Szyfter Z, A Babula. 1995. Ultrastructural study of the female reproductive system of *Saduria entomon* (Linnaeus, 1758) (Isopoda, Valvifera). *Crustaceana*, 68: 720-733.

Ikuta K, T Makioka. 1997. Structure of the adult ovary and oogenesis in *Argulus japonicus* Thiele (Crustacea: Branchiura). *Journal of Morphology*, 231: 29-39.

Ikuta K, T Makioka, R Amikura. 1997. Eggshell ultrastructure in *Argulus japonicas* (Branchiura). *Journal of Crustacean Biology*, 17: 45-51.

Kessel RG. 1968. Mechanisms of protein yolk synthesis and deposition in crustacean oocytes. *Zeitschrift Zellforschung*, 89: 17-38.

Komm BS, GW Hinsch. 1985. Oogenesis in the terrestrial hermit crab *Coenobita clypeatus* (Decapoda, Anomura). I. Previtellogenic oocytes. *Journal of Morphology*, 183: 219-224.

Komm BS, GW Hinsch. 1987. Oogenesis in the terrestrial hermit crab *Coenobita clypeatus* (Decapoda, Anomura). II. Vitellogenesis. *Journal of Morphology*, 192: 269-277.

Krol RM, WE Hawkins, RM Overstreet. 1992. Reproductive components. *In* FW Harrison, AG Humes, eds. Microscopic Anatomy of Invertebrates. Vol. 10 (Decapoda Crustacea). Wiley-Liss, New York, pp. 295-343.

Kung SY, SM Chan, JH Hui, WS Tsang, A Mak, JG He. 2004. Vitellogenesis in the sand shrimp, *Metapenaeus ensis*: the contribution from the hepatopancreas- specific vitellogenin gene (Me Vg2). *Biol.Reprod.*, 71(3): 863-870.

Lepore E, M Sciscioli, M Gherardi. 1993. Some ultrastructural aspects of oocytes during vitellogenesis in two species of *Balanus* (Crustacea, Cirripedia). *Oebalia*, 19: 79-87.

Norrevang A. 1968. Electron microscopic morphology of oogenesis. *International Review of Cytology*, 23: 113-177.

Okumura T, K Yamano, K Sakiyama. 2007. Vitellogenin gene expression and hemolymph vitellogenin during vitellogenesis, final maturation, and oviposition in female kuruma prawn, *Marsupenaus japonicus*. *Comp. Bio. Phys. Part A*, 147: 1028-1037

Papathanassiou E, PE King.1984. Ultrastructural studies on gametogenesis of the prawn *Palaemon serratus* (Pennant). I. Oogenesis. *Acta Zoologica* (*Stockholm*), 65: 17-31.

Quackenbush LS. 1989. Vitellogenesis in the shrimp, *Penaeus vannamei*: in vitro studies of the isolated hepatopancreas and ovary. *Comparative Biochemistry and Physiology*, 94B: 253-261.

Rankin SM, RW Davis. 1990. Ultrastructure of oocytes of the shrimp, *Penaeus vannamei*: cortical specialization formation. *Tissue and Cell*, 22: 879-893.

Rateau JG, C Zerbib. 1978. Étude ultrastructurale des follicules ovocytaires chez le crustacé amphipode *Orchestia gammarellus* (Pallas). *Comptes Rendues de la Academie de Sciences de Paris*, 286D: 65-68.

Sabelli FS, S Tommasini. 1990. Origin and early development of female germ cells in *Eoleptestheria ticinensis* Balsamo-Crivelli, 1859 (Crustacea, Branchiopoda, Conchostraca). *Molecular Reproduction and Development*, 26: 47-52.

Scanabissi FS, M Trentini. 1979. Ultrastructural observations on the oogenesis of *Triops cancriformis* (Crustacea, Notostraca). *Cell and Tissue Research*, 201: 361-368.

Schade ML, RR Shivers. 1980. Structural modulation of the surface and cytoplasm of oocytes during vitellogenesis in the lobster, *Homarus americanus*. An electron microscope-protein tracer study. *Journal of Morphology*, 163: 13-26.

Serrano-Pinto V, C Vazquez-Boucard, H Villarrea-Colmenares. 2003. Yolk proteins during ovary and egg development of mature female freshwater crayfish (*Cherax quadricarinatus*). *Comp. Biochem. Physiol. A Mol. Integr. Physiol.*, 134(1): 33-43.

Suzuki S, K Yamasaki, Y Katakura. 1989. Vitellogenin synthesis by fat body and ovary in the terrestrial Isopod, *Armadillidium vulgare*. *General and Comparative Endocrinology*, 74: 120-126.

Wagele JW. 1992. Isopoda. *In* FW Harrison, AG Humes, eds. Microscopic Anatomy of Invertebrates. Vol. 9 (Crustacea). Wiley-Liss, New York, pp. 529-617.

Yano I, RM Krol, RM Overstreet, WEHawkins. 1996. Route of egg yolk protein uptake in the oocytes of kuruma prawn, *Penaeus japonicus*. *Marine Biology*, 125: 773-781.

Zeni C, F Zaffagnini. 1989. Electron microscopic study on oocytes, nurse cells and yolk formation in *Leptestheria dahalacensis* (Crustacea, Conchostraca). *Invertebrate Reproduction and Development*, 15: 119-129.

Zerbib C. 1979. Étude ultrastructurale de l'ovocyte en vitellogenèse chez les Ecrevisses *Astacus astacus* et *A. leptodactylus*. *International Journal of Invertebrate Reproduction*, 1: 289-295.

Zerbib C. 1980. Ultrastructural observation of oogenesis in the Crustacea Amphipoda *Orchestia gammarellus*. *Tissue and Cell*, 12: 47-62.

Chapter 21

Estrogen Induced Standard Cytotoxic Model in the Uterus of Rat

S. Madhuri[1] and Govind Pandey[2*]

[1]Guest Faculty, Department of Zoology,
Govt. MH College of Home Science for Women, Jabalpur, M.P., India
E-mail: drmadhurig8@yahoo.co.in

[2]Ex-Professor of Pharmacology;
Presently, Officer-in-Charge of Rinder Pest
(Veterinary/AH Department, Govt. of MP),
Jabalpur Division, Jabalpur, M.P., India

ABSTRACT

The present study was conducted to create a standard cytotoxic model in the uterus of rat after administration of ethinyl oestradiol (EO, a highly potent semisynthetic 17 β-oestradiol estrogen). EO was administered @ 750 µg/kg, orally, weekly to the rats of groups 2, 3 and 4 for 8, 12 and 16 weeks, respectively. However, the rats of group 1 were given saline to serve as control. On the 9th week (Group 2), the degeneration and necrosis with infiltration of inflammatory cells in the uterine tissues were observed. On the 13th week (Group 3), marked vascular congestion, epithelial necrosis and extensive fibrosis resulting into compression of endometrial glands were seen.

* Corresponding author: E-mail: drgovindpandey@rediffmail.com

On the 17th week (Group 4), besides above changes, proliferation of endometrium in the form of papillary projections lined with tall columnar cells was also seen. The endometrial glands were necrosed and conspicuous eosinophil infiltration was noticed. Conclusively, estrogen (EO) altered the normal structure of uterus and the duration to develop the standard cytotoxic model in the uterus within 12 to 16 weeks.

Keywords: *Cytotoxic model, Estrogen, Histopathological study, Rat, Uterus.*

Introduction

Estrogen is the most commonly prescribed drug as a component of oral contraceptives (OCs) and hormonal replacement therapy (HRT) in women. The oestrogens are also widely used (misused) in small animals (bitches) for the treatment of misalliance, hypogonadal obesity and hormonal urinary inconsistence. In male dogs, they can be used to treat anal adenoma, excess libido and prostatic hyperplasia (Acke *et al.*, 2003; Cain, 2001). Even being used for various reproductive problems in females of human beings as well as animals, the long-term use of estrogen may cause cytotoxicity, leading to cancers of many organs (Loose and Stancel, 2006, Madhuri, 2008). Stilbestrol (a synthetic estrogen) caused uterine abnormalities such as hyperplasia, carcinoma, adenomyosis and endometriosis in rabbits (Meissner *et al.*, 1957). Excessive estrogen produced the tumours of breast, cervix, endometrium, ovary, pituitary, testicle, kidney and bone marrow in either mice, rats, rabbits, hamsters or dogs (Hertz, 1976). In December 2000, the U.S.A. Government's National Toxicology Program and the National Institute of Environmental Health Sciences listed the estrogen as a carcinogen (Estrogen and cancer website, 2006).

Ethinyl oestradiol (EO), a semisynthetic 17 β-oestradiol, is the highly potent estrogen. Pandey and Madhuri (2008) determined the median lethal dose (LD_{50}) of EO to be more than 1000 µg/kg body weight, orally in female albino rats. EO induced cytotoxicity has been observed in liver (Pandey *et al.*, 2008) and ovary (Madhuri *et al.*, 2007) of rats. Hence, the present study was conducted to create a standard cytotoxic model in the rat uterus by administration of EO.

Materials and Methods

Twenty-four healthy inbred female albino rats (100-150 g) were divided into four groups, each having six rats. The rats were kept in polypropylene cages under standard conditions with 25°±5°C temperature, 45-55 per cent relative humidity and 10 hr light:14 hr darkness in the animal house of the College of Veterinary Science and Animal Husbandry, Jabalpur. The rats were fed on standard pellet diet and drinking water *ad libitum*. The experimental designs and protocols in the study received the approval of Institutional Animal Ethics Committee (IAEC).

The required amount of EO as Lynoral tablets (each tablet containing 0.05 mg of EO only) was purchased, and its suspension was prepared in distilled water mixed with a pinch of *Gum acacia* powder. The EO suspension was administered @ 750 µg/kg, orally, weekly to the rats of groups 2, 3 and 4 for 8, 12 and 16 weeks, respectively. However, the rats of group 1 (to serve as control) were given saline, also mixed with a pinch of *Gum acacia* powder to have uniformity with EO suspension.

For histopathological study, the rats of groups 1, 2, 3 and 4 were sacrificed on the 1st, 9th, 13th, and 17th week, respectively. The uteri of all the rats were collected and preserved in 10 per cent buffered formalin. Later on, the uterine tissues were processed and stained with H and E stain as per the method described by Culling (1963). The histopathological changes in the uterine tissues were observed, microscopically.

Results and Discussion

Microscopic examination of the H and E stained section of the uterus of group 1 showed normal histological picture (Figure 21.1). On the 9th week (Group 2), the vacuolar degeneration and necrosis of columnar epithelial cells in the mucosa and endometrial glands. The blood vessels in the endometrium were dilated and congested. At certain places, proliferation of fibrovascular connective tissues and infiltration of inflammatory cells were distinct (Figure 21.2). On the 13th week (Group 3), marked vascular congestion, epithelial necrosis and fibrous tissue proliferation were seen. The fibrosis was extensive resulting into compression of endometrial glands. Desquamation of glandular epithelium was also noticed (Figure 21.3). On the 17th week (Group 4), proliferation of endometrium in

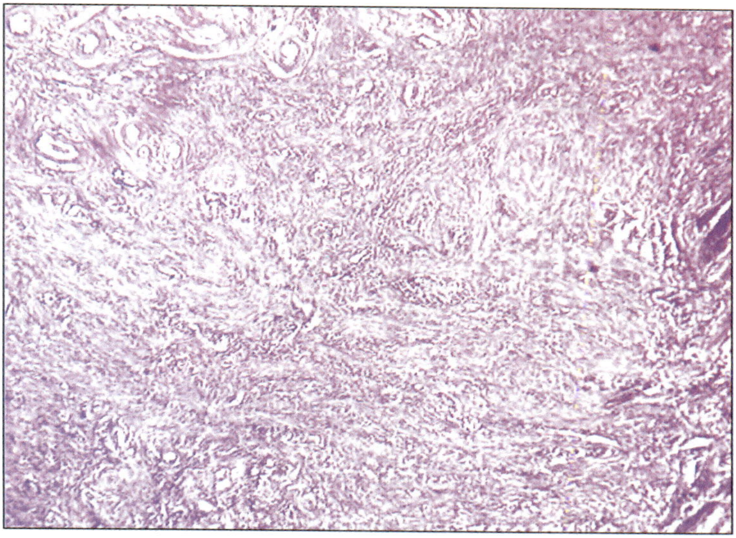

Figure 21.1: Uterus of Rat (Group 1) Showing Normal Structure (H and E, x 100)

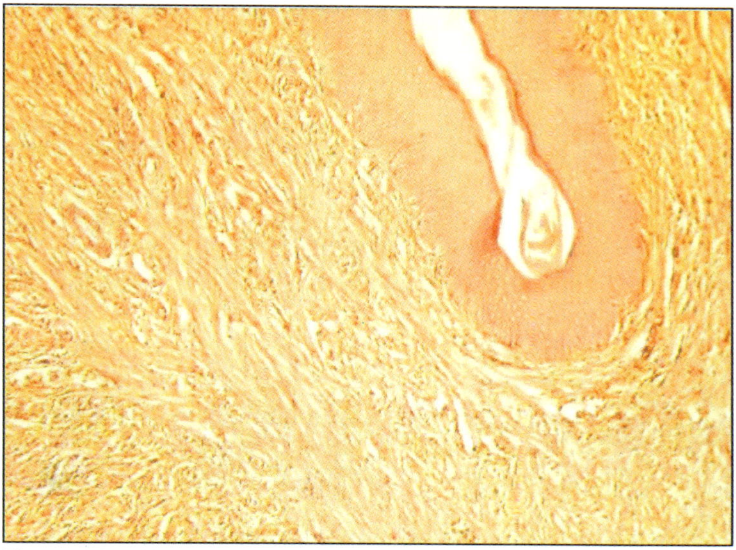

Figure 21.2: Uterus of Rat (Group 2) on 9th Week of EO (750 μg/kg, orally, weekly for 8 wks) Administration Showing Degeneration and Necrosis in the Mucosa and Endometrial Glands with Infiltration of Inflammatory Cells (H and E, x 100)

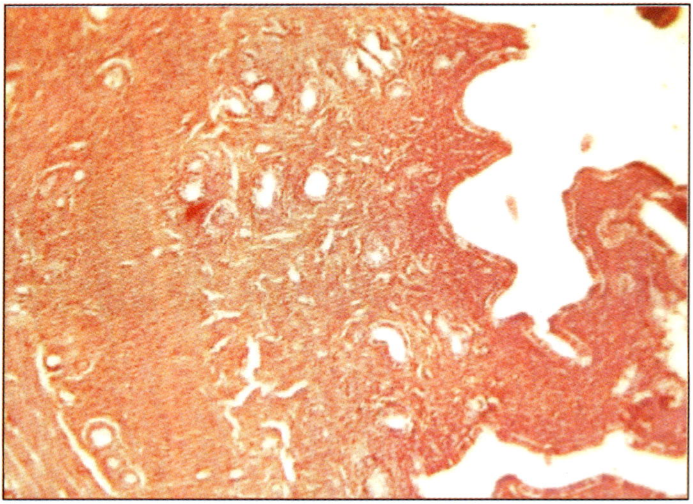

Figure 21.3: Uterus of Rat (Group 3) on 13th Week of EO (750 µg/kg, orally, weekly for 12 wks) Administration Showing Marked Vascular Congestion, Epithelial Necrosis and Extensive Fibrosis Resulting into Compression of Endometrial Glands (H and E, x 100)

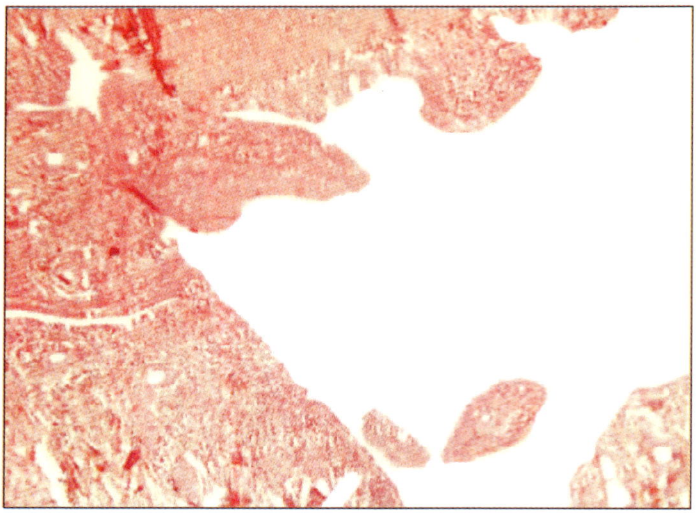

Figure 21.4: Uterus of Rat (Group 4) on 17th Week of EO (750 µg/kg, orally, weekly for 16 wks) Administration Showing Proliferation of Endometrium in the Form of Papillary Projections Lined with Tall Columnar Cells; Necrosis and Eosinophil Infiltration in the Endometrial Glands are also Seen (H and E, x 100)

the form of papillary projections lined with tall columnar cells was also seen. Hyperaemia of blood vessels was also observed. The endometrial glands were necrosed and conspicuous eosinophil infiltration was noticed (Figure 21.4).

The results of the present study may be correlated with the reports of various authors (Hertz, 1976; Loose and Stancel, 2006; Meissner *et al.*, 1957), who cited that the cytotoxic and carcinogenic changes in the uterus of human beings and animals may occur after administration of estrogen. Jabara (1962) reported that the endometrium of bitches after chronic dosing of stilbestrol showed degeneration with marked glandular atrophy. Schwartz *et al.* (1969) observed that high and prolonged dose of quinestrol oestrogen caused uterine enlargement with endometrial hyperplasia and myometrial hypertrophy in bitches. Similar to the present study, EO (250, 500 and 750 μg/kg, orally, weekly for 8 and 12 weeks) induced cytotoxicity has also been reported by Pandey *et al.* (2008) and Madhuri *et al.* (2007) in the liver and ovary of rats, respectively. However, probably no research works, research work on estrogen induced experimental uterine cytotoxicity has been conducted in India as no Indian literature could be available in this regard.

Excessive estrogen is trapped in the uterus, ovary, breast or liver due to stagnation in blood circulation and overstimulates the cell division, leading to cytotoxic changes such as fibroids, cysts or cancers (Estrogen and cancer website, 2006; Madhuri, 2008). In the present study, the cytotoxic changes caused by EO on the 9th, 13th and 17th weeks were indicative of uterine cytotoxicity. The extent and severity of cytotoxicity were time dependent, which suggested that at prolonged period, *i.e.* from 12 to 16 weeks, EO (estrogen) induces the optimum and standard uterine damage, giving rise to a standard experimental cytotoxic model in the uterus of rat.

Acknowledgements

The authors are thankful to the Dean, College of Veterinary Science and Animal Husbandry, Jabalpur for providing laboratory facilities.

References

Acke, E., Mooney, C.T. and Jones, B.R. 2003. Oestrogen toxicity in a dog. *Irish Vet. J.*, 56(9): 465-468.

Cain, J.L. 2001. Rational use of reproductive hormones. In: *Small Animal Clinical Pharmacology and Therapeutics* (Boothe, D.M., ed.). Saunders, Philadelphia. pp. 677-690.

Culling, C.F.A. 1963. *Hand Book of Histological Techniques*, 2nd Edn. Butterworth and Co. Ltd., London. pp. 25-172.

Estrogen and cancer website 2006. www.womenshealt.com; www.amazon.com.

Hertz, R. 1976. The estrogen-cancer hypothesis. *Cancer*, 38(1): 534-540.

Loose, D.S. and Stancel, G.M. 2006. Estrogens and progestins. In: *Goodman and Gilman's The Pharmacological Basis of Therapeutics*, 11th Edn. (Brunton, L.L., ed.). McGraw-Hill Co., New York. pp. 1541-1571.

Meissner, W.A., Sommers, S.C. and Sherman, G. 1957. Endometrial hyperplasia, endometrial carcinoma, and endometriosis produced experimentally by estrogen. *Cancer*, 10(3): 500-509.

Madhuri, S. 2008. Studies on oestrogen induced uterine and ovarian carcinogenesis and effect of ProImmu in rats. PhD thesis, RDVV, Jabalpur, MP, India.

Madhuri, S., Pandey, Govind, Shrivastav, A.B. and Sahni, Y.P. 2007. Ovarian cytotoxicity by oestrogen in rat. *Toxicol. Int.*, 14(2): 143-145.

Pandey, Govind and Madhuri, S. 2008. Median lethal dose, acute and chronic toxicities of ethinyl oestradiol estrogen. *Nat. J. Life Sciences*, 5(2): 291-294.

Pandey, Govind, Madhuri, S., Pandey, S.P. and Shrivastav, A.B. 2008. Hepatic tissue regeneration by OptiLiv in oestrogen induced hepatotoxicity. *Ind. Res. Comm.*, 2(1): 47-52.

Schwartz, E., Tornaben, J.A. and Boxill, G.C. 1969. Effects of chronic oral administration of a long acting estrogen quinestrol to dogs. *Toxicol. Applied Pharm.*, 14: 487-494.

Chapter 22

The Influence of *Acacia nilotica* Flower Phytoestrogenic Extract Administration on Ovulation in Albino Rats

Sharanabasappa A. Patil and Saraswati B. Patil*
Reproductive Endocrinology Laboratory, Department of Zoology, Gulbarga University, Gulbarga – 585 106, Karnataka, India

ABSTRACT

It has been noticed, recently that an ethanol extract from *Acacia nilotica* Linn. flowers induces weak metabolic changes in rat ovary. In the present study *in vivo* influence of an ethanol extract was investigated on morphology and biochemistry of the rat ovary and the uterus. The virgin female rats were orally administered ethanol extract at a dose of 25 mg and 50 mg/100 g body weight for 30 days. The estrous cycle of the extract treated rats was irregular with prolonged estrous and metestrous phases and reduced diestrous and proestrous phases during experimental period. At autopsy on day 31st, ovarian weight was significantly reduced. Histological sections of the ovary indicated a decreased number of developing follicles, Graffian follicles and corpora lutea and increased number of atretic follicles. The ovarian biochemical studies

* Corresponding author: E-mail: sshanu@rediffmail.com

showed increase in cholesterol content and decrease in protein and glycogen contents. The uterine biochemical parameters indicated an increase in glycogen and DNA contents. The weight of the uterus and its morphometric measurements like diameter of uterus, epithelial cell height, endometrial and myometrial thickness were increased significantly in extract treated animals. The extract showed dose dependent increase in its estrogenicity which clearly indicate that ethanol extract inhibits the ovulation in rats.

Keywords: *Acacia nilotica Linn., Ovary, Ovulation, Estrous cycle, Contraception, Phytoestrogen.*

Introduction

Humans have used plants for medicinal and contraceptive purposes for eons. According to modern day analyses, many of the plants historically noted for their ability to prevent pregnancies or cause abortions containing phytoestrogens and other hormonally active substances. Many plants produce chemicals that mimic or interact with hormone signals in animals. At least 20 such phytohormones have been identified in at least 300 plants from more than 16 different plant families (Barret, 1996; Colborn *et al.*, 1996). The estrogen like phytoestrogens are the most studied of all the phytochemicals. In general, phytoestrogens are weaker than the natural estrogen hormones (such as estradiol) found in humans and animals or the very potent synthetic estrogens used in birth control pills and other drugs (Jefferson *et al.*, 2002a). As for adverse health effects, the most likely risks associated with phytoestrogens deal with infertility and developmental problems. For instance, during fourth century BC Hippocrates noted that the wild carrot (now known as Queen Anne's lace) prevented pregnancies (Riddle, 1991). Australian sheep suffered from reproductive problems and infertility after grazing in pasture with the phytoestrogen containing clover *Trifolium subterraneum* (Bennettes and Underwood, 1951). Two phytoestrogens, equol and coumesterol, were identified as the culprits. A group of captive cheetahs experienced infertility while on a diet rich in soy (Setchell and Gosselin, 1987). When the soy was replaced with corn, their fertility was restored.

Phytoestrogen behave like hormones, although they are generally less potent. Like any hormone, too much or too little can alter hormone dependent tissue functions. Taking too much of any hormone may not be good for humans or animals. Similarly, too many phytoestrogens, at the wrong time, may lead to adverse health effects. Experimental animals studies, clarify the possible reproductive and developmental risks associated with phytoestrogens. Some scientists believe that plants make phytoestrogens as a defense mechanism to stop or limit predation by plant eating animals (Ehrich and Raven, 1964; Guillette and Crain, 1995; Hughes, 198?). Instead of protecting themselves with thistles or thorns or tasting bad, these plants use chemicals that affect the predatory animal's fertility. In some countries, phytoestrogenic plants have been used for centuries in the treatment of menstrual and menopausal problems as well as for fertility problems (Muller-Schwarze and Dietland, 2006). The most used plants have later shown higher content of phytoestrogens *i.e. Pueraria mirifica* (Lee *et al.*, 2002) and its close relative, Kudzu (Delmonte and Rader 2006), Angelica (Brown and Waltson, 1999), fennel and anise (Albert-Puleo, 1980). Interestingly, in recent years, the popularity of different kinds of pharmaceutical preparations containing phytoestrogens has been constantly increasing. Therefore, studies of the influence of the above mentioned plant components on female reproduction have become more and more important.

These fertility related activity are attributed to different extracts, compounds and active principles of either the whole or part of plants. Working on the same line, we have undertaken a study on *Acacia nilotica* Linn., which is a traditional medicinal plant. The present experiment is designed to study the effect of ethanol extract of *Acacia nilotica* Linn. on ovulation and estrous cycle in female albino rats.

Materials and Methods

Plant Material

Acacia nilotica Linn. plant was botanically authenticated at the Herbarium, Department of Botany, Gulbarga University, Gulbarga, India. The voucher specimen (HGUG-906) has been deposited in the herbarium. The flowers were collected from the fields, in and around Gulbarga (North Karnataka) in the month of June–July. The flowers were shade dried and powdered. The powdered material was subjected to soxhlation with ethanol. The extract was concentrated to dryness in a flash evaporator (Buchi) under reduced pressure and controlled temperature at 50°c. 200g of flower powder yielded 24g dark gummy ethanol extract. The extract was stored in refrigerator until used for treatment.

Test Animals

Healthy colony bred virgin female Albino Wistar strain rats of weighing 140 -150g with normal estrous cycle (Hariharan, 1980) were used for the experiment. All animals were maintained under standard identical housing condition of temperature, humidity, light and fed with food pellet diet and water *ad libitum* as prescribed by CFTRI, Mysore (India). The study was approved by the institutional ethical committee, which follows the guidelines of CPCSEA (Committee for the Purpose of Control and Supervision of Experiments on Animals, Reg. No. 34800/2001/CPAEA/dated 1/9-08-2001) Government of India, New Delhi, India. All the extracts were prepared in Tween-80 (1 per cent) in distilled water for complete dissolution and were administered orally to the experimental rats by using intragastric catheter at desired doses. The control animals received an equivalent amount of vehicle only.

Experimental Protocol

The animals were divided into three groups consisting of six in each group. Group I: Control, received 0.2 ml Tween-80 (1 per cent). Group II: Received 25mg ethanol extract/100g body weight in 0.2 ml Tween-80 (1 per cent). Group III: Received 50mg ethanol extract/100g body weight in 0.2 ml Tween-80 (1 per cent). All treatments were given for 30 days to cover six regular estrous cycles. Vaginal smear was observed throughout the experimental period.

Histological and Biochemical Examination

All the animals were sacrificed 24 h after the last treatment. The ovaries and uteri were dissected out immediately and separated out from the adherent tissue and weighed to the nearest mg on an electronic balance. Organs from one side of each animal were fixed in Bouin's fluid, embedded in paraffin wax, section at 5μ, stained with Ehrlich's haematoxylene and eosine for histological studies. Number of developing follicles, Graffian follicles, corpora lutea and atretic follicles was counted from stained serial sections of the ovary from each rat. The micrometric measurements like diameter of uterus, thickness of endometrium and myometrium, and height of endometrial epithelium were made from randomly chosen 20 sections from each group using ocular and stage micrometer. Organs from the other side were used for biochemical estimations like cholesterol (Peters and Vanslyke, 1946), protein (Lowry *et al.*, 1951), glycogen (Caroll *et al.*, 1956) and DNA (Burton, 1968).

Statistical Analysis

Data are expressed as the mean ± S.D. After calculation of mean and standard deviation, Duncan test was performed using the software – statistical package for social sciences (SPSS Inc., version 10.0.5) to obtain the significance between the treated groups and the control groups. A value of $P<0.001$ was considered to indicate a significant difference between the groups.

Results

Estrous Cycle (Table 22.1)

The duration of estrus and metaestrus was increased, while diestrus and proestrus was decreased in both doses of extract treated animals. This is significant due to the treatment of ethanol extract in comparison with that of control.

Ovarian Changes (Tables 22.2 and 22.3; Figures 22.1–22.3)

Ovaries of the rats treated with ethanol extract of *Acacia nilotica* Linn. flower extract showed decrease in weight. The ovarian cholesterol increased significantly ($P<0.01$) with 25mg and highly significantly ($P<0.001$) with 50mg ethanol extract treatment in comparison with that of control. The ovarian protein content decreased significantly ($P<0.001$) only with high dose of extract treatment. The glycogen level is decreased highly significantly ($P<0.001$) with both the dose extract treatment. The number of ovarian components such as developing follicles, Graffian follicles and corpora lutea was reduced and that of atretic follicles was increased significantly in comparison with the control values.

Uterine Changes (Tables 22.4 and 22.5)

The uterine weight of rats treated extract of *Acacia nilotica* Linn. increased significantly with ethanol extract treatment. The

Table 22.1: Effect of Ethanol Extract of *Acacia nilotica* Linn. Flower Extract on the Length of Various Phases of Estrous Cycle

Treatment	*Dose (mg/100g body wt.)*	*Duration of Phases of Estrous Cycle (Days)*			
		Estrus	*Metaestrus*	*Diestrus*	*Proestrus*
Control	Tween-80 (1 per cent)	6.80±0.56	6.00±0.51	15.2±0.96	3.80±0.16
Ethanol extract	25	10.34*±0.75	12.82**±0.60	6.80**±0.64	1.65**±0.08
Ethanol extract	50	12.20*±1.37	13.66**±1.34	2.33**±0.42	1.23**±0.25

Duration: 30 days.

Six animals were maintained in each group.

Values are mean ± S.E

*P<0.01, **P<0.001 when compared to control.

Table 22.2: Gravimetric and Histological Changes of the Ovary Due to Administration of *Acacia nilotica* Linn. Flower Extract

Treatment	*Dose (mg/100g body wt.)*	*Weight (mg/100g body wt)*	*Number/Ovary*			
			Developing Follicles	*Graafian Follicles*	*Corpora lutea*	*Atretic Follicles*
Control	Tween-80 (1 per cent)	42.78±0.84	10.42±0.61	6.12±0.28	7.04±0.60	0.86±0.05
Ethanol extract	25	35.03*±1.50	8.49*±2.04	4.92**±1.10	4.43**±0.16	2.98**±0.52
Ethanol extract	50	30.73*±1.52	3.10**±0.30	1.20**±0.30	0.83**±0.30	4.23**±0.30

Duration: 30 days

Six animals were maintained in each group

Values are mean ±S.E

*P<0.01, **P<0.001 when compared to control.

Table 22.3: Ovarian Biochemical Changes Due to Administration of *Acacia nilotica* Linn. Flower Extract

Treatment	*Dose (mg/100g body wt.)*	*Cholesterol (µg/mg)*	*Protein (µg/mg)*	*Glycogen (µg/mg)*
Control	Tween-80 (1 per cent)	15.01±1.44	14.20±0.76	4.80±0.16
Ethanol extract	25	20.98*±1.37	11.89±0.84	2.98**±0.15
Ethanol extract	50	28.16**±1.34	5.20**±1.53	1.38**±0.35

Duration: 30 days.

Six animals were maintained in each group.

Values are mean ±S.E.

*P<0.01, **P<0.001 when compared to control.

Table 22.4: Gravimetric and Histometric Changes in the Uterus Due to Administration of *Acacia nilotica* Linn. Flower Extract

Treatment	*Dose (mg/100g body wt.)*	*Weight (mg/100g)*	*Diameter (μm)*	*Myometrial Thickness (μm)*	*Endometria Thickness (μm)*	*Epithelial Cell Height (μm)*
Control	Tween-80 (1 per cent)	120.98±1.41	2011.37±5.99	105.90±4.09	551.80±4.64	18.02±1.16
Ethanol extract	25	130.55*±3.98	2303.03**±11.39	128.63*±5.98	610.93*±6.91	25.03**±0.75
Ethanol extract	50	148.33**±3.29	2449.60**±9.52	340.00**±7.88	1026.83**±9.85	46.66**±2.46

Duration: 30 days

Six animals were maintained in each group.

Values are mean ±S.E

*P<0.01, ** P<0. 001 when compared to control.

Table 22.5: Biochemical changes in the uterus due to administration of *Acacia nilotica* Linn. Flower Extract

Treatment	*Dose (mg/100g body wt.)*	*Protein (µg/mg)*	*Glycogen (µg/mg)*	*DNA (µg/mg)*
Control	Tween-80 (1 per cent)	4.80±0.20	1.02±0.05	2.33±0.26
Ethanol extract	25	6.70**±0.21	2.28**±0.26	4.01**±2.67
Ethanol extract	50	8.85**±0.36	4.00**±0.25	4.98**±2.72

Duration: 30 days

Six animals were maintained in each group

Values are mean ±S.E.

*P<0.01, **P<0.001 when compared to control.

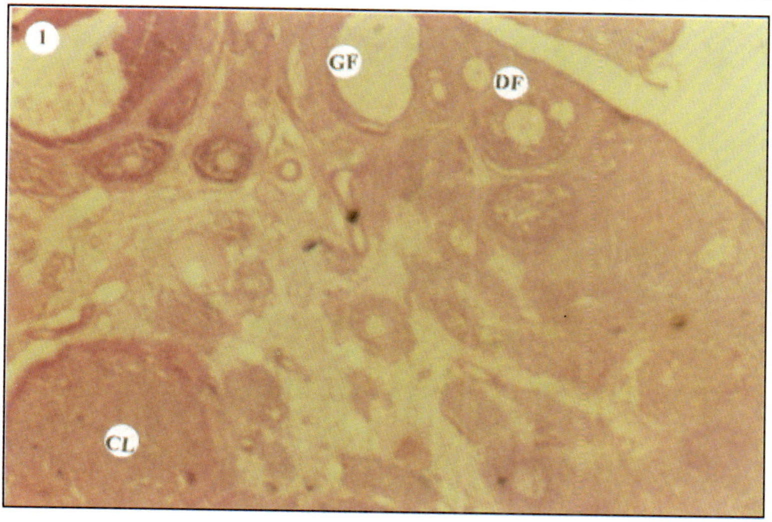

Figure 22.1: Cross Sections of the Ovary of Control Rat Showing Normal Folliculogenesis

micrometric measurements such as diameter of uterus, endometrial and myometrial thickness and epithelial cell height were parallel to that of the weight of uterus. Protein, glycogen and DNA contents of uterus were increased significantly in all the treated rats.

Discussions

The use of plant phytoestrogens/antiestrogens for contraceptive activities especially for prevention/interruption of pregnancy has been in practice since ancient time in India. There are many reviews available on the contraceptive activity of medicinal plants (Bhargava, 1988; Komboj, 1988). The contraceptive effects may be due to antiovulatory, antiimplantation or abortificient activities. To quote a few, seeds of *Nelumbo nucifera* Gaertn., flowers of *Hibiscus rosa sinensis* L., *Momordica charantia* Linn. and many other plants have been reported for antiovulatory activity (Mazumdar *et al.*, 1992; Murthy *et al.*, 1997; Sharanabasappa *et al.*, 2002). Similarly the seed oil of *Daucus carota* L., whole plant of *Striga lutea* Lour., *Striga orabanchioides* Benth., *Striga densiflora* Benth. and *Acalypha indica* L. have shown antiimplantation property (Komboj and Dhawan 1982; Hiremath and Hanumanth Rao, 1990; Hiremath *et al.*, 1994; 1996). The abortificient activity of entire plant of

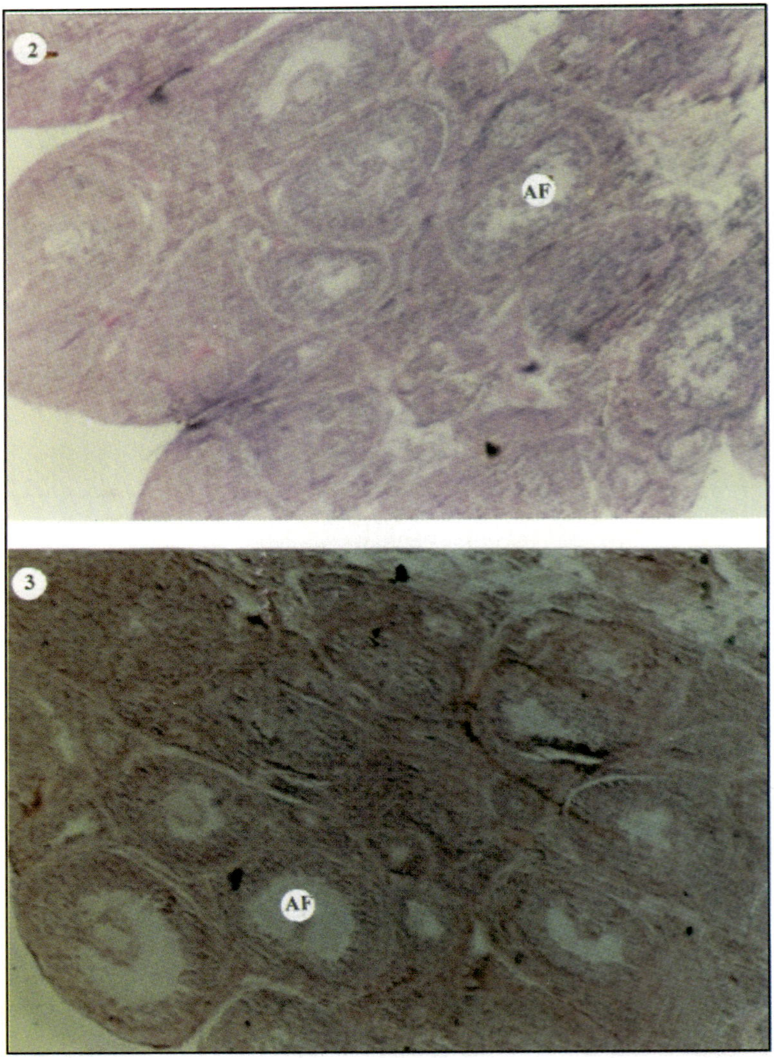

Figures 22.2 and 22.3: Cross Sections of the Ovary of Rats Treated with 25mg and 50mg Ethanol Extract of *Acacia nilotica* Linn. Flower Showing Reduction in the Number of Developing Follicles (DF), Graafian Follicles (GF), Corpora Lutea (CL) and Increase in Atretic Follicles (AF)

Achyranthes aspera L. and *Rubus* species have also been investigated (Garg, 1976; Dhanabal *et al.*, 1999).

In the present experiment, *Acacia nilotica* flower extract administered for 30 days decreased the duration of diestrus and proestrus and increased the duration of estrus and metaestrus significantly during experimental period. The prolonged duration of estrus and metaestrus indicates the mild estrogenecity of the extract, as estrogens are necessary for cornification of vaginal epithelial cells (Murthy *et al.*, 1997). It is well documented that FSH is essential for follicular growth and LH is necessary for ovulation and corpora lutea formation (Marsh, 1975; Richards, 1980), which are responsible for the growth of the ovary and increase in its weight. Therefore, reduction in the observed weight of the ovary after the treatment of *Acacia nilotica* Linn flower extract may be attributed mainly due to non-availability of gonadotrophins. Increase in the cholesterol in the ovaries of treated animals may be due to non-availability of pituitary FSH, LH and prolactin essential for steroidogenesis (Mason *et al.*, 1962; Wang *et al.*, 1979). The lowered protein content of the gonads indicates the retarded ovarian growth, which is dependent on the availability of pituitary FSH, as FSH is essential for protein synthesis in gonads (Means, 1975). The ovarian glycogen, which is an energy source for female reproductive activity, is estrogen dependent (Wallas, 1952).

Therefore, decreased ovarian glycogen observed in the present study may be due to reduced availability of ovarian estrogens. The reduction in the number of developing follicles and Graafian follicles also attributes to the decreased availability of ovarian estrogens as these follicles are the major source of estrogens in the ovary (McNatty *et al.*, 1976). Though the follicular atresia is common in rat ovary, the increased number of atretic follicles in the ovary of experimental animals indicates the non-availability of required amount of gonadotrophins (Friedrich *et al.*, 1975). Uterine growth depends on the availability of ovarian steroid hormones, particularly estrogens (Jalikhani, 1980; Findlay, 1994). As the ovarian activities are impaired in the experimental animals, the increase in the uterine weight may be due to the estrogenic effect of *Acacia nilotica* Linn. flower extract. To further sustain the specificity of the estrogenic nature of extract, we determined other parameters previously described as being modulated by estrogens, namely: the increase in uterine DNA content

(Jenson and De Sombre, 1972), histological changes of uterus and prolonged duration of estrus and metaestrus phases.

In conclusion, this study has shown that the ethanol extract of *Acacia nilotica* flower had inhibited the ovulation in albino rats. The extract had also exhibited phytoestrogenic nature in uterine tissue and further investigations are needed to isolate the active molecules.

Acknowledgements

Financial support from Council of Scientific and Industrial Research (CSIR), New Delhi, India for awarding Research Associate Fellowship to Dr. Sharanabasappa A. Patil is gratefully acknowledged.

References

Albert-Puleo M. 1980. Fennel and anise as estrogenic agents. *J. Ethnopharmacol.,* 2(4):337-44.

Barret J. 1996. Phytoestrogens: Friends or foes? *Environ. Health Persp.,* 104: 478-482.

Bennetts H. W and Underwood E. J. 1951. The oestrogenic effects of subterranean clover (*Trifolium subterraneum*); uterine maintenance in the overiectomised ewe grazing on clover. *Aust. J. Exp. Biol. Med. Sc.,* 29(4): 249-53.

Bhargava S. K. 1988. Antifertility agents for plants. *Fitoterapia, LIX* (3): 163-177.

Brown D. E and Waltson N. J. 1999. Chemicals from Plants: Perspectives on Plant Secondary Products. *World Scientific Publishing,* 141: pp 21.

Burton K. 1968. Determination of DNA concentration with diphenylamine, In methods in Enzymology, (Grossman L. and Moldave K., eds.).12 Part B: 163-166 Academic press New York.

Caroll N. V., Langelly R. W and Row R. H. 1956. Glycogen determination in liver and muscle by use of Anthrone reagent. *J. Biol. Chem.,* 26: 583-593.

Colborn T., Dumanski D and Myers J. P. 1996. Our Stolen Future. New York: Penguin Books, Inc.

Delmonte P and Rader J. I. 2006. Analysis of isoflavones in foods and dietary supplements. *J. AOAC.,* 89(4):1138-46.

Dhanabal S. P., Shibu P., Ramanathan M., Elango K and Suresh B. 1999. Validation of antifertility activity of various *Rubus* species in female albino rats. *Indian J. Pharm. Sc.*, 30: 58-60.

Ehrlich P and Raven P. H. 1964. Butterflies and plants: A study of co evolution. *Evolution.*, 18:586-608.

Findlay J. K. 1994. Molecular biology of the Female reproductive system. Academic Press, California. pp. 457.

Friedrich F., Kemeter P., Salzer H and Breitenecker G. 1975. Ovulation inhibition with human chorionic gonadotrophin. *Acta Endocrinology*, 78: 332-342.

Garg S. K. 1976. Antifertility screening of plants: Effect of four indigenous plants on early pregnancy in female ablino rats. *Indian J. Med. Res.*, 64: 1133-1142.

Guillette Jr. L. J and Crain D. A. 1995. Organization versus activation: the role of endocrine-disrupting contaminants (EDCs) during embryonic development in wildlife. Environ. *Health Persp.*, 103 (suppl 7): 157-64.

Hariharan S. 1980. Laboratory animal's information service center News ICMR, Hyderabad.

Hiremath S. P and Hanumanth Rao S. 1990. Antifertility efficiency of the plant *Striga lutea* (Scrophulariaceae). *Contraception*, 42: 467-477.

Hiremath S. P., Shrishailappa B., Swamy H. K. S., Patil S. B and Londonkar R. L. 1994. Antifertility activity of *Striga orabanchioides*. *Biol. Pharm. Bull.*, 17: 1029-1031.

Hiremath S. P., Swamy H. K. S., Shrishailappa B., Patil S. B and Lodonkar R. L. 1996. Post coital antifertility activity of the plant *Striga densiflora* (Scrophularraceae) on female albino rats. *Int. J. Pharm.*, 34: 48-52.

Hughes Jr. C. L. 1988. Phytochemical mimicry of reproductive hormones and modulation of herbivore fertility by phytoestrogens. *Environ. Health Persp.*, 78: 171-174.

Jafferson W. N., Padilla-Banks E., Clark G and Newbold R. R. 2002a. Assessing estrogenic activity of phytochemicals using transcriptional activation and immature mouse uterotrophic responses. *J. Chromatography B Analyt. Tech. Biomed. Life Sc.*, 777(1-2): 179-189.

Jalikhani B. L. 1980. Ovarian steroids In: Text book of biochemistry and human biology Talawar GP (Ed) Vertical Hall Ind Pri Ltd New Delhi. pp. 805.

Jenson E. V and De Sombre E. R. 1972. In Biochemical actions of Hormones (G.Litwak, Ed.), Vol. II Academic Press New York, pp. 215-255.

Kamboj V. P and Dhawan B.N, 1982. Research on plants for fertility regulation in India. *J. Ethnopharmacol.*, 6: 191-226.

Kamboj V. P. 1988. A review of Indian medicinal plants with interceptive activity. *Indian J. Med. Res.*, 87: 336-355.

Lee Y. S., Park J. S., Cho S. D., Son J. K., Cherdshewasart W and Kang K. S. 2002. Requirement of metabolic activation for estrogenic activity of *Pueraria mirifica*. *J. Vet. Sc.*, 3 (4):273-277.

Lowry O. H., Rosenbrough N. J., Farr N. L and Randoll R. J. 1951. Protein measurement with folin phenol reagent. *J. Biol. Chem.*, 193: 265-275.

Marsh J. M. 1975. The role of cyclic AMP in gonadal function. *Adv. Cycl. Nucl. Res.*, 6: 137-200.

Mason N. R., Marsh J. M and Savard K. 1962. An action of ganadotrophin *In vitro*. *J. Biol. Chem.* 237: 1801-1806.

Mazumdar U. K., Malaya G., Pramanik G., Mukhopadhyay R. K and Swarnali S. 1992. Antifertility activity of seed of *Nelumbo nucifera* in mice. *Indian J. Exp. Biol.*, 30: 533-534.

McNatty K. P., Baird D. T., Bolton A., Chambers P., Corker C. S and MaLean H. 1976. Concentration of oestrogens and androgens in human ovarian venous plasma and follicular fluid throughout the menstrual cycle. *J. Endocrinol.*, 71: 77-85.

Means A. R. 1975. Biochemical effects of follicle stimulating hormone on the testis. *Endocrinology*, 5: 203-218.

Muller-Schwarze and Dietland. 2006. Chemical Ecology of Vertebrate. Cambridge University Press. pp 287.

Murthy D. R. K., Reddy C. M and Patil S.B. 1997. Effect of benzene extract of *Hibiscus rosa sinensis* on the estrous cycle and ovarian activity in albino rats. *Biol. Pharm. Bull.*, 29: 756-758.

Peters J. P and Vanslyke D. D. 1946. Quantitative Clinical Chemistry, Vol I, Co. Baltimore: William and Wilkins.

Richards J. S. 1980. Maturation of ovarian follicles actions and interactions of pituitary and ovarian hormones on follicular cell differentiation. *Physiol. Rev.,* 60: 51-84.

Riddle J. M. 1991. Oral contraceptives and early-term abortifacient during classical antiquity and the Middle Ages. *Past Present No.,* 132: 3-32.

Setchell K. D and Gosselin S. J. 1987. Dietary oestrogens: A probable cause of infertility and liver disease in captive cheetahs. *Gastroenterology.,* 93(2): 225-233.

Shanranbasappa A., Vijayakumar B and Saraswati B.P. 2002. Effect of *Momordica charanatia* seed extracts on ovarian and uterine activities in albino rats. *Pharm. Biol.,* 40: 501- 507.

Wallas O. 1952. Effect of oestrogens in the glycogen contents of the rat uterus. *Acta Endocrinology,* 10: 175-192.

Wang C., Hsuch A. J. W. and Erickson G. F. 1979. Induction of functional prolactin receptors by follicle stimulating hormone in rat granulose cells *In vivo* and *In vitro. J. Biol. Chem.,* 254: 11330-11336.

Chapter 23

Efficacy of Ovaprim and Ovatide in Induced Breeding of *Clarias batrachus*: A Study

A.M. Verma[1], S.K. Prabhakar[2] and K.K. Singh[2]*

[1]Department of Fisheries, Samastipur – 848 101, Bihar, India

[2]Central Institute of Fisheries Education, Salt Lake, Sector – 5, Kolkata – 700 091, W.B., India

ABSTRACT

The Asian Cat fish (locally known as Magur), *Clarias batrachus* is an important air breathing fish in Indian sub – continent. Its increased demand in domestic market is due to its nutritious and therapeutic value. It fetches higher price compared to both Indian and exotic carps. In spite of its immense market potentiality, the culture of this species has not received much attention due to non-availability of seeds. Therefore the artificial seed production in captivity is essential to meet the growing demand for magur culture.

The paper reports the efficacy of ovaprim and ovatide during induced breeding of Asian cat fish, *Clarias batrachus*. A single dose of ovaprim was administered to the female (0.8ml/ kg. body weight) and male (0.4 ml/kg body weight) whereas for ovatide, the dose was 1.0 ml/kg body weight both for male

* Corresponding author: E-mail: verma.anand31@yahoo.com

and female. The number of eggs stripped, percent fertilization, number of hatchlings obtained and the percent hatching were considered during comparison of two inducing agents. The average number of 2500 eggs ($p<0.05$) was stripped while injecting ovaprim. The percent fertilization was 79.89 when ovaprim was used as inducing agent. The percent of hatching was 74.33 and 61.00 in case of ovaprim and ovatide administration respectively. Therefore the synthetic hormone ovaprim could be recommended for induced spawning in *C. batrachus* since it produced better results in terms of breeding performance and hatching ($p<0.05$).

Keywords: *Clarias batrachus, Ovaprim, Ovatide, Efficacy, Induced breeding.*

Introduction

The Asian Cat fish (locally known as Magur), *Clarias batrachus* is an important air breathing fish in Indian sub – continent (Jayaram, 1981). Its increased demand in domestic market is due to its nutritious and therapeutic value. It fetches higher price compared to both Indian and exotic carps. It contains higher percentage of protein (15.0 per cent), Iron (710 mg/100g) and low fat content (1.0 per cent) in muscle compared to other edible fishes (Hossain *et al.*, 2006).

In spite of its immense market potentiality the culture of this species is not picked up. Non-availability of seeds has been attributed as one of the major constraints for this. Presently the seeds required for magur farming are mostly collected from the natural sources. On the other hand, its natural propagation is dwindling due to environmental aberration and habitat stress. Therefore the artificial seed production in captivity is essential to meet the growing demand for magur culture.

Pituitary extract, ovaprim and ovatide have been successfully used for induced spawning of *C. batrachus* (Devraj *et al.*, 1972; Khan and Mukhopadhyay, 1975; Rao and Janakiram, 1991; Saha, 1996; Mahapatra *et al.*, 2000; Sahoo *et al.*, 2005).

In the present study, the efficacy of single application of ovaprim and ovatide have been compared during induced spawning of magur. The study may be relevant to hatchery manager during seed production of this species.

Materials and Methods

Mature healthy magur (90 to 160 g) were collected from Fish farm at Junput for the purpose of present trials (Figure 23.1). Brooders were kept sexwise in plastic drums. Matured females with round and bulging abdomen were selected. The perspective male with pointed papilla was considered for induced breeding trial.

Ovaprim and ovatide were procured prior to the breeding trials. Both stimulants contain Salmon Gonadotropin Releasing Hormone (SGnRH) and Domperidone. Effective doses of ovaprim and ovatide for artificial breeding were considered as reported earlier by Sahu *et al.* (2000) and Sahoo *et al.* (2005) respectively. Three trials were made for each hormone. Details of the stimulants applied their doses and the weight of brood fish are presented in Table 23.1.

The males were dissected and the testes were collected after 18 hr. of injection (Figure 23.2). Testes were cut into pieces and macerated with the help of pestle and mortar. NSS (Normal Saline Solution) was added to make sperm suspension. Simultaneously stripping of ova from female fish was carried out by gently pressing of the abdomen. Eggs were collected in the tray (Figure 23.3). Sperm suspension and eggs were mixed by jerking the tray for 5-10 minutes. The fertilized eggs were washed and transferred to hatchery (Figure 23.4)

One way Analysis of variance (Snedicor and Cochran, 1994) and multiple comparison procedures of the student Newman keuls were used to analyse the significance of the different between the means of treatments ($p<0.05$) by using compare Means (SPSS, 1997).

Results and Discussion

Role of LH-RH analogies in combination with dopamine antagonists like primozide and domparidone in successful spawning is widely acceptable (Peter *et al.*, 1988). Based on this principle, synthetic hormone (ovaprim and ovatide) were developed and were applied for successful induced spawning at magur (Saha, 1996; Mahapatra *et al.*, 2000; Sahoo *et al.*, 2005). Single effective application of ovaprim and ovatide is significant as it saves a considerable amount of time and avoids excessive handling of fish.

The results indicated that the average number of eggs released was 2500 due to ovaprim injection which was significantly ($p<0.05$)

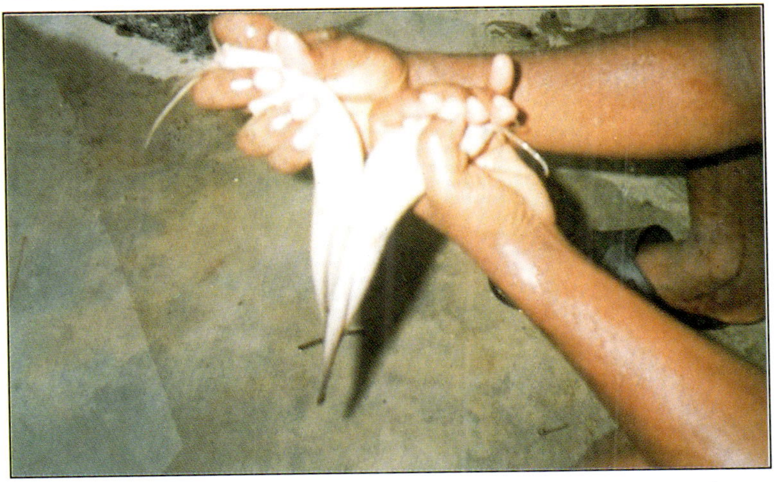

Figure 23.1: Secondary Sexual Characters in *Clarias batrachus*

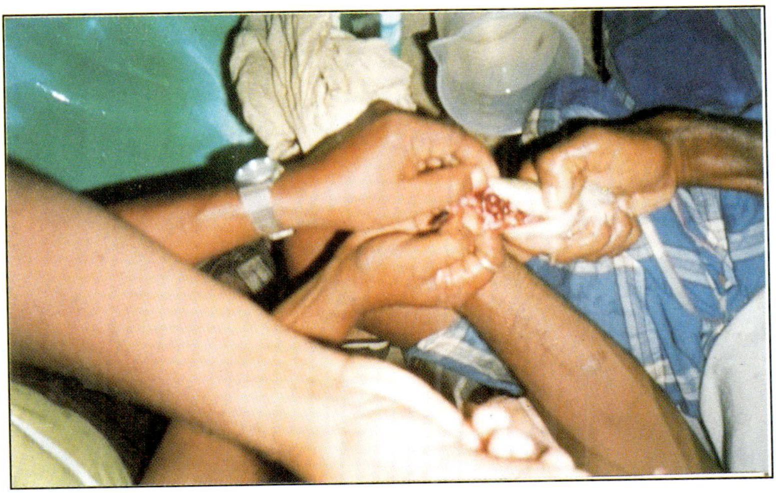

Figure 23.2: Dissection of Male *Clarias batrachus* to take out Testis

higher compared to ovatide injection. The percent fertilization and hatching of egg were also high while using ovaprim as an inducing agent. Our result on fertilization and hatching rate were comparable to earlier reports of Saha (1996) and Das (2002). This might be due to

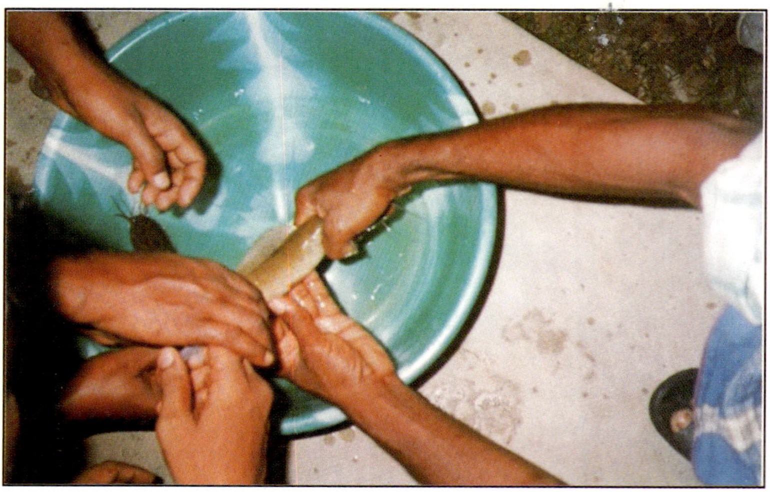

Figure 23.3: Stripping the Eggs of Female *Clarias batrachus*

Figure 23.4: A Plastic Tub Hatchery

better response to ovaprim injection to female fish. The higher number of hatchlings obtained from the females injected with ovaprim was due to higher fertilization and hatching rate.

Table 23.1: Effects of Ovaprim and Ovatide on Induced Breeding in *Clarias batrachus*

Stimulants Applied	*Male*		*Female*		*No. of Eggs Stripped Out*	*Percentage of Fertilization*	*Percentage of Hatching*	*No. of Hatchlings Obtained*
	Body Wt. (g)	*Dose of Stimulants (Ml./kg.)*	*Body Wt. (g)*	*Dose of Stimulants (Ml./kg.)*				
Ovaprim	90±2.28	0.4	142±2.16	0.8	2500^{a}±25.17	79.89±5.16	74.33±7.59	1440^{a}±105.16
Ovatide	102±2.16	1.00	160. ±3.17	1.00	2300^{b}±86.6	72.00±2.71	61.00±2.3	1010^{b}±37.82
p	>0.05	–	>0.05	–	<0.05	>0.05	>0.05	<0.05

Values are means ± SE of triplicate group of fish. Means in the same column with different superscripts differ significantly ($p<0.05$).

The present experimental trials clearly demonstrates that it is better to use ovaprim as the stimulating agent to obtain higher fertilization and hatching rate as well as more number of hatchlings during induced breeding of *C.batrachus*.

References

Das, S.K.2002. Seed production of magur (*Clarias batrachus*) using a rural model portable hatchery in Assam, India – A Farmer Proven technology, *Aquaculture Asia*, VII (2): 19-20.

Devraj, K.V.; Verghese, T.J. and Rac, G.P.S. 1972. Induced breeding of freshwater catfish *Clarias batrachus* (Linn.) by using Pituitary glands from marine catfish. *Curr. Sci.*, 41 (24): 868-870.

Hossain, Q.; Hossain, M.A. and Parween, S. 2006. Artificial breeding and nursery practices of *Clarias batrachus* (Linnaeus, 1958). *Scientific world*, 4(4): 32-37.

Jayaram, K.C. 1981 The freshwater fishes of India, Pakistan, Bangladesh Burma and Sri Lanka – A handbook, Zoological Survey of India, Calcutta 475 pp.

Khan, H.A.. and Mukhopadhyay, S.K. 1975 Production of stocking material of some air breathing fishes by hypophysation. *J. Inland Fish Soc. India*, 7: 156-161.

Mahapatra, B.K., Sengupta, K.K., De, U.K., Rana; U.G., Datta, A.; Basu, A. and Saha, A. 2000 Controlled breeding and larval rearing of *clarias batrachus* (Linn.) for mass scale propagation. *Fishing Chimes*, 19(10 and 11): 97-102.

Peter, R.E. Linn, H.R. and Van Dev Kraak, G. 1988 Induced ovulation and spawning of culture fish in China: Advances in application of Gn RH analogues and dopamine antagonist. *Aquacult.*, 74: 1-10.

Rao, G.R.M. and Janakiram,K. 1991. A effective dose of Pituitary for breeding *Clarias*. *J.Aqua Trop.*, 6(2): 2007-210.

Saha, R. 1996 Effects of various doses of ovaprim for breeding of *clarias* spp. in Tripura. *J. Inland Fish. Soc. India*, 28(2): 75-84.

Sahoo, S.K., Giri S.S. and Sahu, A.K. 2005. Effect on breeding performance and egg quality of *clarias batrachus* (Linn.) at various doses of ovatide during spawning induction *Asian Fisheries Science*, 18(1): 77-83.

Sahu, A. K., Sahoo. S.K. and Ayyappan, S. 2000. Seed production and hatchery management:Asian catfish, *clarias batrachus, Fishing chimes,* 19(10 and 11): 94-96.

Snedecor, G.W. and Cochran,W.G. 1994. Statistical methods 8th edn., Oxford and IBH Publishing Co. Pvt. Ltd., New Delhi.

Statistical packages for Social Sciences. 1997. Base Applications Guide 75, Statistical Packages for Social Sciences, Carry, N.C. USA.

Chapter 24

Photoperiod and Biological Rhythms in Regulation of Avian Reproduction

*Anand S. Dixit**

Department of Zoology, North-Eastern Hill University, Shillong – 793 022, Meghalaya, India

ABSTRACT

In the life history of birds, some changes are ontogenic *i.e.* they happen only once in a life time. Birds hatch into nestlings, and they fledge into juveniles. Then they go through a series of life history changes that repeats every year such as they molt, they may migrate and overwinter. After winter there is spring migration and then breeding which is followed by molt. Breeding consists of nest building, egg laying, incubation and rearing of the young (Figure 24.1). Each stage occupies the whole of time available to it so that the temporal overlap between successive stages tends to be minimized. In fact, each of the stages has evolved to occur at the optimum time and the optimum sequence minimizing overlap between successive stages. Thus, breeding can not start until the spring migration is complete. Molt, migration and breeding are normally temporally separated since all are energy demanding processes (Newton and Rothery, 2005) and birds migrating with

* Corresponding author: E-mail: asdixitnehu@rediffmail.com

incomplete plumage would clearly be at a disadvantage. Breeding activity often starts as soon as the migration to the breeding ground is complete (Wingfield and Farner, 1978; Bauchinger and Klaassen, 2005).

Birds occupy habitats that are different with respect to the amplitude and the predictability of environmental factors.They also differ in the geographical ranges they inhabit over the course of their lives. As a result, they differ greatly in the timing and sequences of life-cycle stages.

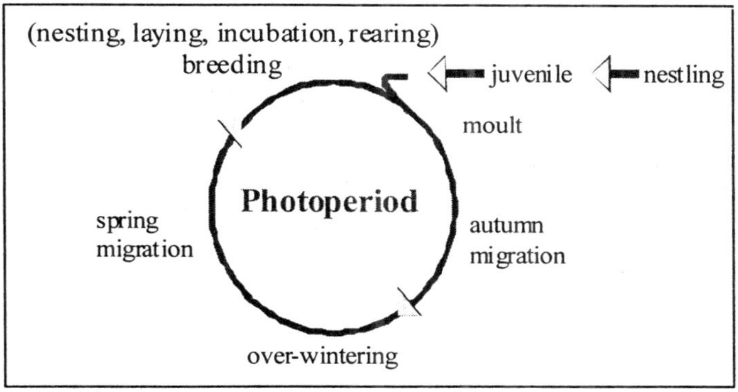

Figure 24.1: Avian Life-History Stages

(Nelson *et al.,* 2001). The evolution of life history strategies in the organisms depends on their response to biotic and abiotic environment. Survival of any species in a non uniform fluctuating environment requires the development of mechanisms that permit an organism to adjust its important functions to changes of the environment. The most critical among these functions is the timing of reproduction, a scheme that causes reproductive activity to occur during the period which assures minimum stress on the adults and maximum probability for survival of young and parents. Reproduction is the part of life cycle with great environmental dependence (Immelmann, 1977). Since physiological preparations for reproduction begins long before the breeding season, the environmental factors selected by an organism for basic timing of its reproductive cycle, must have reliable predictive value, so that the physiological mechanisms controlling reproduction may be activated well in advance of the optimal environmental conditions, necessary for successful rearing of the young. Organisms having annually periodic reproductive season may use periodic changes in the environment as predictive information for the oncoming of favorable season for reproduction.

The birds, like most other wild vertebrates, show a distinct temporal organization in their reproductive activity. Thus, for most bird species a precision in time of reproduction is of interesting occurrence (Farner, 1975). The time and duration of favorable season selected for reproduction differ among different climatic regions and different ecological groups of birds. The annual breeding cycle is most marked at mid and high latitudes but contrary to earlier widely held view, species of tropical and subtropical regions are also seasonal breeders (lack, 1968; Marshall, 1970). Seasonal cycle is represented by the initiation-termination-reinitiation of the physiological processes involved in the expression of an event in the life of the animals *e.g.* reproduction, migration etc. As seasonality insures the occurrence of the events at the most appropriate time of the year, the environmental factors must be involved in its control. Baker (1938) identified such factors into two categories: proximate and ultimate factors. Whereas proximate factors help the organisms to select the most appropriate time window in the year for doing the particular event, the ultimate factors help decide the timing when the actual event would happen within this time window. For a complex phenomenon such as the reproduction, for example, the proximate factors would decide when the breeding season for the species should be scheduled. This will keep both the sexes physiologically ready, should an opportunity in the environment occur for actual reproduction.

Photoperiod as Predictive Information

Organisms live in a complex environment of many interacting components. Environmental factors vary continually and most often predictable from one moment of time to next. The periodic oscillations of geophysical events or factors such as day and night (solar day), lunar day, lunar month, year, season and tidal flow and ebb etc. are regulated by rotation (24 hour) and revolution (365.25days) of earth around sun and that of moon around earth. Life on earth has been subjected to strict regimen of above cyclic changes since its very origin. With little exception, like the organisms that live in underground caves, the depth of ocean, or any similar aperiodic environment, most organisms evolve strategies to counteract or to exploit periodicities in their environment. Through the evolutionary timė, environmental cycles have left their stamps on the basic biology of the organisms. Very early in the course of evolution of life, about three billion years ago, the recurring solar day geophysical pattern served as template upon which were fashioned biological sequences of casually related events. Organisms might have evolved mechanism(s) to make use of highly predictable environment by timing their biological functions for successful survival. Consequently, physiological response of organisms to periodic environment might have been essentially rhythmic. Those that failed to evolve the mechanism(s) to respond and to adapt to ever changing environment probably perished.

The most reliable cue emanating from the earth's rotation on its axis and around sun is the daily cycle of changes in illumination at the earth's surface. The length of light phase (day length) changes with the season, at least away from the equator, and the intensity of any one time of day changes with the weather even at the equator. Many other environmental factors change more or less reliably in parallel with the daily changes in the illumination, such as the temperature and humidity, but these are subjected to considerable frequent short term variation due to local weather conditions. It is not surprising, therefore, that for most organisms the daily cycle of light and dark is the dominant entraining agent for their daily rhythms of biological activities. Daily changes between day and night and annual succession of seasons are the most prominent of all seasonal changes. Synchronization to environmental day-night cycle is key to survival. To affect the anticipation of the favorable season, animal rely on environmental cues. Natural selection ensures that they choose environmental cue that is predictable. Many organisms including birds, therefore, use annual cycle of changes in day length as calendar to time their seasonal physiological and behavioral functions. Photoperiodism is defined as the control of some or all aspects of life cycle in an organism by the timing of light and darkness. According to this definition, photoperiodism affects the long lasting events and it is based on the timing of light, thus darkness, and not on the amount of light or darkness (Kumar, 1997). Since the pioneer discovery of Rowan in 1925, the role of day length in control of seasonal reproduction and premigratory fattening has been confirmed in many high-, mid-, and low-latitude avian species.

As the photoperiod is entirely predictable at given latitude, both within and between years, it is used as a fundamental cue to time the physiological preparations for the three major life-history stages: breeding, molt and migration in birds (Dawson, 2008).

Almost all organisms, which have been investigated, are light sensitive. And in many organisms, especially those living away from the equator, the annual solar cycle has been found influencing various seasonal functions (Murton and Westwood, 1977; Thapliyal, 1981; Hoffman, 1981). At the equator where changes in daily light are small, the changes in daytime light intensity across seasons can influence seasonal responses (Gwinner and Scheuerlein, 1998). Among vertebrates, birds exhibit pronounced seasonal cycles in various behavioral and physiological functions, and several of them are influenced by annual changes in day length. Birds were the first vertebrates in which role of day length (= photoperiod) was demonstrated in control of seasonal functions. That the photoperiod (= day length) could act as a temporal information dates back to eighteenth century when Dutch netters realized that the song associated with breeding activity could be induced in males of several species of passerine birds by keeping them first in darkness from May to August and then returning to natural day lengths. However, nearly for two hundred years there was no scientific investigation. At the end of first-quarter of twentieth century, William Rowan (1925) experimentally demonstrated the role of day length in gonadal development of a passerine bird species, the slate-colored junco (*Junco hyemalis*). In a series of subsequent investigations, he established that the vernal migration and gonadal recrudescence could be induced out of season by exposure of birds in laboratory to increasing day lengths (Rowan, 1926, 1932). Since then, the role of photoperiod in control of seasonal cycles has been demonstrated in about six dozen avian species belonging to 15 different families from both low and high latitudes in the last more than seven decades. It is hoped that the role of day length in control of seasonal responses would be discovered in many more species in future.

The day length contributes the most reliable predictive information for control of annual reproductive cycles of birds inhabiting mid and high latitude (Farner, 1964; Farner and Follett, 1966; Lofts and Murton, 1968). Less is known about the day length in the control of reproduction and other metabolic functions of seasonal breeders at low latitude. In the tropics and subtropics, it is generally believed that that the annual variation in day length, in fact, be too small to provide birds with reliable cue for the timing of seasonal events (Farner and Lewis, 1971) and may be of little use in triggering the reproductive and metabolic activities of the birds of these zones (Immelmann, 1971). Although literature on avian photoperiodism has become voluminous, most of it pertains to mid and high latitude breeders. However, few experimental investigations on low latitude seasonal breeders have been carried out and it has been found that number of tropical and subtropical avian species possess photoresponsive mechanism and are photoperiodic, leading to a tentative conclusion that day length has a more pronounced role in control of reproduction in the tropical and subtropical birds as well than has hitherto been assumed (Thapliyal and Tewary, 1964; Tewary and Kumar, 1982; Tewary and Dixit, 1986; Figure 24.2). Therefore, great caution is needed before minimizing or denying the importance of day length in reproduction and related events in birds inhabiting low latitudes.

Critical Day length

The minimum day length that will induce a physiological response is defined as the critical day length (CD) or threshold photoperiod for the particular response. For example, CD indicates whether an animal is reproductively competent or quiescent at the defined photoperiod. A day length longer than critical day length is described as long day length and a day length shorter than CD is short day length. Thus, a long or short day response is not defined by its own CD, but by whether the response in question will occur under day length longer (long day response) or shorter (short day response) than CD. The minimum photoperiod that induces a photoperiodic response is species specific; varying with the duration of the experiment and may be adapted for breeding at a particular time of the year, at particular latitude. CD may vary between migratory and nonmigratory forms of the same species. The CD is not only species specific but response specific as well, *e.g.* reproduction and migration in a migratory bird may have different CD. Further, CD requirement for a response could be very stringent, *e.g.* a bird with a CD of 12 h will not respond to an 11.5 h photoperiod. However, the concept of critical day length should be treated with caution. In

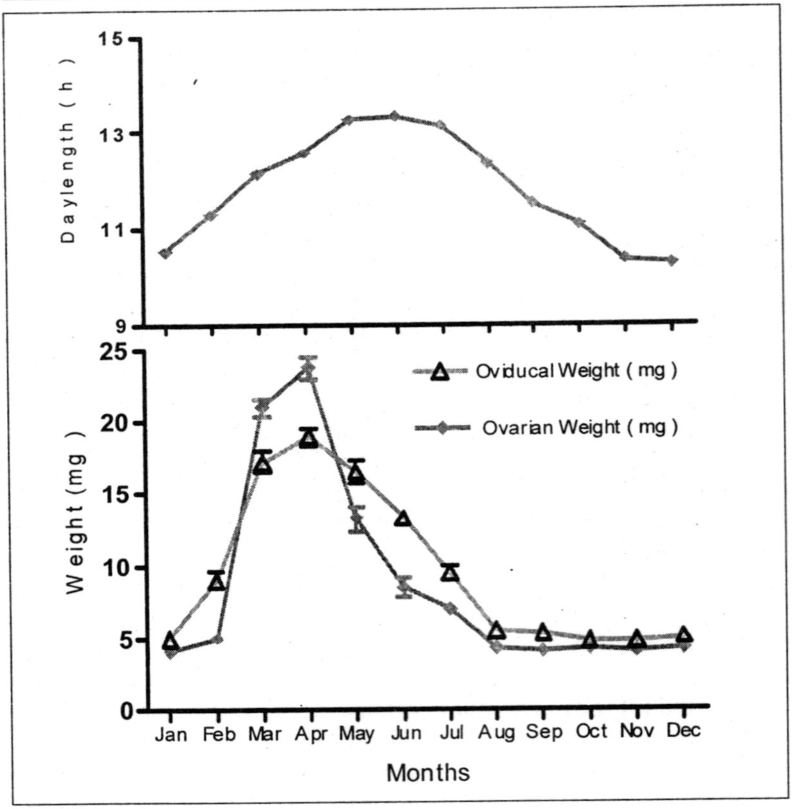

Figure 24.2: Annual Changes in Ovarian and Oviducal Weight of *Gymnorhis xanthocollis* in Relation to Daylength

quails, the gonadal growth rate changes rapidly over a short range of photoperiods and it is easy to define CD while in Starlings (Dawson, 1989) and White-crowned sparrows (Farner and Wilson, 1957), gonadal growth rate varies over a wide range of photoperiods. Moreover, there are many avian species that show substantial gonadal growth under short photoperiods (Wingfield, 1993).

Gonadal growth in most birds at mid and high latitudes occurs during spring when the photoperiod increases. However, the exact time and rate of maturation varies among species, depending on their breeding seasons. It can also vary within the species depending upon latitude and local non photoperiodic factors. In photoperiodic birds, generally the rate of gonadal growth increases with the increasing photoperiods. The relationship between the initial rate of gonadal growth, denoted as k and defined as rate of increase in log gonadal mass (Farner and Wilson, 1957), and photoperiod differs between species. The gonadal growth follows a log linear relationship in Japanese Quails (Farner and Follett, 1966); White - crowned sparrow (Farner *et al.*, 1966); and tree sparrow (Morrison and Wilson, 1972). Unlike birds that breed at high latitude, birds at low latitude are exposed to a limited range of photoperiod and the photoperiodic change over the period between initiation of gonadal growth and active breeding is small. The photoresponse curve of such species must get steeper (within the range of day length change actually seen by the bird) and the CD must be higher if the bird is still to use photoperiod to time its reproductive function. The CD for gonadal growth for some tropical and subtropical birds are reported to lie between 11-12h such as in Weaver bird (Singh and Chandola, 1981); Blackheaded bunting (Kumar and Tewary,1983) and between 12-13h in Redheaded bunting (Prasad, 1983); Rosefinch (Tewary and Dixit,1983) and Yellow-throated sparrow (Tewary and Dixit,1986; Figures 24.3and 24.4).

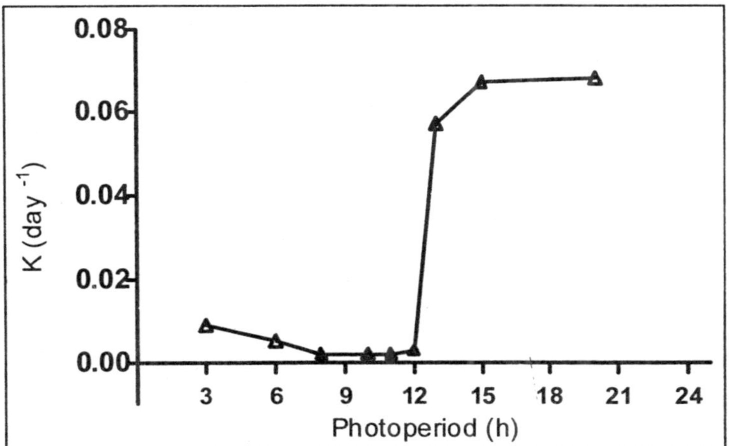

Figure 24.3: Rate of Ovarian Growth (K) in *Gymnorhis xanthocollis* as a Function of the Duration of the Daily Photoperiod. K= (*ln* Wt–*ln* Wo)/t, where Wo and Wt are initial and the final ovarian weights in mg, respectively, and t is time in days.

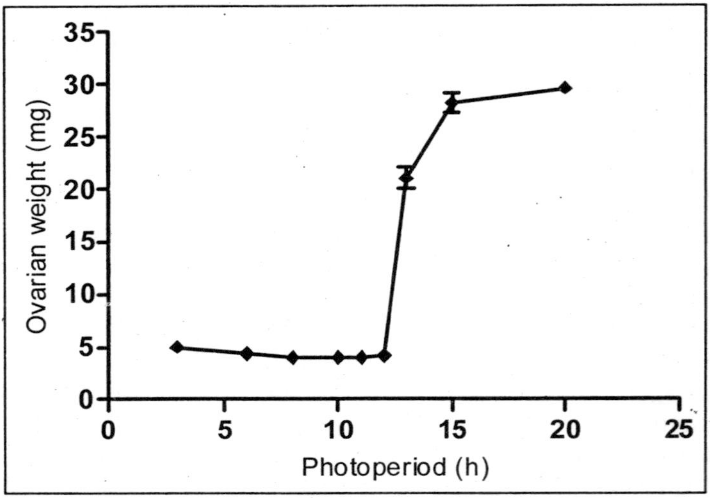

Figure 24.4: Photoperiodic Ovarian Response of *Gymnorhis xanthocollis* Showing the Critical Day Length (Threshold photoperiod)

Long and Short Day Species

Customarily, a bird species that responds to long day lengths is called a long day species and the one that responds to short day lengths is called a short day species. A species which apparently has no relationship between the photoperiod and initiation of its physiological function is called a day- neutral species. However, it may not be possible for a species to be day-neutral when it is exposed to temporally fluctuating photoperiod in the nature. Further, a species is known as absolute photoperiodic if it only responds to a photoperiod longer or shorter than its CD. Thus, a long day qualitative bird remains unstimulated as long as it is maintained under a day length shorter than CD. There are fewer examples of qualitative photoperiodic species in avian photoperiodic literature as most species are not able to maintain themselves in sexually active or inactive state indefinitely under photoperiodic treatments. However, species like White-crowned -sparrow (*Zonotrichea leucophrys gambelii*) can be regarded as absolute photoperiodic species as it maintains its long or short day response for a period up to 130 weeks.

There are many north temperate birds in which long photoperiods induce gonadal growth followed by regression and onset of photorefractoriness, whereas short photoperiods are ineffective (Farner and Follett, 1966; Lofts and Murton, 1968; Farner and Lewis, 1971; Farner and Wingfield, 1980). Although seasonal variation in day length, in tropics and subtropics, is believed to be too small to serve as reliable cue for timing seasonal events, some birds show photoperiodic control of their reproductive and other metabolic functions. It is still difficult to discern general patterns of photoperiodic response among the tropical and subtropical passerines as only few of them have been studied. In Black-headed munia, for example, long as well as short days (ranging from 8-24h) are gonadostimulatory (Thapliyal and Saxena, 1964); Spotted munia responds to unnaturally short photoperiods (ranging from 0.25-6h) but the customary long and short days (ranging from 8-24h) fail to induce the gonad; long days rather retard or inhibit it (Chandola *et al.*, 1975). Although, munias possess photosensitive system, they may not be using photoperiod as a cue in controlling their reproductive function. In contrast, long days (15L/9D) simulate gonadal growth and development while short days (light below 9h/day) fail to do so in Weaver bird (Thapliyal and Saxena, 1964); Red-headed bunting (Prasad, 1983; Tewary and Tripathy, 1983); Black-headed bunting (Tewary and Kumar, 1982); Rosefinch (Kumar, 1981; Tewary and Dixit, 1983) and Yellow-throated Sparrow (Tewary and Dixit, 1985). These birds show gonadal growth for first few weeks under long days followed by regression leading to initial levels. In contrast, they fail to respond to short days even after their exposure for a long period (Figure 24.5).

Photorefractoriness: The Light Lock

An adaptive and advantageous step in importance to the mechanisms that initiate gonadotropic function in the annual gonadal cycle of the birds is the termination of reproductive functions. In many photoperiodic birds each period of gonadal growth and function is followed by rapid collapse of the gonad. Thereafter, majority of birds enter in a state of unresponsiveness to light, during which no known pattern of stimulatory photoperiodic regimes can stimulate their gonad (Farner and Follett, 1966), which has come to be known as photorefractoriness. This post breeding phenomenon was apparently first described by Relay (1936) in

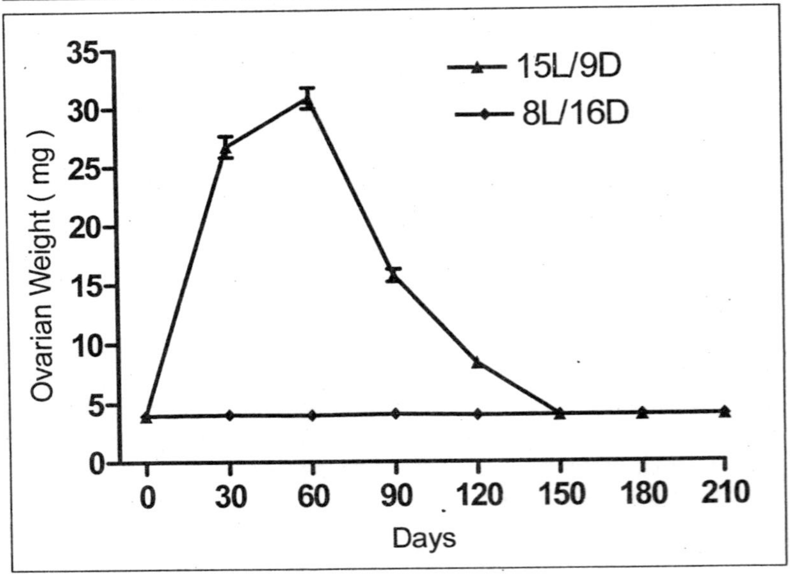

Figure 24.5: Ovarian Response of *Gymnorhis xanthocollis* to Long (15L/9D) and Short (8L/16D) Days

Passer domesticus and by Schildmacher (1939) in *Phoenicurus phoenicurus*. Photorefractoriness has great functional significance. It limits reproduction to the best suited part of the year which ensures maximum reproductive success, permits sufficient time for replenishment of energy stores for post breeding maintenance activity, such as postnuptial molt and preparation for migration. The functional significance of photorefractoriness in all those species in which it occurs appears to be same, *e.g.* mutual exclusion of eargonically intense functions (Farner and Lewis,1971) regardless of the fact that photorefractoriness probably has multiple evolutionary origin.

Photorefractoriness is not a universal phenomenon in the word of avian photoperiodism as in some species, this type of light lock may not operate resulting in the absence of photorefractory period in their annual reproductive cycles. Such birds can be maintained under continuous breeding condition under a stimulatory photoperiod. There are two different mechanisms by which photoperiod controls gonadal regression and terminates the breeding season: absolute and relative photorefractoriness. Some species exhibit "absolute photorefractoriness" which is characterized by the loss of ability to respond to long day length including continuous light. Gonadal regression in such photorefractory birds begin before the day length declines. Birds like House finch, White -crowned sparrow, Blackheaded bunting, Rosefinch etc. show absolute photorefractoriness. On the other hand, some avian species show "relative photorefractoriness" which is characterized by the loss of ability to respond to a stimulatory photoperiod with relatively shorter photophase, but not to the same or longer photophase. For example, individuals of a relative photorefractory species with CD of 12h, if kept under a 15h (15L/9D) photoperiod will become photorefractory to a 13h photoperiod (13L/11D), an otherwise stimulatory photoperiod, but not to the same (15L/9D) or longer photoperiod (20L/4D). Relative photorefractory species can be restimulated on exposure to very long photoperiod following gonadal regression. The gonàdal regression in relative photorefractory species occurs when day length declines but still remains longer than CD. Relative photorefractoriness has only been formally characterized in one domesticated species; the Japanese quail (Robinson and Follett, 1982).Many avian species appear to show elements of both absolute and relative photorefractoriness. Song sparrows (*Melospiza melodia morphna*) exposed to 18h long photoperiod showed spontaneous regression while the birds exposed to 16h photoperiod did not. But a decrease to 14h light induced gonadal regression (Wingfield, 1993). Among three species of Cardualine finches, the degree of absolute to relative photorefractoriness varies with the degree of their reproductive flexibility (Hahn *et al.*, 2004). The House sparrow (*Passer domesticus*) posseses absolute photorefractoriness but a decrease in photoperiod immediately before regression also accelerates regression, as in relative photorefractoriness. Gonadal regression in these birds is induced by the declining day length after the summer solstice (Dawson, 1991).

The development of photorefractoriness is not a sudden but rather termination of physiological processes that have actually commenced much earlier. It has been speculated that photoperiodic stimulation in absolute photoperiodic species results into two opposite effects: firstly, photoperiod stimulates a rapid and complete "switch on", and secondly, it initiates the process that leads to a complete "switch off" of all components of photoneuroendocrine machinery (Nicholls *et al.*, 1988). It is not clear if there is some time lag between the initiations of these two photoperiodic effects. The period of long day exposure required to induce photorefractoriness depends on the daily photoperiod. Generally the photoperiod that causes refractoriness is higher than the photoperiod that induces a photoperiodic response. Further, photorefractory birds are not photoperiodically insensitive as short day exposure for sufficient period can break photorefractoriness as it occurs during autumn-winter months in the nature. Photorefractory birds continue to measure day length. These facts clearly suggest that the photoperiodic relay machinery does not become insensitive to day length in photorefractory birds; instead it becomes unresponsive to the stimulatory effects of day length upon continued exposure. Photorefractoriness has been studied extensively in many avian species and several hypotheses have been proposed to explain its etiology. We do not know precisely at what level in the brain photorefractoriness occurs. Photorefractoriness is believed to occur somewhere at the higher level and not in the peripheral endocrine organs or target sites such as gonad, adipose tissue etc.

Sexual Difference in Photoperiodic Gonadal Response

Photoperiodic stimulation of the gonadal cycle has been less extensively studied in female birds. On comparing photoperiodic gonadal responses of the male with those of the female it is generally found that the ovary does not grow to full breeding condition under artificial photostimulation alone while the testes reach to spermatogenic level in most bird species. Thus, photoperiodically induced gonadal growth in the females of photoperiodic species is less dramatic than in males. Amongst the feral photoperiodic species, especially in passerines, it is general that only partial development of the ovary can be induced by photoperiodic stimulation alone. The females of the species in which the males show strong photoperiodic gonadal response, usually exhibit progressive ovarian and other reproductive changes that are almost equivalent in rate and magnitude to the prenesting changes observed in the feral females. The substantial reduction in the ovarian response in the photoperiodic birds is due to the failure of long daily photoperiods to induce vitellogenesis and the culminative stages of the follicular development (Lofts *et al.*, 1970). These functions are reported to be largely under the control of essential supplementary factors or modifiers, at least, some of which are species specific and may involve environmental information such as presence of active mate, nesting sites or nest material etc. (Steel and Hinde, 1972). Thus, the day length plays a dual role in the development of functional ovary. Long days are essential up to the onset of phase of yolk deposition, and by inducing testes and consequent sexual behavior of the male, long days provide indirectly one of the sources of essential supplementary information for the females.

Photoperiodic Response System in Birds

Photoperiodic response system of birds possesses three principal components: (i) a photoreceptor that interprets photic input, (ii) a clock that measures the photic signal and (iii) a neurosecretary system that translates the photic signal into endocrine secretions.

Photoreception

Photoreception in birds involves nonocular or encephalic receptors and ocular or retinal receptors. Although ocular receptors are not essential for photoreception in domestic mallard, they are functionally complimentary to deep receptors. Putative-opsin expressing deep brain photoreceptors have been identified in avian septum lateralis. Light is perceived by a pigment in photoreceptor and then transduced into a chemical signal. In birds, the photoreception occurs largely through extraretinal photoreceptors located in the hypothalamic area of the brain. The photic information is relayed by deep brain photoreceptors to the reproductive axis. Opsin fibers lie in close apposition with GnRH expressing cells that relay photic information to the reproductive axis. In addition retinal photic input to the circadian system help the animal in fine tuning the timing of photoperiodic induction of reproduction. Photoperiodic information is perceived by specialized photoreceptors that may involve rods, cones and/or specialized photoreceptors in the brain in an undiscovered area. Thus, the light input to the central photoperiodic system could reach through eye, the pineal and one or more kinds of photoreceptors in the brain. At the present state of knowledge, we can say that birds show photoperiodic response in the absence of eye and pineal. It is still to be known whether the eye or pineal gland is unconnected to the photoperiodic response system.

Measurement of Photoperiodic Time

The properties of daily light dark cycle important to time keeping process are its period or cycle length, phase or timing of dawn and/ or dusk, the photoperiod or duration of light per day, and the intensity and wavelength of incident light. Since it was first discovered by Rowan that seasonal changes in day length was primary environmental signal regulating annual reproductive cycles of birds, a great deal of attention has been directed towards answering the question: how do the birds measure the length of day? Although several hypotheses have been forwarded to account the photoperiodic time measurement scheme (Farner, 1975), the discussion has been mostly confined to whether the photoperiodic clock or biochronometer is based on (i) an hour glass phenomena or (ii) endogenous circadian rhythm.

Hour Glass Mechanism

The functioning of this mechanism is based on an hour glass system that starts running by transition of light to darkness and runs for a fixed period, at which point sensitivity to light is generated. The hour glass is not inverted again until the transition of next light to darkness. In birds, this hypothesis has never been accepted widely as the evidence seems to be against it.

Circadian Rhythm and Photoperiodic Responses

Bunning (1963) proposed the involvement of endogenous circadian rhythm in photoperiodic time measurement. According to him, the endogenous circadian rhythm consists of two halves. The first twelve hour is subjective day or photoinsensitive phase while the latter twelve hour is the subjective night or photosensitive phase. The photogonadal stimulation is the result of direct or indirect, repeated (daily or otherwise) illumination of photosensitive phase by external photophase. This device functions as a clock to measure the day length (Figure 24.6). Hamner (1963) experimentally demonstrated this hypothesis in the bird (*Carpodacus erythrinus*). Since then considerable evidence has been accumulated in the literature strongly favoring the involvement of endogenous circadian rhythm in photoperiodic time measurement in many avian species, for example, *Carpodacus maxicanus* (Hamner,1963), *Zonotrichia leucoprays gambelii* (Farner, 1965), *Junco hyaemalis* (Wolfson, 1965), *Coturnix coturnix japonica* (Follett and Sharp, 1969), *Passer montanus*

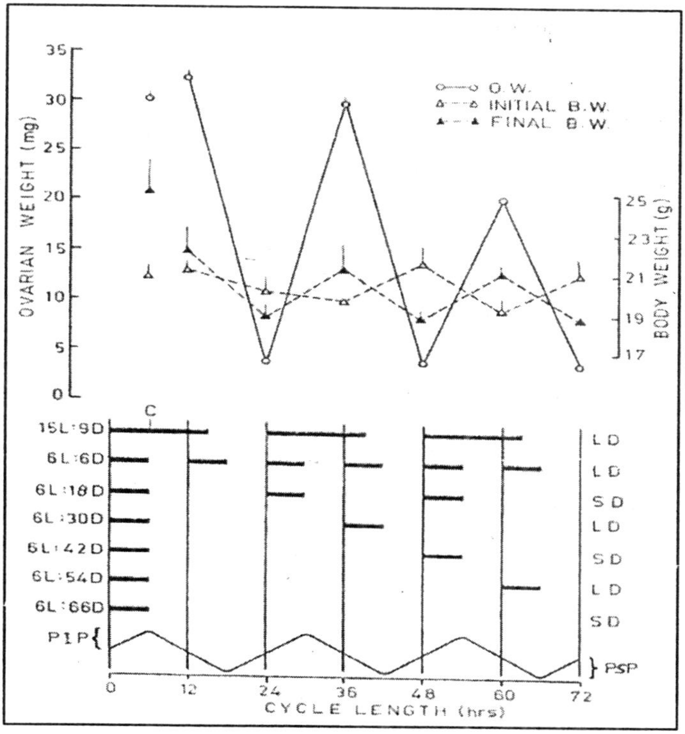

Figure 24.6: A Schematic Representation of Results Obtained with *Carpodacus erythrinus* Under Resonance Light Cycles Showing Involvement of Endogenous Circadian Rhythm in Photoperiodic Time Measurement during Ovarian Growth.

Solid bars represent duration and position of light in hypothetical circadian rhythms, presented at the bottom of the figure. PSP, photosensitive phase; PIP, photoinsensitive phase; LD, long day; SD, short day; OW, ovarian weight; BW, body weight.

(Lofts and Lam, 1973), *Emberiza melanocephala* (Kumar and Tewary, 1984), *Carpodacus erythrinus* (Tewary and Dixit, 1984). As a result, it is now clear that photoperiodic time measurement in majority of birds is based on circadian rhythm of responsiveness to light, but the extent to which the photoinducible phase is the component of the circadian system varies considerably. There have been numerous subsequent refinements of the hypothesis. In a more explicit version (Pittendrigh and minis, 1964) the daily photoperiod has been attributed to play dual role, *i.e.*, of entrainment of circadian rhythm to a 24h cycle and that of producing photoperiodic response if its duration is sufficient to extend into the photosensitive or more precisely photoinducible phase (Pittendrigh, 1966).

Two models those of external and internal coincidence have been advanced to account the possible ways of functioning of the circadian clock in photoperiodic time measurement (Pittendrigh, 1974).

External Coincidence Model

According to this model, a photoperiodic response occurs due to the direct coincidence between the photoinducible phase of circadian rhythm and the environmental photoperiod. First, light entrains the circadian periodicity in photosensitivity which is presumed to be an endogenous free running circadian rhythm and second, when of sufficient duration to extend into photoinducible phase of the entrained circadian periodicity in photosensitivity, it induces the photoperiodic response.

Internal Coincidence Model

This model predicts that external light period brings two or more circadian oscillators in a particular phase relationship with respect to one another this results in a photoperiodic response. According to this model, light serves only as entraining agent without having an active role in inducing a photoperiodic response. This model suggests that the phase relationship between internal circadian oscillators may vary under different photoperiodic conditions (*e. g.* a change in ratio of light to dark) and only under certain conditions (light dark cycle) will the phase relationship be such that photoperiodic induction shall occur. Long day length might establish such a relationship while short photoperiods could maintain the rhythm out of phase with respect to each other.

A common characteristic of both the above models is that the factor determining the stimulatory or inhibitory effect of a particular light dark cycle depends on phase relationship between light and circadian oscillator(s) involved in the transmission of information about day length to the hypothalamo-hypophysial-gonadal axis.

Circannual Rhythm and Seasonal Reproduction

Circannual rhythms are self sustained endogenous rhythms with a period of about a year that affect various physiological and behavioral functions in the organisms. Endogenous circannual clocks are found in many long lived organisms, but are best studied in birds and mammals. Many species living in tropics or close to equator where there is very little or no change in annual photoperiod exhibit clear seasonality in reproductive functions. These observations support the involvement of endogenous circannual rhythm in their control. It has been demonstrated in few bird species that the annual cycles of reproduction are controlled by endogenous circannual rhythmicity (Gwinner, 1975). Under seasonally constant environmental conditions these rhythms persist for several cycles with a period that is usually slightly different from 12 months. Annual changes in day length may act as zeitgeber that entrains circannual rhythms (Farner and Lewis, 1971). Schwab (1971) has shown that male starlings maintained under 12L: 12D for about 2 years show gonadal cycle that corresponds remarkably to the natural testicular cycle of the wild starlings. Similar observations were reported on African weaver finches (Lofts, 1964), Garden warblers (Gwinner, 1996), Spotted munia (Chandola *et al.*, 1983). Endogenous circannual rhythms have been experimentally demonstrated in more than twenty migratory and resident bird species from both tropical and temperate regions (Gwinner and Dittami, 1990; Cadee *et al.*, 1996; Newton 2007). In birds, the overt expressions of circannual rhythms under constant conditions include seasonal changes in reproduction, body mass, molt, and *zugunruhe* or the migratory restlessness. Birds that live in habitats where photoperiod is poor predictor of seasons (equatorial residents, migrants to equatorial/ tropical latitudes) rely more on their endogenous clocks than birds living in the places where there is tight correlation between photoperiod and seasonal events. Such differences on reliance on endogenous clocks may indicate that the annual timing mechanisms are adaptive. The most important temporal cue in the wild and zeitgeber for circannual rhythms is photoperiod. The role of photoperiod in such a temporal scheme is limited where it is used to synchronize the circannual rhythm to the calendar year, but it does not alter the overall temporal course of seasonal programming of the annual event. It has been suggested that the circannual clocks are either based upon (i) a de-multiplication (counting) of circadian days (ii) A sequence of interdependent physiological states, or (iii) one or more endogenous oscillators, similar to circadian rhythms (Wikelski *et al.*, 2007). It takes about 2-3 years to confirm the involvement of circannual rhythm in control of seasonal events like reproduction and that is why only few reports are available on this aspect. The adaptive significance of circannual rhythms has so far been tested by comparing endogenous rhythms in captivity with behavior in wild. The rationale is that endogenous clocks help animals to keep time in the wild (Menaker, 2006).

Photoperiodic Clock

In mammals, SCN generated circadian signal is decoded through daily rhythm in melatonine secretion. The SCN or SCN like structure in non mammalian vertebrates including birds has not been recognized as yet. Some attempts have been made in birds, but the site of SCN homologue is a matter of dispute among avian biologists. Some (Panzica, 1988) consider a nucleus in the medial hypothalamus as the avian SCN whereas the others (Cassone and Moore, 1987) consider a nucleus in the lateral hypothalamus that receives input from retina as the avian SCN. Thus, the sites of photoperiodic clock and its efferent in birds remain in question.

We have no knowledge about anatomical location of the circannual clocks, their neurochemical characteristics, mechanism(s) of generation and their synchronization to environmental light-dark cycles. Further, we are not sure if the circadian and circannual rhythms are physiologically distinct and do not overlap. Whether two circa-rhythms can be separated from each other is yet to be established experimentally.

Hormonal Basis of Photoperiodism

Unlike mammals in which rhythm in melatonin secretion gives photoperiodic message to the neural machinery regulating reproduction, birds may bypass the pineal melatonin in sensing photoperiodic message in regulation of photoperiod induced seasonality. Juss (1993) has discussed in detail the neural aspect of bird photoperiodism. The two hypothalamic areas involved in decoding of photoperiodic message are preoptic and tuberoinfundibular regions. Photoperiod causes secretion of a number of hormones from the anterior pituitary and each of them may be regulated by different neurotransmitter(s). The rate and degree of gonadal maturation is ultimately determined by the rate and secretion of chicken gonadotropin-releasing hormone I (cGnRH-I) and possible antagonistic effects of gonadotropin inhibiting hormone (GnIH; Bentley *et al.*, 2003; Ukena *et al.*, 2003) and/or prolactin (Dawson and Sharp, 1998; Sharp and Blache, 2003). cGnRH is a decapeptide which is synthesized in the cell bodies of specialized neurosecretary neurons. It passes down the axon of these neurons and is secreted in the median eminence at the base of hypothalamus, from where it passes via a blood portal system to the pituitary gland. After reaching to the pituitary, cGnRH-I stimulates the synthesis and release of two gonadotrophic hormones namely, luteinizing hormone (LH) and follicular stimulating hormone (FSH). These two hormones once released into the circulation cause gonadal growth and maturation. The activity of cGnRH-I neurons is primarily controlled by photoperiodic information received from encephalic receptors integrated in some way with a circadian clock. As the photoperiod increases, it causes conversion of thyroxine to triiodothyronine (Yasuo *et al.*, 2005) and this to some extend (Takagi *et al.*, 2007),leads to an increased secretion of cGnRH-I (Yamamura *et al.*, 2006). Photoperiodically induced secretion of cGnRH-I secretion may be modulated by non photoperiodic information from other neural inputs. The conversion of thyroxine to triiodothyronine may act as a long photoperiod signal. This may ultimately downregulate cGnRH-I synthesis as bird become photorefractory.

Reproductive Plasticity

If birds rely entirely on photoperiod to control the time of gonadal growth and regression then these events would occur at the same time every year. In a variety of bird species, the exact timing of breeding differs between years and also within years between different habitats at the same latitude. There can be two possibilities. According to the first possibility, the photoperiod alone controls the time of gonadal growth and regression and thus sets limits to a physiological window within which breeding can occur. Flexibility exists within this window so that non photoperiodic cues can affect the exact timing of egg laying. Female birds often require non photoperiodic cues in the form of essential supplementary factors or modifiers to induce latter stages of ovarian maturation and ovulation. In the second possibility, non photoperiodic cues may themselves modulate photoperiodic regulation of growth and regression and, thus, directly affect the window within which egg laying occurs. In reality, it is possible that both operate. The above possibilities emphasize the importance of non photoperiodic cues. There is a range of non photoperiodic cues that can modulate the timing of breeding, including temperature, rainfall and food availability. There is abundant evidence in many species that the time of egg laying varies with spring temperature (Korpimaki, 1978; Visser *et al.*, 2003;Torti and Dunn, 2005).It is not clear whether this is direct effect of temperature on photoperiodic induction of gonadal growth. Tropical species often have fixed breeding seasons related to predictable periods of rainfall (Wikelski *et al.*, 2000). Although, the annual change in photoperiod is slight in tropics, this is sufficient to cause gonadal maturation (Hau *et al.*, 1998; Dawson, 2008). Rainfall itself may act only as a short term cue to fine tune the timing of breeding. Availability of food is the most important factor that supports the young and parents and controls the timing of breeding (Lack, 1968). Food can also act as proximate factor as the females require sufficient food for production of egg. Egg laying advances when the supplemental food is provided sometimes, but not always in many birds (Meijer *et al.*, 1990). Advances are normally in the range of few days. Yet, restricted food availability has little or no effect on photoperiodically controlled gonadal maturation in either sex of Starlings (Dawson, 1986). The modulating effect of food availability is restricted to the latter stages of ovarian maturation and the effect of food as a proximate factor is to fine- tune the exact date of laying within the physiological window defined by photoperiodically regulated gonadal growth and regression.

Concluding Remarks

This review article examines how birds use annual changes in photoperiod to time their breeding and associated seasonal events at the most suitable time of the year. Seasonality is represented by the initiation-termination-reinitiation of physiological processes. It is a compulsory adaptation for survival in many animals including birds. Most, if not all, species of birds exhibit well-defined seasonality in their several biological processes including gonadal growth and development, molt, body fattening, bill and plumage colors, body mass, hormone levels, song production, nest building, parental care, migration etc. All these cycles remain in a phase relationship and generally centre on reproduction. Therefore, the timing of actual reproduction, which should occur in the year when the survival of young ones is maximum, is critical for the species. Seasonality in reproduction and associated events is species specific and is timed such that when breeding occurs plenty of preferred food is available for both the parents and their offspring. Life on earth has been subjected to strict regimen of cyclic changes since its very origin. Daily changes between day and night and the regular succession of annual seasons are the most prominent of all geophysical changes that the animals experience in nature. Of several factors in the environment, which the birds are exposed to, the day length is most consistent and reliable. Birds tend to adapt to daily light dark cycle using their endogenous time keeping device(s), called "clocks" because of their great precision in the timing of various physiological and behavioral events. It has been found in many studies that the birds exposed to day-to-day variation in light hours in nature adapt to it using their endogenous time keeping machenism(s); day length interacts with endogenous clocks and induces seasonal events. Rhythmic breeding and other physiological processes are evolved through the development of timing mechanisms that are governed by the oscillatory system. These systems are innate (endogenous), inheritable and genetic in origin. In the natural environment, these oscillatory systems are synchronized with day and night, and thus, are expressed as daily overt rhythms. Two mechanisms appear involved in regulation of seasonality in birds. One is the photoperiodism, in which environmental photoperiod times the component events of the seasonality. The other is the circannual rhythm generation, in which a self-sustained endogenous rhythmicity of approximately 1-year times these component events. Photoperiodism is the control of some aspects of life cycle in an organism by the timing of light and darkness. The reproductive cycle of most birds is driven by annual changes in photoperiod and each cycle usually constitutes the physiological state of photosensitivity, photostimulation and photorefractoriness. The minimum photoperiod that will induce a physiological response is known as threshold photoperiod or critical day length for particular response. In many photoperiodic birds, each period of gonadal growth and development is followed by rapid collapse of the gonad. This state of unresponsiveness to stimulatory photoperiodic regime is known as photorefractoriness. A photoperiodic response system has three major components: a photoreceptor that interprets photic input, a clock that measures photic signal, and a neuroseretary system that translates photic signal into endocrine seretions. Photoperiodic photoreception in birds occurs largely through the extra-retinal photoreceptors located in the hypothalamus. Increasing photoperiods of spring stimulate secretion of gonadotropin-releasing hormone from pituitary gland that leads to gonadal maturation. Photoperiodic time measurement in birds involves interaction between endogenous circadian rhythm and deep brain photoreceptors. The photoperiodic signal in the form of cyclic release of melatonin from eye and pineal gland does not seem to be involved in seasonal breeding. In temperate zone species, photoperiod is prominent proximate factor controlling breeding while in tropical breeders endogenous circannual rhythmicity seems to be more important.

Acknowledgements

The financial assistance from the Department of Science and Technology, New Delhi, partly used in preparation of this manuscript is gratefully acknowledged.

References

Baker, J.R., 1938. The evolution of breeding season. In: *Evolution* (Eds. G.R. de Beer) pp. 161-177, Oxford University Press Oxford.

Bauchinger, U. and Klaassen, M., 2005. Long days in spring than in autumn accelerate migration speed of passerine birds. *J. Avian Biol.*, 36: 3-5.

Bently,G. E., Perfito, N., Ukena, K., Tsutsui, K. and Wingfield, J. C., 2003. Gonadotropin-inhibitory peptide in song sparrows (*Melospiza malodia*) in different reproductive conditions, and in house sparrow (*Passer domesticus*) relative to chicken-gonadotropin-releasing hormone. *J. Neuroendocrinol.*, 15: 794-802.

Bunning, E., 1963. Die physiological clock, 3rd ed. University Press Ltd., London, Springer-Verlag, New York, Heidelberg, Berlin.

Cadee, N., Piersma, T. and Daan, S., 1996. Endogenous circannual rhythmicity in a non-passerine migrant, the knot *Calidris canutus*. *Ardea.*, 84: 75-84.

Cassone, V.M and Moore, R.Y., 1988. *J. Comp Neurol.*, 266: 171.

Chandola, A., Saklani, M., Bisht, M., and Bhatt, D., 1983) Adaptation to terrestrial environments, ed. by N.S. Margaris, M. Arianoutsou-Faraggitaki and R.J. Reiter (Plenum Publishing Corparation), 145.

Chandola, A., J. Pavnaskar and J.P. Thapliyal., 1975. Scoto/ photoperiodic responses of a sub tropical finch (Spotted munia) in relation to seasional breeding cycle. *J. Interdiscipl. Cycle Res.*, 6: 189-202.

Dawson, A., 1986. The effect of restricting the daily period of food availability on testicular growth in starlings, *Sturnus vulgaris. Ibis.*, 128: 572-575.

Dawson, A., 1989. Pharmacological doses of thyroxine simulates the effect of increased daylength on the reproductive system of European starlings. *J. Exp. Zool.*, 249: 62-67.

Dawson, A., 1991. Photoperiodic control of testicular regression and moult in male house sparrows, *Passer domesticus. Ibis.*, 133: 312-316.

Dawson, A., 2008. Control of the annual cycle in birds: endocrine constraints and plasticity in response to ecological variability. *Phil. Trans. R. Soc. B.*, 363: 1621-163.

Dawson, A. and Sharp, P.J., 1998. The role of prolactin in the development of reproductive photorefractoriness and postnuptial molt in the Europian starling (*Sturnus vulgaris. Endocrinology,* 139: 485-490.

Farner, D.S., 1964. The photoperiodic control of reproductive cycles in birds. *Amer. Sci.*, 52: 137-156.

Farner, D.S., 1975. Photoperiodic controls and reproductive cycles in *Zonotrichia.* In: Proc. XVI Intern. Ornithol. Congr. Pp. 369-382. Australian Academy Science, Canberra.

Farner, D.S. and Follett, B.K., 1966. Light and other environmental factors affecting avian reproduction. *J. Anim. Sci.*, 25: 90-118.

Farner, D.S. and Lewis, R.A., 1971. Photoperiodism and reproductive cycles in birds. In: *Photophysiology* (Ed. Giese, A.C.) pp. 325-370, Academic Press, New York and London.

Ferner, D.S., 1965. Circadian system in the photoperiodic responses of Vertebrates. In: Circadian Clocks, pp. 357-369. Ed. J. Aschoff. North-Holland, Amsterdam.

Ferner, D.S. and Follett, B.K., 1966. Light and other environmental factors affecting avian reproduction. *J. Anim. Sci.*, 25: 90-118.

Ferner, D.S. and Wilson. A.C., 1957. A quantitative examination of testicular growth in the white-crowned sparrow. *Biol. Bull.*, 133: 254-267.

Ferner, D.S. and Wingfield, J. C., 1980. Reproductive endocrinology of birds. *Ann. Rev. Physiol.*, 42: 457-472.

Ferner, D.S., Follett, B.K. King, J.R. and Morton, M.L., 1966. A quantitative examination of ovarian growth in the white-crowned sparrow. *Biol. Bulletin*, 130: 67-75.

Follett, B.K. and Sharp, P.J., 1969. Circadian rhythmicity in photoperiodically induced gonadotropin release and gonadal growth in the Quail. *Nature*, 223: 968-971.

Gwinner, E and Dittami, J., 1990. Endogenous reproductive rhythms in a tropical birds. *Science*, 249: 906-908.

Gwinner, E., 1975. Effect of season and external testosterone on the free-running circadian activity rhythm of European Starlings (*Sturnus vulgaris*). *J. Comp. Physiol.*, 103: 315-328.

Gwinner, E., 1996. Circadian and circannual programmers in avian migration. *J. Exp. Biol.*, 199: 39-48.

Gwinner, E. and Scheuerlein, A., 1998. Seasonal changes in daylight intensity as a potential zeitgeber of circannual rhythm in equatorial Stonechats. *J. Ornithol.*, 139: 4 07-412.

Hahn, T.P., Pereyra, M. E., Sharbaugh, S.M. and Bentley, G.E., 2004. Physiological responses to photperiod in three cardueline finch species. *Gen. Comp. Endocrinol.*, 137: 99-108.

Hamner, W.M., 1963. Diurnal rhythm and photoperiodism in testicular recrudescence of the House Finch. *Science*, 142: 1294-1295.

Hau, M.,Wikelski, M. and Wingfield, J.C., 1998. A neotropical forest bird can measure the slight change in tropical photoperiod. *Proc.R. Soc. B.*, 1391: 89-95.

Immelmann, K., 1971. Ecological aspect of periodic reproduction as source of predictive information. In: *Breeding Biology of Birds* (Eds. Farner, D.S. and King, J.R.). pp. 341-389, Academic Press, New York and London.

Juss, T. S., Meddle, S.L., Servent, R.S. and King, V. M., 1993. Melatonin and photoperiodic time measurement in Japanese quail (*Coturnix coturnix japonica*. *Proc. R. Soc. B.*, 254: 21-28.

Korpimaki,E., 1978. Breeding biology of the Starling *Sturnus vulgaris* in western Findland. *Ornis Fenn.*, 55: 93-104.

Kumar, V., 1981. Photoperiodic response of some migratory birds. Ph.D. dessertation, Banaras Hindu University, Varanasi, India.

Kumar, V., 1997. Photoperiodism in higher vertebrates: An adaptive strategy in temporal environment. *Indian J. Exp. Biol.*, 35: 427-437.

Kumar, V. and Tewary, P.D., 1984. Circadian rhythmicity and the termination of photorefractoriness in the Black-headed Bunting. *Conndor.*, 86: 27-29.

Kumar, V. and Tewary. P.D., 1983. Response to experimental photoperiods by a migratory bunting, *Emberiza malanocephala*. *Ibis*, 125: 305-312.

Lack, D., 1968. Bird migration and natural selection. *Oikos*, 19: 1-9.

Lack, D., 1968. Ecological adaptation for breeding in birds. London, UK: Methuen.

Lofts, B., 1964. Evidence of an autonomous reproductive rhythm in an equatorial bird (*Quelea quelea*. *Nature*, 201: 523-524.

Lofts, B. and Lam, W.L., 1973. circadian regulation of gonadotrophin secretion. *J. Repod. Fert. Suppl.*, 19: 19-34.

Lofts, B. and Murton. R.K., 1968. Photoperiodic and physiological adaptations regulating avian breeding cycles and their ecological significance. *J. Zool.*, 155: 32.

Lofts, B., Follett,B.K. and Murton, R.K., 1970. Temporal changes in the pituitary-gonadal axis. *Mem. Soc. Endocrinol.*, 18: 545-575.

Marshall, A.J., 1970. Environmental factors other than light involved in the control of sexual cycles in birds and mammals. *Colloq. Internat. Centre. Nat. Rech. Sci.*, 172: 53-69.

Meijer, T., Daan, S. and Hall, M.R., 1990. Family planning in the Kestrel (*Falco tinnunculus*): the proximate control of laying date and clutch size. *Behaviour*, 114: 117-136.

Menaker, M., 2006. Circadian organization in the real world. *Proc. Natl Acad. Sci.* USA, 103: 3015-3016.

Mirrison,J.V. and Wilson, F.E., 1972. Ovarian growth in Tree sparrow, *Spizella arborea*. *Auk.*, 89: 146-155.

Murton, R.K., and Westwood. N.J., 1977. Avian breeding cycles. Clarendon Press, Oxford.

Nelson, R.J., Demoas, G., Klein, S.L. and Kriegsfeld, L.J., 2001. Seasonal cycles in immune function and disease proscesses. New York, NY: Cambridge University Press.

Newton, I., 2007. The ecology of birds migration. London, UK: Academic Press.

Newton, I. and Rothery, P., 2005. The timing, duration and pattern of molt and its relationship to breeding in a population of the European greenfinch *Carduelis chloris*. *Ibis*, 147: 667-679.

Panzica, G.C., 1988. *Cell Tissue Res.*, 242: 371.

Pittendrigh, C.S., 1966. The circadian oscillation in *Drosophila pseudoobscura* pupae: a model for the photoperiodic clock. 2. *Z. Pflanzenphysiol.*, 54: 275-307.

Pittendrigh, C.S., 1974. Circadian oscillation in cells and the circadian organization of multicellular systems. In: The Neurosciences: Third Study program, pp. 437-458. Eds. F.O. Schmitt and F. G. Worden. MIT Press, Cambridge.

Pittendrigh, C.S. and Minis, D. H., 1964. The entrainment of circadian oscillations by light and their role as photoperiodic clocks. *Am. Nature*, 98: 261-294.

Prasad, B.N., 1983. Photoperiod; Gonadal growth and premigratory fattening in the Redheaded Bunting, *Emberiza bruniceps*. *Environ. Control. Biol.*, 21: 53-59.

Robinson, J. E. and Follett, B.K., 1982. Photoperiodism in Japanese quail: the termination of seasonal breeding by photorefractoriness. *Proc. R. Soc. B.*, 215: 95-116.

Rowan, W., 1925. Relation of light to bird migration and developmental changes. *Nature*, 115: 494-495.

Rowen, W., 1926. On photoperiodism, reproductive periodicity and the annual migrations of birds and certain fishes. *Proc. Boston Sco. Natur. Hist.*, 38: 147-189.

Rowen, W., 1932. Experiments in bird migration III. The effects of artificial light, castration and certain extracts on the autumn movements of the American Crow (*Corvus brachyrhynchos*). *Proc. Nat. Acad. Sci. USA*, 18: 639-654.

Schildmacher, H., 1939. Uber die kunstliche Aktivierung der Hoden einiger Vogalarten in Herbst durch Belichtung und Verderlappen hormone. *Biol. Zentralbl. Leipzig*. 59: 653-657.

Schwab, R.G., 1971. Circadian testicular periodicity in the European Starling in the absence of photoperiodic change. In: Biochronometry, pp. 428-447. ed. M. Menaker. U. S. Nat. Acad. Sci. Washington.

Sharp, P.J. and Blache, D., 2003. A neuroendocrine model for prolactin as the key mediator of seasonal breeding in birds under long- and short-day photoperiods. *Can. J. Physiol. Pharmacol.* 81: 350-358.

Singh, S. and Chandola, A., 1981. Photoperiodic control of seasonal reproduction in tropical weaver bird. *J. Exp. Zool.* 216: 293-298.

Steel, E.A. and Hinde, R. A., 1972. Influence of photoperiod on oestrogenic induction of nest building in cararies. *J. Endocrinol.* 17: 105-114.

Takagi, T., Yamamura, T., Anraku, T., Yasuo, S., Nakao, N., Watanabe, M., Iigo, M., Ebihara, S. and Yoshimura, T., 2007. Involvement of transforming growth factor alpha in the photoperiodic regulation of reproduction in birds. *Endocrinology.* 145: 2788-2792.

Tewary,P.D. and Anand S. Dixit., 1986. Photoperiodic regulation of reproduction in subtropical female yellow-throated sparrow (*Gymnorhis xanthocollis. Condor.* 88: 70-73.

Tewary, P. D. and Anand S. Dixit, 1983) Photoperiodic control of ovarian cycle in Rosefinch, *Carpodacus erythrinus. J. Exp. Zool.* 228: 537-542.

Tewary, P.D. and Kumar, V., 1982. Photoperiodic responses of a subtropical migratory finch, The blackheaded bunting, *Emberiza melanocephala. Condor.*, 84: 168-171.

Thapliyal, J.P., 1981. Endocrinology of avian reproduction, Presidential Address. *Indian Sci. Congr.*, Varanasi.

Thapliyal, J.P. and Saxena. R.N., 1964. Gonadal development of the male Blackheaded Munia under constant nine-hour (short) days. *J. Exp. Zool.* 156: 153-156.

Thapliyal, J.P. and Tewary, P.D., 1964. Effect of light on pituitary, gonad and plumage pigmentation in the Avadavat, *Estrilda amandava* and baya weaver, *Ploceus Pphilippines. Proc. Zool. Soc., London*, 142: 67-71.

Thapliyal, P.D. and Tripathi, B.K., 1983. Photoperiodic control of reproduction in a female migratory Bunting, *Emberiza brunuceps. J. Exp. Zool.* 226: 269-272.

Torti, V. M. and Dunn, P.O., 2005. Variable effects of climate change on six species of North American birds. *Oecologia.*, 145: 486-495.

Ukana,K., Ubuka, T. and Tsutsui, K., 2003. Distribution of a novel avian gonadotropin-inhibitory hormone in the quil brain. *Cell Tissue Res.*, 312: 73-79.

Visser, M.E. *et al.*, 2003. Variable responses to large-scale climate change in European *Parus* population. *Proc. R. Soc. B.*, 270: 367-372.

Wiikelski, M., Hau, M. and Wingfield, J.C., 2000. Seasonality of reproduction in a neotropical rain forest bird. *Ecology*, 81: 2458-2472.

Wikelski, M., Martin, L.B., Scheuerlein, A.,Robinson, M.T., Robinson, N.D., Helm, B., Hau, M., Gwinner, E., 2007. Avian circannual clocks: adaptive significance and possible involvement of energy turn over in their proximate control. *Phil. Trans. R. Soc. B.*

Wingfield, J.C., 1993. Control of testicular cycles in the song sparrow, *Melospiza melodia*: interaction of photoperiod and an endogenous prograsm? *Gen. Comp. Endocrinol.*, 92: 388-401.

Wingfield, J.C. and Farner, D.S., 1978. The annual cycle of plasma irLH and steroid hormones in feral populations of the white-crowned sparrow, *Zonotrichia leucophrys gambelii. Biol. Reprod.*, 19: 1046-1056.

Wolfson, A., 1965. Circadian rhythm and the regulation of the annual reproductive cycle in birds. In: Proceeding of the feldefing Summer School; pp. 370-378. ed. J. Ascoff. North-Holland Publ. Co., Amsterdam.

Yamamura, T., Yasuo, S., Hirunagi, K., Ebihara, S. and Yoshimutra, T., 2006. T-3 implantation mimica photoperiodically reduced encasement of nerve terminals by glial processes in the median eminence of Japanese quail. *Cell Tissue Res.*, 324: 175-179.

Yasuo, S., Watanabe, M., Nakao, N., Takagi, T., Follett, B.K., Ebihara, S. and Yoshimura, T., 2005. The reciprocal switching of two thyroid hormone-activating and –inactivating enzyme genes is involved in the photoperiodic gonadal response of Japanese quail. *Endocrinology*. 146: 2551-2554.

Chapter 25

Environmental Modulations and Reproduction: The Favourable and Unfavourable Paradigms

Urmi Chatterji*

Department of Zoology, University of Calcutta,
35 Ballygunge Circular Road, Kolkata – 700 019, W.B., India

ABSTRACT

Environmental cues are known to control and synchronize the reproductive physiology of various classes of animals. Both asexual and sexual modes of reproduction are intensely dependent on environmental factors, be it natural phenomena such as photoperiod, temperature, availability of food or to adverse environmental conditions, either natural or man-made. Such conditions have been shown to modulate reproduction, either individually or in combination. The environmental cues affect a number of reproductive parameters, including sex determination, gametogenesis, spawning on one hand and structural changes of reproductive units, fecundity, spontaneous abortions and stillbirths on the other. Although such changes may not always lead to extinctions, they may cause local extirpations, depending on the speed of adaptation or ability to overcome the changing environmental conditions. Seasonal

* Corresponding author: E-mail: urmichatterji@gmail.com

reproduction and environmental sex determination are intricately associated with epigenetic factors and are extremely essential for propagation of species under adverse conditions. Although most organisms have adopted expedient mechanisms to cope with changes in the environment, the increase in environmental pollutants has been a constant threat to the reproductive potency of both invertebrates and vertebrates. Such environmental pollutants range from increase in global temperature and ultraviolet radiations to chemicals that are either found naturally on the earth's crust or those that are added to the environment as agricultural or industrial effluents. Some of these chemicals can mimic the activity of naturally occurring hormones and lead to disastrous, often irreversible, alterations in the reproductive physiology. This review is an attempt to unravel how divergent reproductive strategies in various animal species are affected by natural or adverse alterations in the environment.

Keywords: *Environmental cues, Reproductive physiology, Reproductive potency, Reproductive strategies, Environmental pollutants, Ultraviolet radiations.*

Introduction

Reproduction is the culmination of individual transcendence. Every existing animal or living organism is the result of biological reproduction, either asexually or sexually. Common forms of asexual reproduction include binary fission (amoeba), budding (hydra), gemmules (sponges), fragmentation (flatworms), regeneration (echinoderms) and parthenogenesis (aphids). Sexual reproduction, on the other hand, involves union (fertilization, both internal and external) of two distinct gametes, in both aquatic and terrestrial conditions. Whatever the case may be, several environmental cues are known to have profound effect on the process of sex determination and reproduction and are known to influence the outcome accordingly. Such environmental factors may involve specific conditions of temperature, photoperiod, humidity and availability of food on one hand, and the presence of different naturally-occurring or man-made pollutants on the other, which may disrupt the reproductive physiology and fecundity of different organisms.

Animals that face difficulties in encountering sexual partners, because of low mobility or recurrently low population densities have evolved self-fertilization as a reproductive assurance strategy, allowing reproduction when mates are not available, even at the cost of inbreeding depression (Jarne and Charlesworth, 1993; Kalisz *et al.*, 2004). Delayed selfing allows self-fertile organisms to reproduce even in the absence of mate, and is thought to have evolved as a reproductive assurance strategy. In animal species with strong inbreeding depression, the time during which selfing is postponed in the absence of mates (waiting time) is predicted to have evolved as a function of inbreeding depression, resource reallocation and survival (Pierre-Yves *et al.*, 2006).

Most habitats worldwide exhibit temporal fluctuations in biotic and abiotic conditions. The degree of environmental predictability has important consequences for the reproductive physiology of organisms. On one hand, opportunistic breeders must keep their reproductive organs in a near-functional state in order not to miss the narrow time window during which reproduction is possible (Emerson and Hess, 1996; Hahn, 1998). This is usually a costly strategy, because being prepared is energetically demanding (Murton and Westwood, 1977). In contrast, since reproduction is energetically expensive, animals living in variable environments breed only when conditions are suitable for the production of viable offspring (Wallen and Schneider, 1999). During the non-breeding season, most animals allow the total regression of reproductive organs and allocation of vital resources to other organismal functions, like the buildup of flight muscles in migrating birds (Gwinner, 1996). While the complete inactivation of the reproductive system during the non-breeding season conveys energetic benefits, it also imposes a temporal constraint because the reactivation of the reproductive system can require weeks or even months (Dawson *et al.*, 2001). In order to counteract this temporal constraint, animals living in seasonally predictable habitats generally prepare for the next reproductive season long in advance using photoperiod (daylength) as a reliable signal for the ensuing favorable conditions (Dawson *et al.*, 2001). Increasing photoperiod in spring stimulates the hypothalamo-pituitary-gonad axis, leading to the secretion of reproductive hormones and the growth and maturation of reproductive organs. Photoperiod also controls the termination of reproduction, shutting down reproductive processes in summer or fall (Nicholls *et al.*, 1988).

Although most animals have the capability of adjusting to and overcoming naturally-occurring environmental fluctuations, most, if not all species, often succumb to the deleterious effects of environmental contaminants. Such contaminants, which may occur in the form of xenoestrogens or toxic chemicals, are often irreversibly detrimental to the reproductive health of animal species across the animal kingdom. Hence, modulations of the environment and reproductive fecundity are intensely interrelated features and has been studied in great detail throughout the world to (i) provide an insight into conservation of species that are facing the danger of extinction; (ii) improve breeding conditions of economically important animals; and (iii) devise improved strategies of overcoming reproductive failures in human beings.

Environmental Cues that Sustain Reproduction

Seasonal reproduction is commonly seen in animals at all latitudes, even in the deep tropics. The two environmental factors of most concern when considering seasonal breeding are (i) foraging conditions, as they determine energy balance, and (ii) predictive cues like photoperiod. Food availability and the ambient temperature encountered during foraging determine energy balance and that, in turn, is the ultimate cause of seasonal breeding in all mammals and the proximate cause in many. Low temperatures experienced during foraging are particularly important for small mammals. Photoperiodic cueing is common among long-lived mammals from the highest latitudes to the mid-tropics; it is much less important for shorter-lived mammals, many of which exhibit great individual variations in photo-responsiveness and thus mixed strategies for adapting to winter conditions. It has been suggested that as the climate changes, the small rodents of the world will probably reproduce successfully wherever they can survive the higher temperatures, since their short generation time is an asset. The situation may be quite different for longer-lived mammals, particularly those whose reproduction in rigidly controlled by photoperiod.

Although relatively little is known about possible sex differences in response to cues from the environment that control the timing of seasonal breeding, it has been established that the sexes can be expected to differ in the cues they use to time reproduction (Ball and Ketterson, 2008). Sex differences in the control of reproduction could be regulated via the response to photoperiod or in the relative importance and action of supplementary factors (such as temperature, food supply, nesting sites and behavioural interactions) that adjust the timing of reproduction, so that it is in agreement with local conditions. A wealth of information has accumulated for vertebrate and invertebrate species about how cues such as variations in photoperiod as well as temperature, food availability, nest sites and social interactions can affect the timing of the onset and the end of reproduction (Goldman *et al.*, 2004). In seasonally unpredictable habitats, organisms cannot anticipate and prepare for a regular yearly reproductive period. Therefore, they maintain an activated reproductive system all year around and opportunistically breed at any time of year (Hau *et al.*, 2004). Seasonal reproduction, on the other hand, synchronizes breeding with food abundance to maximize reproductive success (Lack, 1968). In addition, day length or photoperiod has long been known to be crucial to the timing of seasonal processes (Gwinner, 2003). Photoperiodicity has been especially well studied in temperate-zone birds where the ultimately predictable change in the length of the day is the initial predictive cue used to time reproduction (Wingfield and Farner, 1993). Supplementary cues, such as food availability, temperature, and rainfall, are then used to fine-tune the timing of breeding to the local environment (Wingfield and Kenagy, 1991). The use of photoperiodic cues by animals increases with latitude and results in synchronized breeding, both within and among populations at the same latitude (Wingfield *et al.*, 1997). For example, in Siberian hamsters (*Phodopus sungorus*), long summer days stimulate, whereas short winter days inhibit, reproductive physiology and behavior (Paul *et al.*, 2009).

In Polychaetes, two main reproductive strategies are known: (i) semelparity, which is characteristic of species in which breeding occurs once during the lifetime during a spawning crisis, and (ii) iteroparity, that can be observed in species in which breeding occurs several times during the lifetime. Although most polychaete species are iteroparous, nereids breed only once during their lifetime (Andries, 2001). In addition, their sexual reproduction is often accompanied by a transition to an epitokous form that transforms them to enable them to swim and spawn in surface waters.

Reproduction in marine invertebrate species requires the synchronous spawning of neighbouring individuals in a population. In some species, entire populations can spawn abruptly, sometimes within a few days, suggesting that spawning is coordinated by external environmental factors. Such factors include temperature and day length which may interfere with the reproductive cycle without triggering spawning. Although how a threshold temperature or day length would be a reliable means of synchronizing the liberation of gametes has not yet been elucidated, a certain threshold must be reached for their release, as seen by the fact that spawning is delayed or suppressed at temperatures below 10°C in *Nereis diversicolor*, (Olive, 1981) and 12°C in *Perinereis nuntia* (Hardege *et al.*, 1994).

The basic principles of reproductive physiology necessitates delineation of the environmental factors such as photoperiod (daylength), temperature, rainfall and other biotic and abiotic factors, which promote gametogenesis in several species of fish. Studies on thermoperiod relationship with gonadal weight gain has been given a high priority since gonadal weight gain can be stimulated or inhibited depending on the time of day when the thermocycle is commenced. This chronobiological approach is of considerable importance to aquaculture and may hold the key to producing precociously gravid fish. Timed application of heat by releasing or recirculating thermal waters into fish ponds can help optimize growth and reproduction, since environmental factors favouring gonadal recrudescence are entirely different from those stimulating maturation and ovulation.

The temporal organization of life history events depends on how predictable seasonal changes in an environment are. In environments where good conditions can occur at any time, many organisms are quite flexible in their life histories. Under such conditions, a quantification of the predictability can only be achieved by determining the reaction of organisms to environmental changes (Wikelski *et al.*, 2000). It is common for non-tropical avian species to limit reproductive activity to the time of year when temperatures are relatively mild and the necessary food resources are present (Wingfield, 1983). In order to coordinate gonadal recrudescence and the associated increase in endocrine-mediated behaviour with conditions that are favourable for breeding, birds use specific cues in the environment to time their reproduction. Environmental factors that influence timing of reproduction are typically referred to as ultimate causes of variation in timing (Wingfield and Kenagy, 1991), and food is the common example. In contrast, environmental factors that initiate reproduction by stimulating the neural substrates that lead to reproductive cascades are usually referred to as proximate causes, and here variation in day length is the common example (Wingfield and Kenagy, 1991). This distinction between factors as ultimate or proximate based on whether they influence fitness or act to stimulate the neuroendocrine system has proved quite useful. It has allowed investigators to ask not only why animals breed when they do–because they are more successful than they would have been at other times–but also how the animal 'knows' when it is best to breed, that is, what cues it uses to initiate and maintain reproductive readiness.

For many animals, strategies for optimally timing reproduction involve monitoring not only the physical environment, but also the social context. With a few exceptions, most bat species studied to date display strong seasonality and synchrony in their reproductive cycles such that lactation coincides with maximum food availability (Heideman, 2000). To explore the potential for social factors to modulate reproductive seasonality, the influence of social and environmental cues on birth timing was examined in greater spear-nosed bats *Phyllostomus hastatus* (Porter and Wilkinson, 2001). It was found that the timing of births within caves and in captivity indicated that social cues also affect the timing and synchrony of births within female social groups. Bats brought into captivity and maintained without seasonal cues initially exhibited less birth synchrony than wild groups.

Environment-Dependent Sex Determination

Sex determination is of central importance to the propagation of a species. During this step, differentiation of males and females is initiated in the embryo. There is a remarkable diversity of mechanisms by which sex determination can be accomplished, two of which includes genetic sex determination (GSD) and environmental sex determination (ESD). GSD is dependent on the genetic factors that are located on sex chromosomes, while ESD is dependent on epigenetic factors. ESD can be dependant on a wide range of influences, including visual cues, population cues, hormonal activities, and temperature (Yao and Capel, 2005). Mechanisms of ESD not only allow for the possibility of rapid adaptation of sex ratios within a changing environment, but can also make these species vulnerable to extrinsic changes that might lead to significant changes in the sex ratio.

Environmental sex determination occurs in a species when the sex of offspring is dictated by prevailing environmental conditions Species exhibiting this form of sex determination can be found among rotifers, nematodes, polychaetes, crustaceans, insects, fish, and reptiles (Schröder, 2005). In all cases of ESD, animals interpret cues from the environment that indicate whether male or female progeny would maximize population sustainability (Olmstead and LeBlanc, 2003). Hormonal or metabolic pathways are altered in receptive individuals in response to the environmental stimuli that lead to the production of offsprings of the desired sex. Human activity, including the introduction of xenobiotics into the environment, can disrupt this process. For example, exposure of turtle eggs to some polychlorinated biphenyls (PCB) can skew sex ratios of offspring in favor of females (Turk *et al.*, 1999).

A form of ESD where the temperature influences the differentiation between sexes is called temperature-dependent sex determination (TSD). The differentiation of gonads in species with TSD is sensitive to the incubation temperature of the eggs during a critical period of embryonic development (Figure 25.1). In species with TSD, the environmental condition of temperature experienced during incubation stimulates the determination of one sex over another and the levels and types of hormones that individuals are exposed to can be established (Judith *et al.*, 1999). Janzen (1995) determined the specific temperature ranges of the common snapping turtle (*Chelydra serpentina*) that produced fully male and fully female clutches. Nests incubated at 26°C produced 100 per cent males, at 30°C produced 100 per cent females. However, when incubated at 28°C, they produced equal numbers of females and males. It has been shown that sex reversal can be stimulated by shifting the temperatures from a male-inducing temperature to a female-inducing temperature, only during the middle third stage of embryogenesis (Turk *et al.*, 1999).

Mechanisms whereby discrete phenotypes arise from a single genotype as a result of differing environmental conditions are known

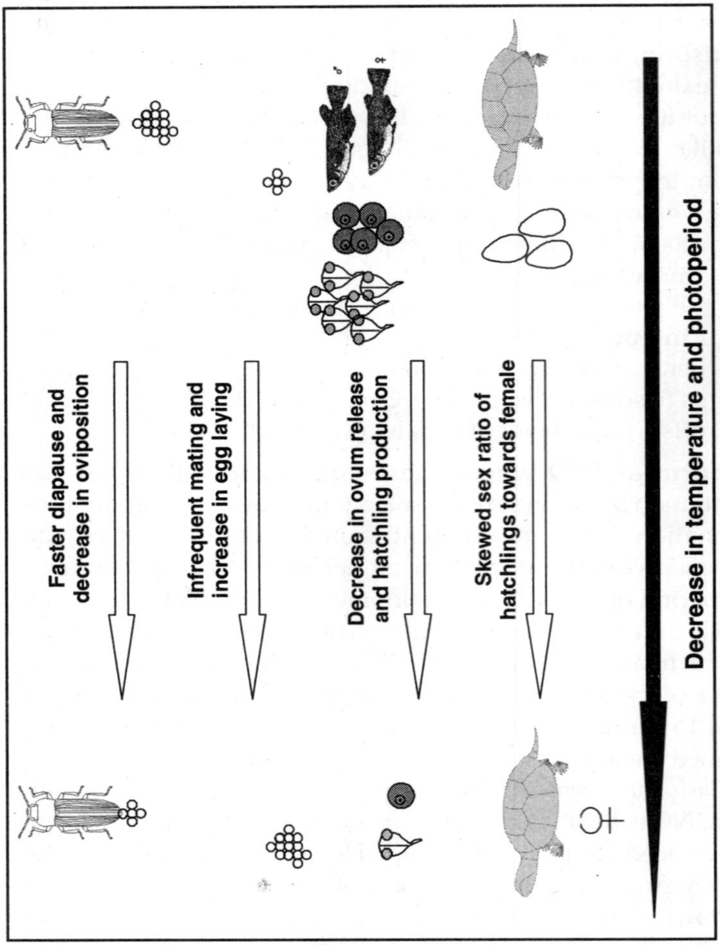

Figure 25.1: Effect of Temperature and Photoperiod on Reproduction of some Invertebrates and Lower Vertebrates

as polyphenism. For example, crocodiles possess a sex-determining polyphenism, and therefore, their gender is a polyphenic trait. When polyphenic forms exist at the same time in the same panmictic (interbreeding) population, they can be compared to genetic polymorphism. With polyphenism, the switch between morphs is environmental; but with genetic polymorphism, the determination of morph is genetic. The discrete nature of polyphenic traits differentiates them from traits like weight and height, which are also dependent on environmental conditions but vary continuously across a spectrum. When a polyphenism is present, an environmental cue causes the organism to develop along a separate pathway, resulting in distinct morphologies. The nature of these environmental conditions includes seasonal cues like temperature and moisture, pheromonal cues, kairomonal cues (signals released from one species that can be recognized by another), and nutritional cues. Sex-determining polyphenisms are beneficial to the species because a large female-to-male ratio maximizes reproductive capacity. However, temperature-dependent sex determination (as seen in crocodiles) makes the species susceptible to endangerment by changes in weather pattern. Interestingly, TSD has been proposed as a plausible explanation for the extinction of the dinosaurs.

The caste system of insects enables eusociality, the division of labour between non-breeding and breeding individuals. A series of polyphenisms determines whether larvae develop into queens, workers, and in some cases soldiers. In the case of the ant, *P. morrisi*, an embryo must develop under certain temperature and photoperiod conditions in order to become a reproductively-active queen, and thus allows for control of the mating season but limits the spread of the species into certain climates. In bees, royal jelly provided by worker bees causes a developing larva to become a queen. Royal jelly is only produced when the queen is aging or has died. This system is less subject to influence by environmental conditions, yet prevents unnecessary production of queens.

The importance of social interaction in the sexual maturation of cichlid fish is also well documented. Francis (1993) showed that sex differentiation in the Midas cichlid is mediated by relative body size during the juvenile stage of development. Removal of large fish from a tank of juveniles dramatically increases the growth rate of the largest remaining fish, indicating social determination of maturation rate. It has been hypothesized that such a system would evolve where there is intense competition for mates and only a few males are able to breed successfully (Fraley, 1982).

In addition to social factors, hormones are crucial for sex determination. Generally, estrogen feminizes embryos while aromatase converts testosterone to estradiol and makes females (Crews, 2003). It has been postulated that progesterone works with estrogen to make females receptive to males, but in males small amounts of progesterone work with testosterone to increase male sexual behavior as well (Crews, 1993). While it is clear that hormones and social/behavioural interactions are important in sex determination it is unclear how the two are related. It may be hypothesized that growth pattern, hormone secretion and behavior are expressions of brain activity, which could be the key to connecting the three (Wingfield *et al.*, 1997; Crews, 2003). Scientists have observed that electrical stimulation of appropriate areas of the brain changes gonadal steroid secretion, and gonadotropin-releasing hormone containing cells in the brain are rendered sensitive to changes in social status (Crews, 1993; White *et al.*, 2002).

Reproduction Under Adverse Environmental Conditions

Organisms cope with harsh environmental conditions either by tolerating environmental stress (through physiological adaptations), or by avoiding it in space (through migration) or time (diapause). Some species rely on a fixed mode of defense, while others may choose from an array of options as they face different environmental stressors. Having a choice, organisms should employ that mechanism which maximizes their reproductive success.

The rotifer *Brachionus plicatilis* is a zooplanktonic invertebrate typically found in water bodies where environmental factors restrict population growth to short periods lasting days or months. The adverse conditions for growth include evaporation of water in temporary habitats leading to desiccation, unfavorable temperatures, scarcity of food and appearance of predators. Survival of the population in such a hostile environment is ensured via the production of resting eggs, which show a remarkable tolerance to unfavorable conditions and may be stored for decades (Kotani *et al.*, 2001; Schröder, 2005). Their high reproductive rates facilitate colonization of vacant niches with extreme rapidity, converting primary production (algal and bacterial) into a form usable for secondary consumers with remarkable efficiency (Nogrady *et al.*, 1993). Consequently, this rotifer has been developed as an essential food source for raising marine fish larvae in marine fish hatcheries (Lubzens and Zmora, 2003). In absence of males, parthenogenesis dominates their life cycle, but following certain environmental cues, sexual reproduction takes place. Females that reproduce asexually are termed "amictic" and females that reproduce sexually are "mictic". Amictic females produce diploid eggs that develop by ameiotic pathenogenesis into females. The factors inducing this signal are largely unknown, although population density and environmental factors such as salinity, presence of pheromones and food availability have been shown to play a role (Hagiwara *et al.*, 2005; Snell *et al.*, 2006). In addition, real-time PCR confirmed that small heat shock proteins and some antioxidant genes were upregulated in resting eggs, therefore, suggesting that desiccation tolerance is a characteristic feature of resting eggs even though they do not necessarily fully desiccate during dormancy (Denekamp *et al.*, 2009).

Bivalves spawn in response to both environmental cues and internal chemical cues. Spawning may be influenced and regulated by water temperature, environmental chemicals (like those released by phytoplanktons), and by chemical pheromones released with sperm and eggs when neighbouring animals spawn. These cues ensure that both sperm and eggs are present so that fertilization can occur, and that temperature is appropriate and food is available for larval development. It was observed that in western Lake Erie, the zebra mussel spawn peaked following a late summer increase in phytoplankton (Figure 25.2), ensuring adequate food for the developing larvae (Ram *et al.*, 1992). However, if the late summer phytoplankton levels were much smaller than usual, the spawning peak never occurred.

Within the last decade research indicates that there are reproductive effects in freshwater gastropods exposed to steroids and steroid-mimicking compounds. Oehlmann *et al.* (2006) reported that exposure to bisphenol-A (an oestrogenic compound in vertebrates) increased the reproductive rate of the tropical prosobranchs when exposures were made at 20°C. Snails exposed

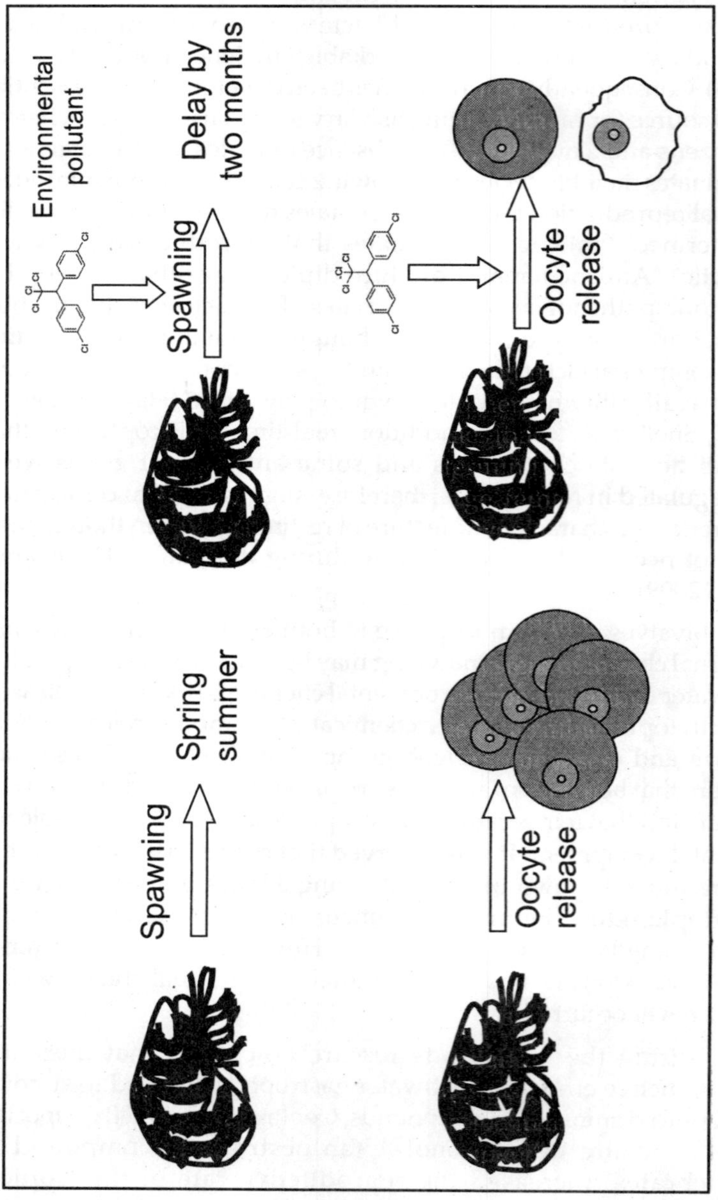

Figure 25.2: Effect of Environmental Pollutants on Reproduction of Zebra Mussel

to bisphenol-A produced significantly more eggs at certain times of the year, most notably at the onset of autumn (Clarke *et al.*, 2009). This suggests that the observed effect is steroid driven, and occurs in snails with widely divergent reproductive strategies and only under certain (autumnal) conditions.

Reproductive physiology can mediate between environmental signals, such as poor-quality food, and adaptive responses. Insects have the ability to resorb oocytes that are not oviposited. This 'oosorption' is proposed to be an adaptive mechanism to optimize fitness in hostile environments, recouping resources that might otherwise be lost and reinvesting them into future reproductive potential. Milkweed bug (*Oncopeltus fasciatus*) females that feed on a poor-quality diet of pumpkin seeds had higher levels of ovarian apoptosis (oosorption), lower reproductive output, but no change in lifespan. Such bugs have adapted to both food-abundance and food-limited environments.

Recent studies have shown that the chemical serotonin (5-hydroxytryptamine, 5-HT), when externally applied to zebra mussels, overrides the natural cues of the mussels and artificially stimulates the reproductive organs of males and females, causing the untimely release of sperm and eggs (Fong *et al.*, 1993). However, the degree of response to serotonin was affected by water temperature. Animals did not respond to serotonin at 4°C, but did respond at 20°C and 27°C; approximately 387 g/l of serotonin were required to stimulate mussels to spawn (Fong *et al.*, 1993). Researchers are investigating other substances which may elicit the same response in mussels at lower concentrations.

Dissolved gases also have profound effects on the reproductive efficiency of organisms living in an aquatic environment. In a series of experiments, it was shown that water environment, saturated with hypoxic oxygen-nitrogen mixture, has low stimulation effect on hydra budding. The water environment, saturated with hypoxic oxygen-argon mixture has high stimulation effect on budding, not only increasing the amount of young buds, but reducing the bud development period in the comparison with control groups. Thus, saturating water environment with gas mixture, containing 85 per cent of argon, significantly increases hydra budding activity at standard temperature and under normal ambient pressure.

The annual cycles of birds, including breeding, are typically coordinated with environmental events, such as the advent of spring in temperate and Arctic and Antarctic regions, and with periodic rainfall or dry periods in tropical regions. Some species of birds, however, breed irregularly or at times of specific food supplies. The timing of annual cycles has been under stringent natural selection to ensure that young are being raised at times of optimal food supplies. While migration and breeding of many birds are controlled by photoperiod, their food supplies are usually temperature-dependent, since the growth of insect larvae, leaves, and other food supplies in the spring is caused by rising air and soil temperatures. Other climate-driven changes can also affect food supplies during the post-breeding season when birds are molting and fattening in preparation for fall migration. In some areas, early growth of plants in the spring can lead to frost damage, causing failure to set seed. Birds reliant on these seeds for molting and fattening may find restriction in food supplies required for appropriate preparation for migration (Bauchinger *et al.*, 2009).

The environment in which a breeding female lives prior to conception and during the early stages of her pregnancy has striking effects on oocytes developing in the ovarian follicle and on early embryos in the reproductive tract. Of the various environmental factors known to affect oocyte and embryo development, altered nutrition during this critical period has been particularly well studied. Alterations in the quantity of food consumed or the composition of the diet imposed solely during the pre-mating period affect oocyte maturity, blastocyst yield, prenatal survival and the number of offspring born alive. Importantly, nutrition at this time also affects the quality of embryos and resultant offspring, with increasing evidence from a variety of species showing that periconception nutrition can alter behaviour, cardiovascular function and reproductive function throughout post-natal life. In livestock species it is important to devise nutritional strategies that improve reproductive efficiency and the quality of offspring but which do not add to the environmental footprint of the production system and which recognise likely changes in food availability arising from predicted changes in climate (Carey, 2009).

Heat stress can have large effects on most aspects of reproductive functions in mammals, which include disruptions in spermatogenesis and oocyte development, oocyte maturation, early embryonic development, fetal and placental growth and lactation. These deleterious effects of heat stress are the result of either the hyperthermia associated with heat stress or the physiological adjustments made by the heat-stressed animal to regulate body temperature. Many effects of elevated temperature on gametes and the early embryo involve increased production of reactive oxygen species (ROS). Artificial selection for production of milk, meat and fiber has increased the susceptibility of some breeds of farm animal species to heat stress. Thus, the impact of global warming on body temperature regulation and reproduction may be more severe for domestic animals than for wild mammalian species. Nonetheless, there is allelic variation in genes controlling body temperature regulation and cellular resistance to heat shock, so that genetic adaptation to increasing global temperature may be possible for many wild and domesticated species (Ashworth *et al.*, 2009).

Effect of Environmental Pollutants on Reproduction

Of the several chemical contaminants that are found in nature 2,3,7,8-tetrachlorodibenzo-p-dioxin (2,3,7,8-TCDD) has been studied extensively with regard to its effects on the reproductive physiology of different animal species. Bivalve molluscs are as sensitive to the effects of 2,3,7,8-TCDD on gonad development, embryonic development, and epithelial lesion occurrence as are higher vertebrates. 2,3,7,8-TCDD alters normal development of reproductive organs and early development in bivalve molluscs and preferentially accumulates into the gonads (Clarke *et al.*, 2009). The sensitivity of gonad maturation is most likely due to disruption of cross-talk between highly conserved steroid, insulin, and metabolic pathways that are involved in gonad differentiation (Cooper and Wintermyer, 2009).

Annelids are also susceptible to adverse environmental conditions. The earthworm *Eisenia fetida* was exposed to different concentrations of cadmium, copper, lead and zinc in artificial soil. Mortality, growth and cocoon production were measured for eight weeks to determine the LC_{50} values. Furthermore, the percentage of viable cocoons and number of juveniles emerging per cocoon was recorded. The results indicated that cocoon production was more sensitive than mortality for all the metals, particularly cadmium and copper (Spurgeon *et al.*, 1994). However, there was no significant effect of metals on the viability of cocoons. The weights of earthworms declined in all treatments during the experiment, probably due to the lack of suitable food in the soil medium used.

The number of amphibians like frogs, toads and salamanders has been dropping in many areas of the world. The causes range from destruction of their local habitats to global depletion of the ozone layer, in addition to pollutants. Amphibians are found in many ecosystems and habitats, including deserts, grasslands and forests, from sea level to high mountaintops and vary in physical size, reproductive capacity and population density (Pounds and Crump, 1994). Initially it was believed that eggs had perished due to pollution or excess acidity of lake waters. However, when such eggs were cultured in laboratory conditions in the same lake water, it was concluded that not pollution but increased exposure to ultraviolet radiation could explain the reproductive problems that were seen. Many amphibian species known to be in jeopardy were mountain dwellers that lay their eggs in open, often shallow, water. Such eggs undergo prolonged exposure to sunlight and thus to any ultraviolet radiation that passes through the ozone shield (Blaustein *et al.*, 1994a). Researchers speculated that excessive exposure to ultraviolet radiation could be contributing to the problems that cause abnormal development of amphibian embryos by causing DNA breaks and consequently disrupt the functioning of cells and may even kill them. In addition, since ultraviolet rays can impair immune function, it seemed plausible that some amount of egg damage in amphibians is caused by an ultraviolet-induced breakdown in the ability of amphibian embryos to resist infection by fungus found in the waters (Blaustein *et al.*, 1994b).

Certain synthetic compounds can mimic the activity of naturally occurring hormones and have drastic consequences, such as a reduction in sperm count and the alteration of male genitalia. Diseases–possibly related to environmental pollution–seem to jeopardize some amphibians as well. Amphibian eggs are susceptibility to fungal infection, probably increased by exposure to excessive ultraviolet radiation. Bacterial contamination is also highly contagious and has been implicated in the death of adult frogs, toads and salamanders (Wake, 1991). Several environmental pollutants have been identified as anti-androgenic endocrine disrupting chemicals (EDC), interfering with sexual differentiation and reproduction in amphibians. In a study with flutamide (FLU), adult male *Xenopus laevis* were injected with human chorionic gonadotropin (hCG) to initiate mate calling behavior. After one day hCG-stimulated frogs were treated via aqueous exposure over three days without and with FLU. Androgen-controlled mate calling behavior was recorded during the 12h dark period. FLU caused a significant decrease in calling activity starting at the second day of exposure, in addition to elevated levels of testosterone, indicating the adverse reproductive effects of the anti-androgenic FLU (Behrends *et al.*, 2010).

p-tert-Octylphenol (OP) is an endocrine disruptor known to bind to the estrogen receptor; however, the consequences of OP on males are controversial. Effects of chronic exposure of OP on rodent male reproduction revealed significant decrease in body weight and total sperm motility in rats exposed to OP (Gregory *et al.*, 2009). Rapid industrialization has increased the burden of chemicals in the environment. These chemicals may be harmful to development and reproduction of any organism. Analysis of the adverse effects of leachates from a tannery solid waste was studied on development and reproduction using *Drosophila* as a model (Siddique *et al.*, 2009). Sub-organismal analyses revealed Hsp70 expression and tissue damage in a sex-specific manner. Refractoriness of Hsp70 expression in accessory glands of male flies and ovaries of females was concurrent with tissue damage. Genes encoding certain seminal proteins from accessory glands were significantly down-regulated at higher concentrations of the leachates, suggesting that waste leachates cause adverse effects on the expression of genes encoding seminal proteins that facilitate normal reproduction by causing cellular damage to reproductive organs.

There is extensive range of evidence for the effects of endocrine disrupting chemicals (EDCs) in wild populations of several animal phyla, including reptiles, amphibians, fish, birds, mammals and a range of invertebrates. Endocrine disrupting chemicals (EDCs) include a wide range of organic and inorganic chemicals derived from many different anthropogenic sources. They are present in several everyday items, including plastics, detergents, electrical equipment, food-can linings, paints, adhesives and pesticides and are also produced as a result of incomplete combustion of hydrocarbon fuels. Unlike more conventional pollutants which tend to be found at high concentrations close to the source, they are largely invisible, highly dispersed and present throughout the environment. Physiological processes in animals of all taxa can be affected by exposure to EDCs, and particularly to mixtures, even when environmental concentrations are very low. Effects of prolonged exposure to multiple EDCs at environmental concentrations have been investigated using sheep exposed to pastures fertilised with sewage sludge, which contains a mixture of anthropogenic pollutants. The consequences of EDC exposure range from developmental abnormalities to sexual dysfunction, often leading to the occurrence intersex phenotypes. For example, disruption of kisspeptin expression in the fetal hypothalamus and pituitary has been demonstrated and may contribute to the observed perturbations of fetal ovary and testis structure, and gene and protein expression (Figure 25.3). Transcriptomic investigations are starting to reveal the complex molecular endocrinology underlying the occurrence of the intersex phenotype and may provide novel insight into the targets of EDCs in arthropods and show whether these molecules are mechanistically related to those identified in vertebrates. The consequences of intersex for individual fecundity have been demonstrated both by direct perturbations in sperm density and motility, and through the reduced proportion of eggs that result in viable offspring. Competitive breeding studies have also demonstrated that severely intersex individuals have a reduced competitive success. The intersex phenotypes can be induced through controlled exposure to acute and chronic water effluents which correlate with the uptake of estrogen mixtures entering the fish. Harnessing genome wide transcript and informatic data has facilitated the characterisation of the global processes underlying gonadal differentiation and of novel EDC targets within gonadal tissues. Exploitation of microsatellites to determine parentage and dominance in the presence of a known EDC and an approach involving brain-area specific transcriptomic analysis may reveal the role of the endocrine system in controlling aggressive behaviour patterns (Hansen *et al.*, 2009).

Certain natural agents can induce changes in the physiology of pregnant mammals, leading to alterations in DNA methylation patterns of the developing fetus and to the emergence of new phenotypes. Nevertheless, in order for this process to occur and to lead to evolutionary changes it would require (i) certain key periods

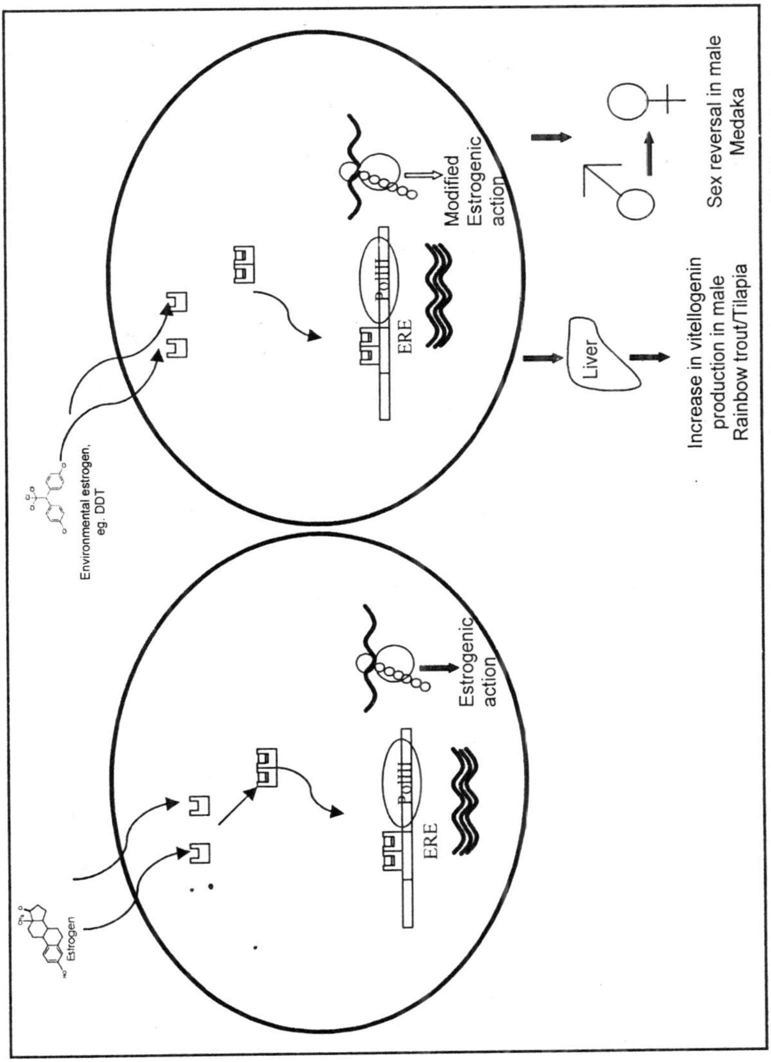

Figure 25.3: Effect of Environmental Estrogens on Reproduction in Aquatic Vertebrates

in the ontogeny of the organism where the environmental stimuli could produce effects, (ii) particular environmental agents to act as such stimuli, and (iii) that a genomic persistent change be consequently produced in a population. These persistent changes can be achieved by exposing animals to EDCs. This class of compounds can alter DNA methylation patterns in mammals in two ways. One is by repeated exposures inducing an altered DNA methylation pattern in each generation (extrinsic process). This may occur, for example, after exposure to nutritional phytoestrogens, which can produce both changes in DNA methylation and in reproductive features. Another way in which EDCs can produce transgenerational changes is by affecting intrinsic features in the DNA that allow for epigenetic transgenerational transmission of characters. For example, exposure of the germ line to the endocrine disruptor vinclozolin during the critical period of sex differentiation produces changes in DNA methylation patterns in several future generations of rats. In addition to transgenerational reproductive alterations, this transient exposure to vinclozolin can also cause transgenerational behavioral changes. The implications for the action of endocrine disrupting chemicals are usually limited to physiological, toxicological or epidemiological studies, but this recent epigenetic transgenerational evidence poses new questions about the role of environment as an inducer of changes that are relevant from an evolutionary perspective (Rhind, 2009).

Arsenic is a potent non-steroidal environmental estrogen and may mimic an estrogenic mechanism to induce lesions in the rat uterus, disrupt the estrogen signaling pathway and consequently lead to reproductive failures. It is known that the estrogen receptor is a hormone-activated transcription factor which mediates the biological effects of estrogen in the target tissue by stimulating the expression of estrogen-regulated genes. The sensitivity of a given tissue to estrogen thus varies with the level of estrogen receptors present in it (Chatterjee and Chatterji, 2010). Studies have revealed that arsenic treatment significantly down regulated the expression of ERa and its downstream element VEGF (vascular endothelial growth factor) in the uterus, indicating that arsenic either suppresses the bonafide action of estradiol on the uterus by decreasing the expression of specific receptors, at both the mRNA transcript and protein levels, or acts via a parallel mechanism in the rat uterus, that eventually disrupts the estrogen signaling pathway and the G1 cell cycle proteins responsible for cell proliferation. Estradiol-regulated VEGF is chiefly responsible for modulating *in vivo* angiogenesis in the uterus, and its down regulation by arsenic may be a primary cause for spontaneous abortions, still-births and other reproductive failures. The D-type cyclins are rate-limiting for the progression through the G1 phase of the cell cycle. In fact, a strong correlation between the expression of increased levels of cyclin D1 mRNA and ER over-expression has been reported (Dickson and Stancel, 2000). Arsenic, on the other hand, decreases the expression of uterine estrogen receptors, and consequently suppressed cell cycle progression and reduced the proliferation-promoting effects of estradiol in the rat uterus. Arsenic exposure also reduces the serum levels of gonadotropins in female rats, which in turn leads to degeneration of the uterine tissue architecture (Chatterjee and Chatterji, 2010). Interestingly, liver toxicity assays did not indicate significant differences in the SGPT or SGOT levels in the experimental animals as compared to the control ones. Hence, it may be emphasized that level of arsenic which failed to reduce the general body weight of the rats or even affect the liver toxicity enzymes, were capable of bringing about such severe detrimental changes in the uterine physiology and steroid signaling pathway. Finally, it has been demonstrated that arsenic at low but chronic doses, relevant to the exposure level in different parts of the world, is a major endocrine disruptor (Chatterjee and Chatterji, 2010) and thus, may be responsible for the different reproductive failures seen in women exposed to such levels of arsenic.

The Intracellular Mediators

Nereidae are major components in the diets of over-wintering birds and commercial fish and are thus vital to the functioning of estuarine ecosystems. They use environmental cues like photoperiodism to synchronize reproduction and initiate vitellogenesis in a temperature compensated process. Although in Nereidae the prevailing paradigm is of a single "juvenile" hormone controlling growth and reproduction, a new multi-hormone model has been reported, which integrates the environmental and endocrine control of reproduction. The juvenile hormone is shown to be heat stable and cross reactive between species. In addition, a second neurohormone, present in mature females, is found to promote oocyte growth. Furthermore, dopamine and melatonin appear to switch off the juvenile hormone while serotonin and oxytocin promote oocyte growth. Genotypic adaptation of the photoperiodic response may be possible, but significant impacts on fecundity, spawning success and recruitment are likely in response to short-term extreme events. EDCs may also have an impact on putative steroid hormone pathways and have significant implications for the functional role of Nereidae (Skinner *et al.*, 2010).

Many environmental pollutants can cause cells in the body to work differently than normal. Reproduction is a target for a number of chemicals because of the synchronization of neuronal and hormonal cues. The mechanisms by which this occurs are largely unknown, but a ligand activated transcription factor known as the aryl hydrocarbon receptor (AhR) plays an important role in some toxic effects of environmental pollutants (Pocar *et al.*, 2005). In addition to pollutants, many natural compounds that are ingested or inhaled are also known to bind to AhR. Although the normal function of AhR in the body is relatively unknown, it has been shown that many pollutants can bind to the receptor and turn on different genes in the body. Interestingly, some pollutants may work through epigenetic mechanisms, where rather than changing the sequence of DNA, they make modifications to DNA sequences, like addition of a methyl group, that result in differential expression of genes (Barnett *et al.*, 2007).

AhR is a ligand-activated transcription factor responsive to both natural and man-made environmental compounds. AhR and its nuclear partner ARNT are expressed in the female reproductive tract in a variety of species and play a pivotal role in the physiology of reproduction. Benedict *et al.* (2003) suggested that AhR deletion impairs follicular growth and, concomitantly, reduces the number of follicles that ovulate and become corpus luteum, thus implicating AhR as a novel regulator of ovulation. In addition, the ovulatory gonadotropin surge has been shown to induce the expression of AhR gene activity in the macaque ovary (Chaffin *et al.*, 1999) and AhR-mediated induction of CYP1A1 in the absence of exogenous ligands, suggesting a direct role of the AhR signaling pathway in the resumption of meiosis in mammalian oocytes (Pocar *et al.*, 2004). However, AhR has been implicated as the mediator of most of the adverse effects on reproduction of potent environmental pollutants, like TCDD, which are collectively referred to as aryl hydrocarbons.

Studies by Cooper and Wintermyer (2009) and Rhodes *et al.* (1997) indicate that TCDD preferentially accumulates into the gonads of *M. arenaria*, not solely by lipid-mediated uptake but TCDD binding to a tissue-specific receptor (Wintermyer *et al.*, 2005). However, studies indicate that it is unlikely that the highly conserved AhR and ARNT evolved solely as a means of interacting with polycyclic aryl hydrocarbons (Denison *et al.*, 2002). Exposure to TCDD in animals or cells with non-binding AhRs results in abnormal reproduction and developmental effects, altered estrogenic responses, activation of cell proliferation pathways (Oikawa *et al.*, 2001), interference with the transcription of the estrogen receptor (ER) or binding, and accelerated steroidal metabolism or conjugation reactions (Barnett *et al.*, 2007). Hence, AhR independent pathways can contribute to TCDD toxicity by cross talk, competition for ARNT with other independent pathways, or possibly direct membrane receptor activation (Wintermyer and Cooper, 2003, 2007). In addition, in molluscs AhR/ARNT activation is though a temperature sensitive mechanism that could change the configuration to allow binding (Cooper and Wintermyer, 2009).

The involvement of AhR in reproductive processes may be attributed to an intricate network of direct and indirect interactions between sex steroid receptors and the AhR–ARNT system which can modify reproductive processes (Pocar *et al.*, 2005). AhR-mediated actions affect all levels of a reproductive system, from the hypothalamo-pituitary-gonadal axis to the reproductive organs themselves, causing adaptive short-term and irreversible long-term effects on the reproductive system. Developmental and transgenic mouse studies have clearly demonstrated that the AhR transcription factor is more than just a xenobiotic sensor but potentially an integral key component of normal reproductive physiology.

Conclusions

Reports of Florida alligators with shrunken penises, of female birds not reproducing normally, and of feminized male birds have been leading to the idea that environmental pollutants were contributing to these reproductive problems. However, most of the studies were being conducted on the possible effects of environmental pollutants on reproduction at the cellular and molecular levels.

It seemed that if pollutants were causing reproductive problems in animals, they were probably interfering with the normal mechanisms of steroid hormone action, since steroids are necessary for the development and function of the male and female reproductive systems. The first step in the mechanism of steroid hormone action is its binding a receptor or binding protein followed by gene activation or suppression. If environmental pollutants were acting through steroid-dependent mechanisms, they must be interacting with the estrogen receptor, androgen receptor or with certain steroid binding proteins like the androgen-binding protein (ABP), which is produced by the testes, or sex hormone-binding globulin (SHBG), which is produced by the liver.

Environmental contamination by agricultural chemicals and industrial waste disposal results in adverse effects on reproduction of exposed birds too (Fry, 1995). The diversity of pollutants results in physiological effects at several levels, including direct effects on breeding adults as well as developmental effects on embryos. The effects on embryos include mortality or reduced hatchability, failure of chicks to thrive (wasting syndrome), and teratological effects producing skeletal abnormalities and impaired differentiation of the reproductive system through mechanisms of hormonal mimicking of estrogens. Estrogenic organochlorines represent an important class of toxicants to birds because differentiation of the avian reproductive system is estrogen-dependent.

Evidence of demasculinizing effects of pollutants in wildlife species led investigators to assume that pollutants were primarily estrogenic, or worked through the estrogen receptor. A lot of scientists think that the concentration of environmental pollutants is not high enough to cause problems except to animals living in contaminated lakes or to animals that eat animals that live in those lakes and that had acquired high concentrations of pollutants in them (Thorpe *et al.*, 2009). However, there have also been cases of reproductive problems in humans that have eaten fish from polluted lakes. There are also some clinical observations indicating that there may be a decrease in sperm counts in human males resulting from exposure to environmental pollutants. While animals or humans might not be exposed to a high enough concentration of a single pollutant in their general surroundings, there are also many different pollutants around that might interfere with reproductive functions. Adult reproductive tracts have been shown to be sensitive to pollutants. However, pollutants seem to be more effective in disrupting the reproductive systems of developing males and females because the systems are much more sensitive to steroid hormones during early development.

Infertility issues now affect 15-20 per cent of couples as opposed to 7-8 per cent fifty years ago. Average sperm counts among adult men have decreased by 50 per cent since 1938, with a decline of 2 per cent every year from 1973 (http://www.yourfuturehealth.com// resources_stations.htm). This decline in reproductive health has been linked to an increased presence in the environment of man-made chemical contaminants in the form of pesticides and other pollutants. Rapid and unplanned industrialization caused large amounts of these synthetic compounds and their by-products to be released in the environment (air, soil, water and food). Studies have shown that occupational exposure to pesticides caused neonatal deaths, congenital defects, testicular dysfunction and infertility. Many of these chemicals found in our environment and households have estrogenic properties and are toxic because they affect the endocrine system. Hence, a thorough evaluation and understanding of the environmental modulators is an absolute necessity in order to design strategies which would enhance reproductive success in both animals and humans.

References

Andries, J-C., 2001. Endocrine and environmental control of reproduction in Polychaeta. *Can. J. Zool.*, 79: 254–270.

Ashworth, C.J., Toma, L.M. and Hunter, M.G., 2009. Nutritional effects on oocyte and embryo development in mammals: implications for reproductive efficiency and environmental sustainability. *Philos. Trans. R. Soc. Lond. B. Biol. Sci.*, 364(1534): 3351-61.

Ball, G.F. and Ketterson, E.D., 2008. Sex differences in the response to environmental cues regulating seasonal reproduction in birds. *Philos. Trans. R. Soc. Lond. B. Biol. Sci.*, 363(1490): 231-46.

Barnett, K.R., Tomic, D., Gupta, R.K., Miller, K.P., Meachum, S., Paulose, T. and Flaws, J.A., 2007. The aryl hydrocarbon receptor affects mouse ovarian follicle growth via mechanisms involving estradiol regulation and responsiveness. *Biol. Reprod.*, 76: 1062–1070.

Bauchinger, U., Van't Hof, T. and Biebach H., 2009. Food availability during migratory stopover affects testis growth and reproductive behaviour in a migratory passerine. *Horm. Behav.*, 55(3): 425-33.

Behrends, T., Urbatzka, R., Krackow, S., Elepfandt, A., Kloas, W., 2010. Mate calling behavior of male South African clawed frogs (*Xenopus laevis*) is suppressed by the antiandrogenic endocrine disrupting compound flutamide. *Gen. Comp. Endocrinol.* (In Press).

Benedict, J.C., Miller, K.P., Lin, T.M., Greenfeld, C., Babus, J.K., Peterson, R.E. and Flaws, J.A., 2003. Aryl hydrocarbon receptor regulates growth, but not atresia, of mouse preantral and antral follicles. *Biol. Reprod.*, 68: 1511–1517.

Blaustein, A.R., Hoffman, P.D., Hokit, D.G., Kiesecker, J.M., Walls, S.C. and Hays, J.B., 1994b. UV repair and resistance to solar UV-B in amphibian eggs: A link to population declines? *Proc. Natl. Acad. Sci.*, 91(5): 1791-1795.

Blaustein, A.R., Wake, D.B. and Sousa, W.P., 1994a. Amphibian declines: Judging stability, persistence and susceptibility of population to local and global extinction. *Conversation Biol.*, 8(1): 60-71.

Carey, C., 2009. The impacts of climate change on the annual cycles of birds. *Philos. Trans. R. Soc. Lond. B. Biol. Sci.*, 364(1534): 3321-30.

Chaffin, C.L., Stouffer, R.L. and Duffy, D.M., 1999. Gonadotropin and steroid regulation of steroid receptor and aryl hydrocarbon receptor messenger ribonucleic acid in macaque granulosa cells during the periovulatory interval. *Endocrinol.*, 140: 4753–4760.

Chatterjee, A. and Chatterji, U., 2010. Inorganic arsenic abrogates the estrogen signaling pathway in the rat uterus. *Reprod. Biol. Endocrinol.*, 8: 80–90.

Clarke, N., Routledge, E., Garner, A., Casey, D., Benstead, R., Walker, D., Watermann, B., Knass, B., Tomsen, A. and Jobling, S., 2009. Exposure to Treated Sewage Effluent Disrupts Reproduction and Development in the Seasonally Breeding Ramshorn Snail (Subclass: Pulmonata, *Planorbarius corneus*). *Environ. Sci. Technol.*, 43: 2092–2098.

Clarke, N., Routledge, E.J., Garner, A., Casey, D., Benstead, R., Walker, D., Watermann, B., Gnass, K., Thomsen, A. and Jobling, S., 2009. Exposure to treated sewage effluent disrupts reproduction and development in the seasonally breeding Ramshorn snail (subclass: Pulmonata, Planorbarius corneus. *Environ. Sci. Technol.*, 43(6): 2092–2098.

Cooper, K.R. and Wintermyer, M.J., 2009. A critical review: 2,3,7,8 - tetrachlorodibenzo-p-dioxin (2,3,7,8-TCDD) effects on gonad development in bivalve mollusks. *Environ Sci Health C Environ Carcinog Ecotoxicol Rev.*, 27(4): 226-45.

Crews, D., 1993. The organizational concepts and vertebrates without sex chromosomes. *Brain Behav. Evol.*, 42: 202-214.

Crews, D., 2003. Sex determination: where environment and genetics meet. *Evol. Dev.*, 5: 50-55.

Dawson, A., King, V.M., Bentley, G. E. *et al.*, 2001. Photo-periodic control of seasonality in birds. *J. Biol. Rhythms*, 16: 365-380.

Denekamp, N.Y. Thorne, M.A.S., Clark, M.S., Kube, M., Reinhardt, R. and Lubzens, E., 2009. Discovering genes associated with dormancy in the monogonont rotifer *Brachionus plicatilis. BMC Genomics*, 10: 108.

Denison, M.S., Pardini, A., Nagy, S.R., Baldwin, E.P. and Bonati, L., 2002. Ligand binding and activation of the Ah receptor. *Chemico-Biological Interac.*, 141: 3–24.

Dickson, R.B. and Stancel, G.M., 2000. Estrogen receptor-mediated processes in normal and cancer cells. *J. Natl. Cancer Inst. Monogr.*, 27: 135–145.

Emerson, S. B. and Hess, D.L., 1996. The role of androgens in opportunistic breeding, tropical frogs. *Gen. Comp. Endocrinol.*, 103: 220–230.

Fong, G., Wall, D. and Ram, J., 1993) Characterization of Serotonin Receptors in the-Regulation of Spawning in the Zebra Mussel Dreissena polymorpha (Pallas). *J. Exp. Zool.*, 267: 475-482.

Fraley, B. (1982. Social control of developmental rate in the African cichlid, *Haplochromis burtoni. Z. Tierpsychol.*, 60: 66-82.

Francis, R., 1993. Social control of primary sex differentiation in the Midas cichlid. *Proc. Natl. Acad. Sci., USA*, 90: 10673-10675.

Fry, D.M., 1995. Reproductive effects in birds exposed to pesticides and industrial chemicals. *Environ Health Perspect.*, 103 Suppl. 7: 165-71

Goldman, B. D., Gwinner, E., Karsch, F. J., Saunders, D., Zucker, I. and Ball, G. F., 2004. Circannual rhythms and photoperiodism. *In* Chronobiology: biological time keeping (eds. J. C. Dunlap, J. J. Loros and P. J. DeCoursey), pp. 107–142. Sunderland, MA: Sinauer Associates, Inc.

Gregory, M., Lacroix, A., Haddad, S., Devine, P., Charbonneau, M., Tardif, R., Krishnan, K., Cooke, G.M., Schrader, T. and Cyr, D.G., 2009. Effects of chronic exposure to octylphenol on the male rat reproductive system. *Toxicol Environ Health A.*, 72(23): 1553-60.

Gwinner, E., 1996. Circadian and circannual programmes in avian migration. *J. Exp. Biol.*, 199: 39–48.

Gwinner, E., 2003. Circannual rhythms in birds. *Curr. Opin. Neurobiol.*, 13: 770–778.

Hagiwara, H., Kadota, Y. and Hino, A., 2005. Maternal effect by stem females in *Brachionus plicatilis*: Effect of starvation on mixis induction in offspring. *Hydrobiologia*, 546: 275–279.

Hahn, T. P., 1998. Reproductive seasonality in an opportunistic breeder, the Red Crossbill (*Loxia curvirostra*). *Ecol.* 79: 2365–2375.

Hansen, P.R., Taxvig, C., Christiansen, S., Axelstad, M., Boberg, J., Kiersgaard, M.K., Nellemann, C. and Hass, U., 2009. Evaluation of endocrine disrupting effects of nitrate after in utero exposure in rats and of nitrate and nitrite in the H295R and T-screen assay. *Toxicol. Sci.*, 108(2): 437-44.

Hardege, J.D., Bartels-Hardege, H.D., Yu, Y., Zhu, M.Y., Wu, B.L., and Zeeck, E., 1994. Environmental control of reproduction of *Perinereis nuntia* var. *brevicirrus*. *J. Mar. Biol. Assoc., U.K.*, 74: 903–918.

Hau, M., Wikelski, M., Gwinner, H. and Gwinner, E., 2004. Timing of reproduction in a Darwin's finch: temporal opportunism under spatial constraints. *OIKOS*, 106: 489-500, 2004

Heideman, P. D., 2000. Environmental regulation of reproduction. *In* The reproductive biology of bats: 469-499. Crichton, E. G. and Krutzsch, P. H. (Eds.) London: Academic Press.

Janzen, F.J., 1995. Experimental Evidence for the Evolutionary Significance of Temperature Dependent Sex Determination. *Evolution.*, 49: 864-873.

Jarne, P. and Charlesworth, D., 1993. The evolution of the selfing rate in functionally hermaphrodite plants and animals. *Annu. Rev. Ecol. Syst.*, 24: 441-466.

Judith, B., Willingham, E., Todd Osborn III, C., Turk Rhen, T. and Crews, D., 1999. Developmental Synergism of Steroidal Estrogens in Sex Determination. *Environ. Health Perspect.*, 107: 93-97.

Kalisz, S., Vogler, D. W. and Hanley, K. M., 2004. Context-dependent autonomous self-fertilization yields reproductive assurance and mixed mating. *Nature*, 430: 884-887.

Kotani, T., Ozaki, M., Matsuoka, K., Snell, T.W. and Hagiwara, A., 2001. Reproductive isolation among geographically and temporally isolated marine *Brachionus* strains. *Hydrobiologia*, 446/447: 283–290.

Lack, D., 1968. Ecological adaptations for breeding in birds. London: Methuen.

Lubzens, E. and Zmora, O., 2003. Production and nutritional value of rotifers. *In* Stottrup JG, McEvoy LA, eds. Live Feeds in Marine. Aquaculture. Oxford: Blackwell Publishing. pp. 300–303.

Murton, R. K., and Westwood, N.J., 1977. Avian breeding cycles. Clarendon Press, Oxford, UK.

Nicholls, T.J., Goldsmith, A.R. and Dawson, A., 1988. Photorefractoriness in birds and comparison with mammals. *Physiol Rev.*, 68(1): 133-76.

Nogrady, T., Walalce, R.L. and Snell, T.W., 1993. Rotifera. *In* Guides to the identification of the micro-invertebrates of the continetal waters and the world. Dumont HJF, editor. The Hague: SPB Academic Publishing; pp. 1–142.

Oehlmann, J., Schulte-Oehlmann, U., Bachmann, J., Oetken, M., Lutz, I., Kloas, W. and Ternes, T., 2006. Bisphenol A Induces Superfeminization in the Ramshorn Snail Marisa cornuarietis (Gastropoda: Prosobranchia) at Environmentally Relevant Concentrations. *Environ. Health Perspect.*, 114: 127–133.

Oikawa, K., Ohbayashi, T., Mimura, J., Iwata, R., Kameta, A., Evine, K., Fujii-Kuriyama, Y., Kuroda, M. and Mukai, K., 2001. Dioxin suppresses the checkpoint protein, MAD2, by an aryl hydrocarbon receptor-independent pathway. *Cancer Res.*, 61: 5707–5709.

Olive, P.J.W., 1981. Environmental control of reproduction in Polychaeta: experimental studies of littoral species in N.E. England. *In* Advances in invertebrate reproduction. W.H. Clark, Jr. and T.S. Adams. (eds) Elsevier/North Holland, Amsterdam. pp. 37–52.

Olmstead, A.W. and LeBlanc, G.A., 2003. Insecticidal Juvenile Hormone Analogs Stimulate the Production of Male Offspring in the Crustacean Daphnia magna. *Environ. Health Perspect.*, 111 (7): 919-924.

Paul, M.J., Galang, J., Schwartz, W.J. and Prendergast, B.J., 2009. Intermediate-duration day lengths unmask reproductive responses to nonphotic environmental cues. *Am J Physiol Regul Integr. Comp. Physiol.*, 296.

Pierre-Yves, H., Vimond, L., Lenormand, T. and Jarne, P., 2006. Is delayed selfing adjusted to chemical cues of density in the freshwater snail *Physa acuta*? *OIKOS*, 112: 448-455.

Pocar, P., Augustin, R. and Fischer, B., 2004. Constitutive expression of CYP1A1 in bovine cumulus oocyte-complexes in vitro: mechanisms and biological implications. *Endocrinol.* 145: 1594–1601.

Pocar, P., Fischer, B., Klonisch, T. and Hombach-Klonisch, S.· 2005. Molecular interactions of the aryl hydrocarbon receptor and its biological and toxicological relevance for reproduction. *Reprod.* 129: 379-389.

Porter, T.A. and Wilkinson, G.S., 2001. Birth synchrony in greater spear-nosed bats (*Phyllostomus hastatus*) *J. Zool., Lond.* 253: 383-390.

Pounds, J.A. and Crump, M.L., 1994. Amphibians declines and climatic disturbance: The case of the golden toad and the harlequin frog. *Conversation Biol.*, 8(1): 72-85.

Ram, J.L., Crawford, G.W., Walker, J.U., Mojares, J.J., Patel, N., Fong, P. and Kyozuka, K., 1993. Spawning in the Zebra Mussel

(*Dreissena polymorpha*): Activation by Internal or External Application of Serotonin. *J. Exp. Zool.*, 265: 587-598.

Ram, J.L., Fong, P., Croll, R.P., Nichols S.J. and Wall, D., 1992. The Zebra Mussel (*Dreissena polymorpha*), A New Pest in North America: Reproductive Mechanisms As Possible Target of Control Strategies. *Invert. Reprod. Dev.*, 22: 1-3.

Rhind, S.M., 2009. Anthropogenic pollutants: a threat to ecosystem sustainability? *Philos. Trans. R. Soc. Lond. B. Biol. Sci.*, 364(1534): 3391-401.

Rhodes, L., Gardner, G. and Van Beneden, R., 1997. Short-term tissue distribution, depuration and possible gene expression effects of (H3)-2,3,7,8-TCDD exposure in soft-shell clams (*M. arenaria*), *Environ. Toxicol. Chem.*, 16: 1888–1894.

Schröder, T., 2005. Diapause in monogonont rotifers. *Hydrobiologia.* 546: 291–306.

Siddique, H.R., Mitra, K., Bajpai, V.K., Ravi Ram, K., Saxena, D.K. and Chowdhuri, DK., 2009. Hazardous effect of tannery solid waste leachates on development and reproduction in *Drosophila melanogaster*: 70kDa heat shock protein as a marker of cellular damage. *Ecotoxicol Environ Saf.* 72(6): 1652-62.

Skinner, M.K., Manikkam, M. and Guerrero-Bosagna, C., 2010. Epigenetic transgenerational actions of environmental factors in disease etiology. *Trends Endocrinol Metab.* [Epub].

Snell, T.W., Kubanek, J., Carter, W., Payne, A.B., Kim, J., Hicks, M.K. and Stelzer, C.P., 2006. A protein signal triggers sexual reproduction in Brachionus plicatilis (rotifera). *Mar. Biol.*, 149: 763–773.

Spurgeon, D.J., Hopkin, S.P. and Jones, D.T., 1994. Effects of Cadmium, Copper, Lead and Zinc on growth, reproduction and survival of the earthworm *Eisenia fetida* (Savigny): Assessing the environmental impact of point-source metal contamination in terrestrial ecosystems. *Environ. Pollu.*, 84: 123-130.

Thorpe, K.L., Maack, G., Benstead, R and Tyler, C.R., 2009. Estrogenic wastewater treatment works effluents reduce egg production in fish. *Environ. Sci. Technol.*, 43(8): 2976-82.

Turk, R., Willingham, E., Sakata, J. and Crews, D., 1999. Incubation Temperature Influences Sex-Steroid Levels in Juvenile Red-Eared Slider Turtles, *Trachemys scripta*, a Species with Temperature-Dependent Sex Determination. *Biol. Reprod.*, 61: 1275-1280.

Wake, D.B., 1991. Declining amphibian populations. *Science*, 253: 860.

Wallen, K. and Schneider, J. E., 1999. Reproduction in context: social and environmental influences on reproduction. MIT Press.

White *et al.*, 2002. Social regulation of gonadotropin-releasing hormone. *J. Exp. Biol.*, 205: 2567-2581.

Wikelski, M., Hau, M. and Wingfield, J.C., 2000. Seasonality of reproduction in a neotropical rain forest bird. *Ecol.*, 81(9): 2458–2472.

Wingfield, J.C., 1983. Environmental and endocrine control of avian reproduction: an ecological approach. In Avian endocrinology (eds. S. I. Mikami, K. Homma and M. Wada), pp. 265–288. Berlin, Germany; Tokyo, Japan: Springer; Japan Science Society Press.

Wingfield, J. C. and Kenagy, G.J., 1991. Natural regulation of reproductive cycles. In Vertebrate endocrinology: fundamentals and biomedical implications (eds P.K.T. Pang and M.P. Schreibman), pp. 181–241. San Diego, CA: Academic Press.

Wingfield, J.C. and Farner, D.S., 1993. Avian endocrinology-field investigations and methods. *Condor.*, 78: 571–573.

Wingfield, J.C., Jacobs, J. and Hillgarth, N., 1997. Ecological constraints and the evolution of hormone-behavior interrelationships. *In* The integrative neurobiology of affiliation (C. S. Carter, Lederhendler II, B. Kirkpatrick, eds). New York: New York Academy of Sciences; 22–41.

Wintermyer, M. and Cooper, K., 2003. Dioxin/furan and PCB concentrations in eastern oyster (*Crassostrea virginica*) tissues and the effects on egg fertilization and development. *J. Shellfish Res.*, 22: 737-746.

Wintermyer, M., Skaidas, A., Roy, A., Yang, Y., Georgapoulos, P., Burger, J. and Cooper, K.R., 2005. The development of a physiologically-based pharmacokinetic (PBPK) model using the distribution of 2,3,7,8-tetrachlorodibenzo-p-dioxin (2,3,7,8-TCDD) in the tissues of the eastern oyster (*Crassostrea virginica*), Mar. *Environ. Res.*, 60: 133–152.

Wintermyer, M.L. and Cooper, K.R., 2007. The development of an aquatic bivalve model: Evaluating the toxic effects on gametogenesis following 2,3,7,8-tetrachlorodibenzo-p-dioxin (2,3,7,8-TCDD) exposure in the eastern oyster (*Crassostria virginica*). *Aquatic Toxicol.*, 81: 10–26.

Yao, H. H-C. and Capel, B., 2005. Temperature, Geners, and Sex: a Comparative View of Sex Determination in *Trachemys scripta* and *Mus musculus*. *J. Biochem.*, 138: 5-12.

Chapter 26

Assessment of Reproductive Toxicity of Orally Administered Technical Phosphamidon in Male Albino Mice

T.S. Shreelakshmi and B.B. Kaliwal*

Post-Graduate Department of Studies in Biotechnology and Microbiology, Karnatak University, Dharwad – 580 003, Karnataka, India

ABSTRACT

The effect of organophosphate insecticide phosphamidon at four graded dosage levels (1.3, 2.6, 3.9 and 5.2 mg/kg/day) for 30 consecutive days and an effective dose 5.2 mg/kg/day for 5, 10 and 20 days duration treatments on male reproduction in albino mice was studied. The phosphamidon was given orally by gavage below their acute LD_{50} level of intoxication to male mice and the animals were sacrificed on 31st day or 24 hrs after the terminal exposure. The blood was collected from cardiac puncture and serum was separated and stored at -20°C for determination of hormone levels. The reproductive and accessory reproductive organs were removed, weighed and processed for biochemical and histopathological studies. The results of the present study showed significant decrease in the body weight, testis and accessory reproductive organs weight

* Corresponding author: E-mail: b_kaliwal@yahoo.com

($P < 0.05$) in 3.9 and 5.2 mg/kg/day dosage and at 20 days duration with 5.2 mg phosphamidon treatment groups. Phosphamidon with higher dosage and prolong duration of treatment significantly decreased the serum testosterone levels, the biochemical parameters such as DNA, RNA, total proteins, glycogen and the activities of SDH, 3βHSD, 17βHSD, ACP, Na^+-K^+, Mg^{++}, Ca^{++}ATPases and ACP enzymes whereas cholesterol, LDH and alkaline phosphatase enzymes activity were increased significantly ($P < 0.05$) both in testis and epididymis. However there was no significant change in the organs weight, biochemical parameters and testosterone levels found at lower dosage groups and shorter duration treatment with phosphamidon when compared with those of corresponding parameters of control mice. The histometric and histological changes in mice revealed that the number and diameter of spermatogenic (spermatogonia, primary spermatocytes, secondary spermatocytes and spermatids) and Leydig cells were decreased significantly ($P < 0.05$) in the mice treated with 3.9 and 5.2 mg dosage and 10 and 20days duration exposure to phosphamidon. The results of the present study suggest that a sub-acute oral exposure to phosphamidon affects the testicular structure and functions of reproductive and accessory reproductive organs and also the biochemical parameters in the testis and epididymis.

Keywords: *Phosphamidon, Spermatogenesis, Enzymes, Testosterone, Histology, Mice.*

Introduction

For centuries, pesticides have been used in agriculture to enhance food production by eradicating unwanted insects and controlling disease vectors. The widespread use of pesticides in public health and agricultural program has been severe environmental pollution and potential health hazards, including severe acute and chronic cases of human poisonings (Abdollahi *et al.,* 2004). The complex process of reproduction in mammals is influenced by various physical, chemical and nutritional factors (Ellis *et al.,* 1970; Mann *et al.,* 1982). Reproductive tissues are vulnerable to attack by toxic chemicals. The exposure of environmental chemicals and drugs varying degrees of infertility may result (Meistrich *et al.,* 1986; Georgellis *et al.,* 1989). Each of the spermatogenic line cells may be the target for the action of a toxic agent (Foster *et al.*, 1986). It has been reported that the germinal epithelium of testis of rodents is sensitive to a wide variety of internal and external factors (Bockelhide *et al.*, 1989). Disorder of development and function of the male reproductive tract have increased in incidence over the past 30-50 years (Giwercman *et al.*, 1992). Atrophy of the reproductive structures associated with spermatogenic arrest has been reported in animals as a result of exposure to pesticides and other chemicals (Waulkes *et al.*, 2003; Sarkar *et al.*, 2003; Pant *et al.*, 2004).

Studies on organophosphorus (OP) compounds, altered male reproductive function, particularly semen quality and hormone balance in men and animals have revealed that (Pandungtod *et al.*, 2000; Sanchez-Pena *et al.*, 2004; Pina-Guzman *et al.*, 2005; Narayan *et al.*, 2005) sub lethal doses of these pesticides lead to alterations in reproductive performance in birds and mammals (Maitra and sarkar, 1996 a, b). Mathew (1992) has reported that methyl parathion acts as a germ cell mutagen in inducing sperm shape abnormalities in mice. Few studies have reported that organophosphorus pesticides induce oxidative stress in humans (Ranjbar *et al.*, 2002) and animals (Sharma *et al.*, 2005). Meeker *et al.*(2004 *a*, *b*) have found environmental exposure to OPs can be associated with increased sperm DNA fragmentation. Debanth and Mandal (2000) have observed that testicular damage caused by quinolphos was due to free-radical mediated LPO. Pina-Guzman *et al.*(2006) have reported that genotoxic potential of methyl parathion to male germinal cells at two critical stages of spermatogenesis (meiosis and epididymal maturation) and that oxidative stress caused in sperm by acute methyl-parathion exposure may be associated with DNA damage as a possible mechanism of action. Dichlorovos affects the testis leading male reproductive toxicities like degeneration of spermatogenic cells, sperm shape abnormalities and decrease in the sperm count (Krause and Homola, 1974). Toxicants bring about pathological changes by killing cells through necrosis or apoptosis. The mechanisms differ intensively in terms of the mechanism of action as well as the manifestation of the microscopically discernible organization and the molecular biology of the affected cell (Averal *et al.*, 1996). Phosphamidon (Dimecron) is one such insecticide widely used in agricultural practices among Indian formers. There are few reports on the specific toxic effects of phosphamidon on male reproduction in experimental animals. Therefore, the present study was aimed to investigate the effects of phosphamidon with emphasis on the biochemical changes, enzymatic and hormonal milieu, histologic and histometric evaluation of the testis and epididymis in albino mice.

Materials and Methods

Pesticide

Technical grade (purity 94.25 per cent) Phosphamidon was obtained from Cheminova India Ltd., Gujarat (India). The graded doses of phosphamidon were prepared in distilled water vehicle to obtain desired concentrations as 1.3, 2.6, 3.9 and 5.2 mg/kg/day dose of $1/10^{th}$ of LD_{50} (Jacques and Bein, 1960). The doses were given orally below their acute LD_{50} level of intoxication according to their body weight for 30 consecutive days and an effective dose 5.2 mg for 5, 10 and 20 days for durational study to respective groups.

Animals

Laboratory bred 90 days old healthy fertile male mice weighing about 25 to 30 g were used under standard animal housing condition (temperature controlled 22 ± 3°C facility and maintained with 12 hrs light/dark cycle) with unlimited access to pellet diet "Gold Mohar" (Hindustan Lever Ltd., Mumbai) and water *ad libitum* throughout the study in the animal house, P.G. Department of Studies in Zoology, Karnatak University, Dharwad. Animals were randomly divided into control and seven treatment groups (Distilled water vehicle was served as control). Each group consists of 10 mice and were housed in separate polypropylene cages containing sterile paddy husk as bedding material. Body weights were recorded daily throughout the experiment.

All the experimental animals were autopsied on 31st day or 24 hrs after terminal exposure by cervical dislocation. The blood was collected from cardiac puncture using sterile syringe and serum was separated and stored at -20°C for hormone assays. The testes and accessory reproductive organs were dissected out weighed and were used for histologic and biochemical estimations such as DNA, RNA, protein, glycogen, cholesterol and activity of SDH, LDH, 3β-HSD, 17β-HSD, Na^{+}-K^{+}ATPase, Ca^{++}ATPase, Mg^{++}ATPase, ACP, AKP enzymes in testis and epididymis.

Biochemical Studies

Freshly removed testis and epididymis were weighed to the nearest milligrams and used for the biochemical analysis such as nucleic acids (DNA and RNA) following the methods of Schneider (1957) and Plummer (2001), protein estimation was performed as per the method described by Lowry *et al.*(1951), glycogen by Sciffer *et al.* (1950) and Carrol *et al.*(1956), and cholesterol by Abell *et al.* (1952) and activity of enzymes such as succinic dehydrogenase (SDH) and lactate dehydrogenase (LDH) as described by Nachlas *et al.*(1960) and King (1965) respectively, Acid phosphatase (ACP) and Alkaline phosphatase (AKP) as described by Bergermeyer and Bernt (1963) 3β-Hydroxysteroid dehydrogenase (3βHSD) and 17-Hydroxysteroid dehydrogenase was measured biochemically according to the method of Shivanandappa and Venkatesh (1997) and Jarabak *et al.*(1962) respectively and sodium, potassium, calcium and magnesium dependent ATPases were assayed as described by Jinna *et al.*(1989). Serum testosterone concentrations were measured using testosterone kit from Monobind Ltd., Lakeforest, USA, following the immunoenzymatic method and according to the standard protocol given by National Institute of Health and Family Welfare (NIHFW, New Delhi) Srivastava (2002). The cross-reaction of the testosterone antibody to dehydrotestosterone was 10.0 per cent and intra-assay run precision had a co-efficient of variation of 6.2 per cent. There was no inter-assay variation as all samples were assayed at a time. Results of testosterone are expressed as ng/ml serum.

Histology and Histometry

The testis were fixed in aqueous bouin's fixative and processed for paraffin wax sectioning (5 µm) and stained with heamatoxylin-Eosin (Humason, 1979). Randomly chosen 10 good sections from each testis in each group were observed under the microscope. The seminiferous tubules were examined for counting the different spermatogenic cells and leydig cells lying around them. The diameter of spermatogenic cells and Leydig cells were determined after 1000 observations of particular cell types/testis from each animal of control and treated groups. Spermatogonia, primary spermatocytes, secondary spermatocytes, spermatids, and Leydig cells were identified based on the findings of earlier investigators (Dym and Fawcett, 1971; Fawcett *et al.*, 1973) as reviewed by De Krester and

Kerr (1994). The values were expressed as number and diameter of spermatogenic and Leydig cells per seminiferous tubules.

Statistical Analysis

Statistically significant difference between treatments and controls were determined by one-way ANOVA together with Dunnett's test ($P<0.05$). Values are expressed as mean ±SE.

Results

Body Weight, Testis and Accessory Reproductive Organs Weight

The findings of the present study on testes and accessory reproductive organs weight revealed that treatment with 3.9 and 5.2 mg dosage and 20 days duration of treatment with 5.2 mg phosphamidon caused significant decrease in the weight of testes and accessory reproductive organs (epididymides, vasa deferentia, seminal vesicles, prostate gland, coagulatory glands and Cowper's glands). However, treatment with 1.3 and 2.6 mg dosage and 5 and 10 days treatment duration with 5.2 mg phosphamidon caused no significant change in the weight of testes and accessory reproductive organs, except the epididymis weight where it was a significant decreased with 2.6 mg phosphamidon treatment when compared with those of control mice (Table 26.1).

Histometric and Histological Changes in the Testis

The findings of the present study on the number and diameter of spermatogenic cells (spermatogonia, primary spermatocytes, secondary spermatocytes, spermatids and leydig cells revealed that there was a significant decrease in the number and diameter of spermatogenic and Leydig cells with 3.9 and 5.2 mg dosage groups and 20 days duration exposure with 5.2 mg phosphamidon treatment. However, treatment with 1.3 and 2.6 mg dosage and 5 and 10 days duration exposure with 5.2 mg phosphamidon showed no significant change in the number and diameter of spermatogenic and Leydig cells when compared with those of the corresponding parameters of the control mice (Table 26.2).

The testis of the control mouse shows normal spermatogenesis which includes spermatogonia, primary spermatocytes, secondary spermatocytes and spermatids. Lumen of seminiferous tubules

Table 26.1: Dose and Durational Effect of Phosphamidon on Testes and Accessory Reproductive Organs Weight in Mice

Group	*Treatment*		*Relative Organs Weight (mg/100 g body weight)*						
	Dose (mg)	*Duration (days)*	*Testes (mg)*	*Epididymis (mg)*	*Vasa Deferentia (mg)*	*Seminal Vesicles (mg)*	*Prostate Gland (mg)*	*Coagulatory Gland (mg)*	*Cowper's Gland (mg)*
I	Control	30	836.8±1.00	395.0±2.00	202.2±1.02	856.0±2.01	92.57±1.02	169.09±0.65	159.30±1.65
II	1.3	30	813.6±1.53	351.4±2.02	192.0±2.02	812.0±1.09	84.25±1.09	158.90±0.91	141.90±1.44
III	2.6	30	785.4±1.25	300.3± 3.07*	184.2±2.00	764.9±2.02	70.95±1.04	140.05±1.05	133.35±0.94
IV	3.9	30	705.4±1.32*	286.0±2.90*	161.4±1.53*	702.1±2.42*	62.35±0.95*	132.95±1.00*	122.95±0.89*
V	5.2	30	618.4±2.91*	245.4±0.37*	152.0±1.04*	654.9±2.07*	50.90±1.01*	122.90±0.78*	119.29±1.03*
VI	5.2	20	743.0±1.33*	264.0±2.37*	160.4±1.58*	680.5±2.00*	64.32±1.14*	136.92±0.59*	126.91±1.64*
VII	5.2	10	794.0±1.22	302.0±2.41*	180.2±2.04	752.9±2.01	78.92±1.05*	147.12±0.66	138.58±1.89
VIII	5.2	5	804.2±1.06	364.0±2.21	196.0±2.11	821.0±2.24	88.21±1.01	159.90±0.77	145.18±1.54

Values are mean±SEM of 10 animals.

* Significant $P \leq 0.05$ compared to control.

Table 26.2: Dose and Durational Effect of Phosphamidon on Number and Diameter of Spermatogenic and Leydig Cells in Mice

Group	Treatment		Number of Spermatogenic and Leydig Cells				
	Dose (mg)	Duration (days)	Spermatogonia	Primary Spermatocytes	Secondary Spermatocytes	Spermatids	Leydig Cells
I	Control	30	78.13±3.34	90.13±3.12	98.87±4.56	161.53±2.41	57.61±2.31
II	1.3	30	73.41±3.10	84.41±3.14	90.[illegible]3±4.45	156.74±3.21	51.96±2.28
III	2.6	30	65.84±3.34	77.84±2.41	81.58±4.00*	152.35±3.31	45.94±2.83
IV	3.9	30	57.33±2.89*	70.59±2.81*	73.46±3.13*	148.56±2.78*	38.51±2.19*
V	5.2	30	49.53±3.57*	61.13±3.47*	61.43±3.16*	140.31±3.15*	30.18±3.05*
VI	5.2	20	59.23±2.12*	70.61±2.53*	72.59±3.11*	145.83±3.08*	38.98±3.15*
VII	5.2	10	67.54±2.67	77.58 ±3.00	86.81±2.81*	150.61±2.83	47.89±3.05*
VIII	5.2	5	75.56±3.01	86.13±2.13	92.57±3.18	157.78±3.15	53.61±2.15

Contd...

Table 26.2–Contd...

Group	*Treatment*		*Diameter (μm) Spermatogenic and Leydig Cells*				
	Dose (mg)	*Duration (days)*	*Spermatogonia*	*Primary Spermatocytes*	*Secondary Spermatocytes*	*Spermatids*	*Leydig Cells*
I	Control	30	8.59±0.09	8.98±0.34	7.58±0.28	5.91±0.36	8.94±0.38
II	1.3	30	7.86±0.11	8.15±0.28	7.05±0.34	5.21±0.34	8.68±0.28
III	2.6	30	6.80±0.31	7.54±0.18	6.55±0.22	4.45±0.28	7.94±0.31
IV	3.9	30	5.90±0.21*	6.75±0.23*	5.95±0.31*	3.61±0.21*	7.13±0.27*
V	5.2	30	4.70±0.26*	5.83±0.31*	5.08±0.35*	2.88±0.24*	6.87±0.34*
VI	5.2	20	5.81±0.25*	6.65±0.28*	5.83±0.31*	3.49±0.28*	7.05±0.26*
VII	5.2	10	6.75±0.10	7.41±0.31	6.52±0.29	4.43±0.31	7.81±0.34*
VIII	5.2	5	7.91±0.05	8.35±0.25	7.13±0.21	5.32±0.26	8.57±0.41

Values are mean±SEM of 5 animals.

* Significant $P \leq 0.05$ compared to control.

filled with sperms. Interstitial tissue contains clusters of Leydig cells (Figure 26.1). Phosphamidon affected the testis of the mouse treated with 3.9 and 5.2 mg dosage and 20 days duration treatment with marked reproduction in spermatogenic cells, formation of giant cells and vacuoles. Lumen with tissue debris and clumped sperms and leydig cells with deformed condition (Figures 26.4, 26.5 and 26.8). However, treatment with 1.3 dose and 5 days duration with 5.2 mg treatment showed normal spermatogenesis and includes normal spermatogenic cells and sperms in the lumen of seminiferous tubules and interstitial cells contains clusters of Leydig cells (Figures 26.2 and 26.6). Whereas treatment with 2.6 dose and 10 days duration exposure with 5.2 mg phosphamidon exhibited decrease in the number of spermatogenic cells, formation of giant cells, vacuoles and less number of sperms in the lumen of the seminiferous tubules and Leydig cells are in deformed condition (Figures 26.3 and 26.7). The tubular atrophy was more in higher dose and longer duration treatment groups (Figures 26.4, 26.5 and 26.8).

Biochemical Studies

Study on biochemical contents of the testis and epididymis revealed that, mice treated with 3.9 and 5.2 mg dosage and 20 days duration treatment with 5.2 mg phosphamidon caused significant decrease in the level of DNA, RNA, total proteins, glycogen, whereas cholesterol content increased significantly. Whereas, the levels of DNA and RNA were not changed significantly in 3.9 mg dose and 20 days duration treatments groups. However, with 1.3 and 2.6 mg dosage and 5 and 10 days duration treatment with 5.2 mg phosphamidon showed no significant change in the level of the biochemical contents of the testis and epididymis, except with 2.6 mg dose and 10 days duration treatment with 5.2 mg phosphamidon caused significant decrease in the glycogen content and with significant increase in the cholesterol content when compared with those of control mice (Tables 26.4 and 26.5).

Enzymes Activity

Phosphamidon decreased the activity of SDH, 3βHSD, 17βHSD, Na^+-K^+ATPase, Mg^{++}ATPase, Ca^{++}ATPase and ACP and increased the activity of LDH and AKP in the mice treated with 3.9 and 5.2 mg dosage and 20 days duration treatment groups. Whereas the mice treated with 1.3 and 2.6 mg dosage and 5 and 10 days duration treatments with phosphamidon caused no significant change in the

Table 26.3: Dose and Durational Effect of Phosphamidon on Biochemical Contents of Testis in Mice

Group	*Treatment*		*DNA*	*RNA*	*Protein*	*Glycogen*	*Cholesterol*
	Dose (mg)	*Duration (days)*					
I	Control	30	2.96±0.06	4.97±0.01	167.01±1.28	6.65±0.40	8.91±0.33
II	1.3	30	2.58±0.08	4.56±0.05	158.09±2.01	6.04±0.46	9.53±0.37
III	2.6	30	2.04±0.06	4.28±0.04	150.00±2.21*	5.41±0.5*	9.96±0.37*
IV	3.9	30	1.90±0.05	3.93±0.08	144.21±2.28*	4.93±0.41*	10.65±0.39*
V	5.2	30	1.79±0.05*	3.23±0.04*	128.95±2.23*	3.85±0.32*	11.90±0.51*
VI	5.2	20	2.00±0.06	3.64±0.10	137.52±1.41*	4.82±0.51*	10.84±0.40*
VII	5.2	10	2.51±0.01	4.16±0.04	149.09±1.31	5.42±0.45*	10.00±0.32*
VIII	5.2	5	2.80±0.04	4.84±0.09	160.04±1.45	6.32±0.37	9.45±0.36

Values are mean±SEM of 5 animals.

* Significant $P \leq 0.05$ compared to control.

Effect on Testes of the Albino Mice After Exposure to Phosphamidon

Photographs original exposure at 200 X. SG: Spermatogonia; PS: Primary spermatocytes; SS: Secondary spermatocytes; SP: Spermatids; SM: Sperms; ST: Seminiferous tubule; IN: Interstitial tissue; A: Autropic tubule

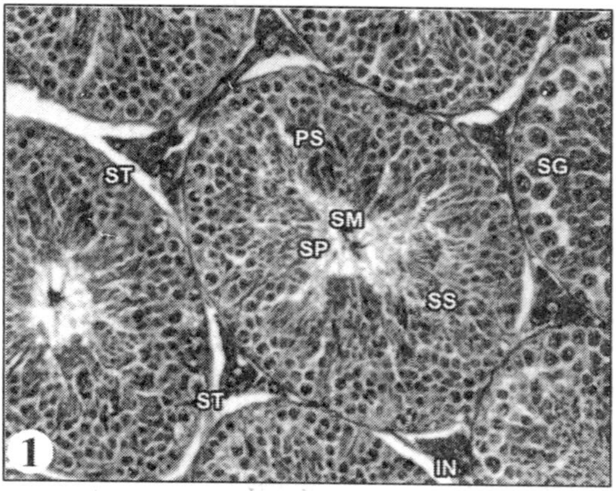

Figure 26.1: T.S. of the Testis of the Control Mouse Showing Stages of Spermatogenesis Includes Spermatogonia, Primary Spermatocytes, Secondary Spermatocytes and Spermatids. Lumen of seminiferous tubules filled with sperms. Interstitial tissue contains clusters of Leydig cells.

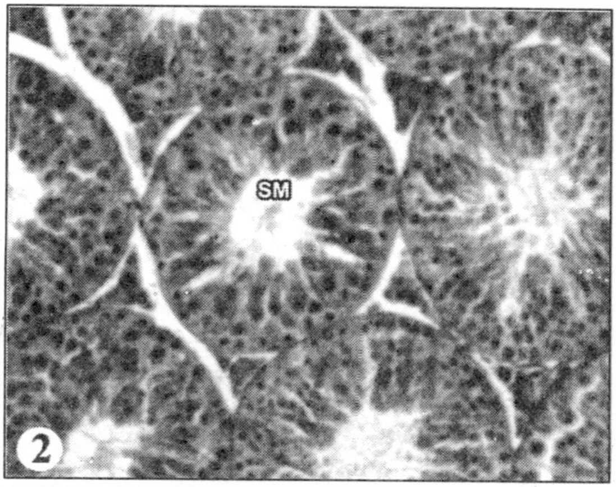

Figure 26.2: T.S. of the Testis of the Mouse Treated with 1.3 mg/kg b.wt./day Phosphamidon for 30 Days Showing Decreased Number of Spermatocytes, Spermatids and Sperms in Lumen. Interstitial tissue contains clusters of Leydig cells.

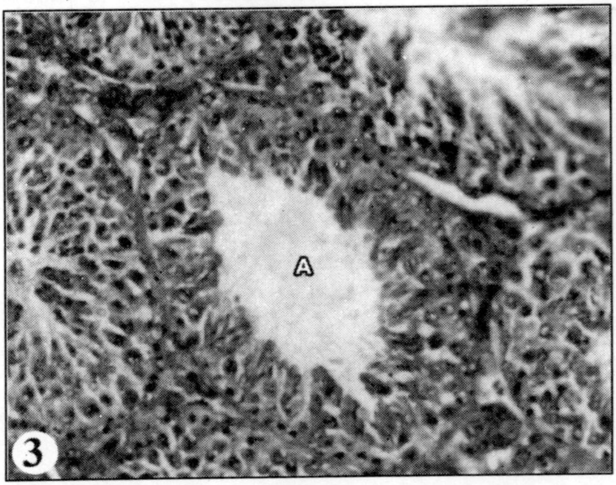

Figure 26.3: T.S. of the Testis of the Mouse Treated with 2.6 mg/kg b.wt./day Phosphamidon for 30 Days Showing Decreased Number of Spermatogenic Cells, Formation of Giant Cells, Vacuoles and Less Number of Sperms Showing Autropic Tubules in the Seminiferous Tubules. Leydig cells are in deformed conditions.

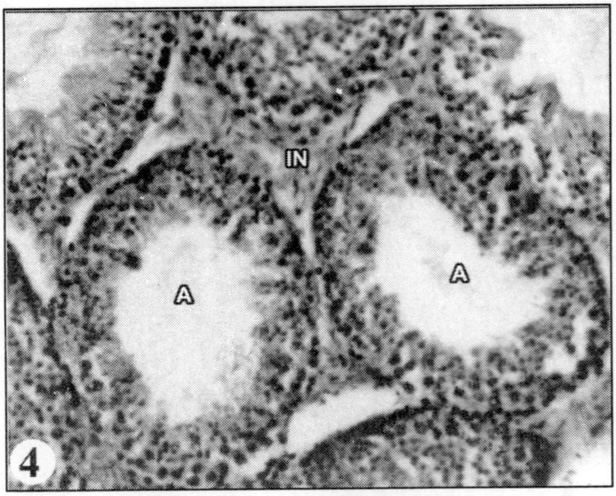

Figure 26.4: T.S. of the Testis of the Mouse Treated with 3.9 mg/kg b.wt./day Phosphamidon for 30 Days Showing Formation of GiantCells, Decreased Number of Spermatogenic Cells and Autropic Tubules with Loss of Sperms. Leydig cells are in deformed condition.

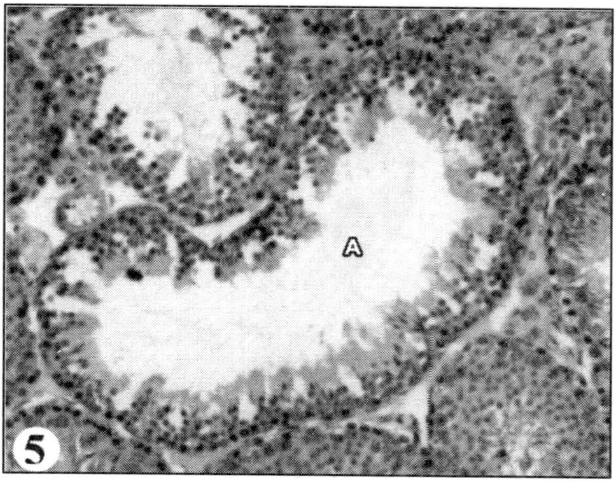

Figure 26.5: T.S. of the Testis of the Mouse Treated with 5.2 mg/kg b.wt/day Phosphamidon for 30 Days Showing Gormation of Giant Cells, Vacuoles and Marked Teduction in Spermatogenic Cells. Autropic tubules with tissue debris and clumped sperms. Leydig cells are in deformed condition.

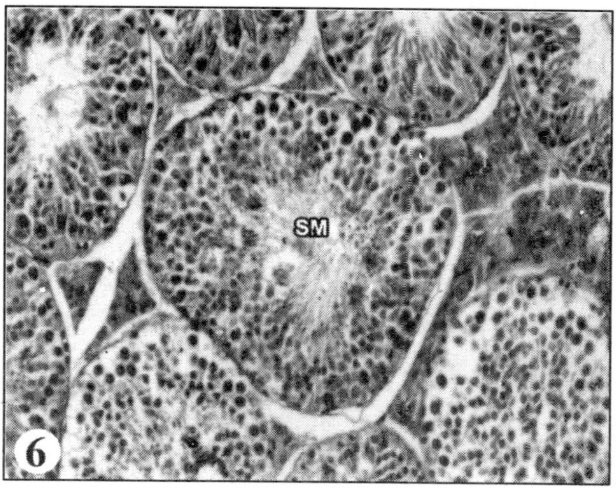

Figure 26.6: T.S. of the Testis of the Mouse Treated with 5.2 mg/kg b.wt./day Phosphamidon for 5 Days Showing Normal Spermatogenesis. The seminiferous tubules are closely packed. The tubular space is packed with interstitial tissue, containing clusters of Leydig cells.

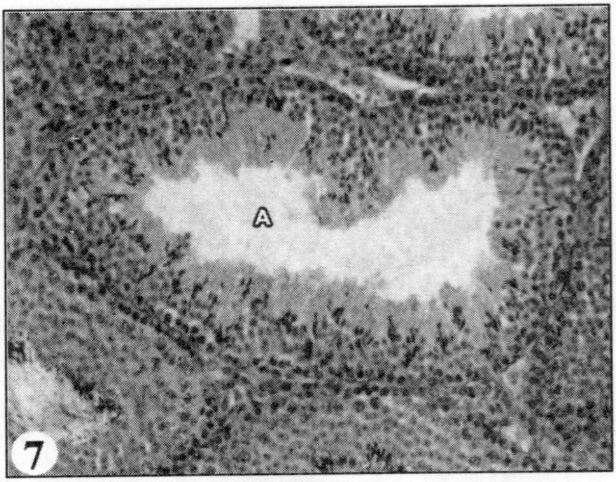

Figure 26.7: T.S. of the Testis of the Mouse Treated with 5.2 mg/kg b.wt./day Phosphamidon for 10 Days Showing Seminiferous Tubules with Different Stages of Spermatogenesis. Autropic tubules with loss of sperms.

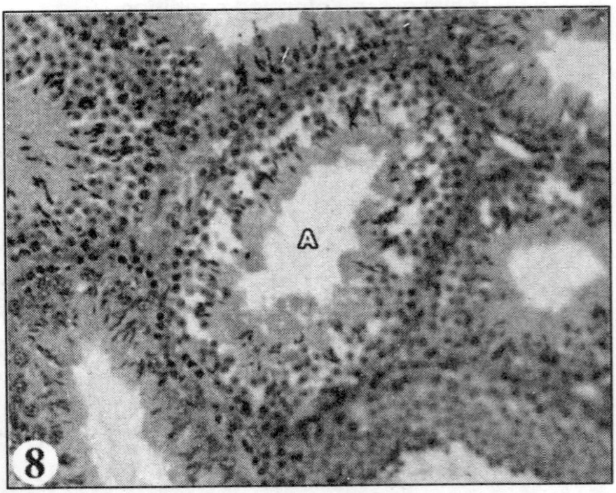

Figure 26.8: T.S. of the Testis of the Mouse Treated with 5.2 mg/kg b.wt./day Phosphamidon for 20 Days Showing Formation of Giant Cells, Decreased Number of Spermatogenic Cells and Autropic Tubules with Loss of Sperms. Leydig cells are in deformed condition.

Table 26.4: Dose and Durational Effect of Phosphamidon on Biochemical Contents of Epididymis in Mice

Group	Treatment		DNA	RNA	Protein	Glycogen	Cholesterol
	Dose (mg)	Duration (days)					
I	Control	30	3.09±0.14	5.84±0.38	215.29±2.01	7.40±0.10	7.98±0.76
II	1.3	30	2.87±0.16	5.21±0.52	202.59±2.28	7.21±0.16	8.66±0.53
III	2.6	30	2.40±0.14	4.81±0.31	197.49±2.21	7.02±0.12*	9.08±0.50*
IV	3.9	30	2.21±0.18	4.40±0.50	181.69±2.06*	6.83±0.09*	9.72±0.39*
V	5.2	30	1.84±0.14*	4.17±0.37*	169.23±2.04*	6.51±0.19*	10.54±0.64*
VI	5.2	20	2.12±0.11	4.33±0.41*	176.43±2.09*	6.64±0.16*	9.88±0.54*
VII	5.2	10	2.34±0.18	4.92±0.42	190.56±2.12	7.00±0.18	9.15±0.65*
VIII	5.2	5	2.93±0.14	5.44±0.43	209.41±2.08	7.35±0.13	8.38±0.81

Values are mean±SEM of 5 animals.

* Significant $P \leq 0.05$ compared to control.

Table 26.5: Dose and Durational Effect of Phosphamidon on Testis Dehydrogenase and Phosphatase Enzymes Activity in Mice

Group	Treatment		Enzyme Activity mmoles/min/g tissue								
	Dose (mg)	*Duration (days)*	*LDH*[a]	*SDH*[b]	*3βHSD*[c]	*17βHSD*[d]	*Na⁺-K⁺ ATPase*[e]	*Mg⁺⁺ ATPase*[e]	*Ca⁺⁺ ATPase*[e]	*ACP*[f]	*AKP*[f]
I	Control	30	9.38±0.19	12.74±1.30	0.52±0.05	0.96±0.08	8.16±0.53	9.14±0.56	6.83±0.23	18.69±0.63	14.98±0.23
II	1.3	30	9.45±0.17	12.11±1.28	0.47±0.08	0.91±0.05	7.68±0.41	8.73±0.51	6.22±0.22	17.87±0.58	15.17±0.20
III	2.6	30	9.73±0.21	11.32±0.89	0.41±0.03	0.85±0.06*	6.83±0.43	7.56±0.43	5.64±0.19	16.76±0.53	16.08±0.16
IV	3.9	30	10.08±0.17*	10.31±0.74*	0.34±0.05*	0.78±0.08*	5.65±0.51*	6.54±0.37*	4.89±0.23*	15.63±0.58*	16.93±0.18*
V	5.2	30	10.58±0.19*	9.12±1.13*	0.25±0.04*	0.60±0.05*	4.47±0.45*	5.43±0.54*	3.86±0.21*	14.45±0.56*	17.96±0.22*
VI	5.2	20	10.19±0.18*	10.09±0.78*	0.31±0.05*	0.73±0.06*	5.41±0.58*	6.31±0.50*	4.54±0.24*	15.30±0.54*	17.16±0.20*
VII	5.2	10	9.80±0.16	11.12±0.84	0.43±0.06*	0.83±0.07*	6.65±0.53*	7.44±0.51*	5.38±0.18	16.62±0.50	16.11±0.23
VIII	5.2	5	9.32±0.18	12.19±1.34	0.48±0.07	0.94±0.06	7.73±0.44	8.83±0.53	6.31±0.21	17.93±0.54	15.09±0.20

a: mmoles of pyruvate formed/min/g tissue; b: mmoles of formazon formed/min/g tissue; c: mmoles of NAD converted to NADH/min/g tissue; d: mmoles of NADPH converted to NADP/min/g tissue; e: mmoles of inorganic phosphorus formed/min/g tissue; f: mmoles of p-nitrophenyl formed/min/g tissue.

Values are mean±SEM of 5 animals.

* Significant $P \leq 0.05$ compared to control.

Table 26.6: Dose and Durational Effect of Phosphamidon on Epididymis Dehydrogenase and Phosphatase Enzymes Activity in Mice

Group	Treatment		Enzyme Activity mmoles/min/g tissue						
	Dose (mg)	Duration (days)	LDH[a]	SDH[b]	Na^+-K^+ ATPase[c]	Mg^{++} ATPase[c]	Ca^{++} ATPase[c]	ACP[d]	AKP[d]
I	Control	30	13.51±0.31	12.62±0.43	8.75±0.28	10.46±0.35	7.08±0.22	13.83±0.55	14.18±0.26
II	1.3	30	14.03±0.26	11.88±0.36	8.12±0.23	9.97±0.33	6.61±0.25	12.98±0.47	14.86±0.30
III	2.6	30	14.98±0.25	11.03±0.32	7.34±0.26	9.22±0.30	6.02±0.24	12.18±0.34	15.43±0.28
IV	3.9	30	16.14±0.24*	10.13±0.42*	6.43±0.22*	8.36±0.28*	5.42±0.23*	11.41±0.30*	16.10±0.26*
V	5.2	30	17.32±0.28*	9.08±0.44*	5.28±0.26*	7.24±0.26*	4.10±0.28*	10.43±0.30*	17.04±0.30*
VI	5.2	20	16.23±0.31*	10.10±0.41*	6.16±0.22*	8.16±0.28*	5.14±0.22*	11.21±0.48*	16.28±0.26*
VII	5.2	10	15.04±0.30	10.94±0.36*	7.14±0.23	9.06±0.33*	5.98±0.26	12.00±0.33	15.54±0.28*
VIII	5.2	5	13.98±0.28	11.98±0.34	8.04±0.26	10.94±0.32	6.54±0.23	12.96±0.54	14.81±0.30

a: mmoles of pyruvate formed/min/g tissue; b mmoles of formazon formed/min/g tissue; c: mmoles of inorganic phosphorus formed/min/g tissue; d: mmoles of p-nitrophenyl formed/min/g tissue.

Values arc mean±SFM of 5 animals.

*: Significant $P \leq 0.05$ compared to control.

activity of dehydrogenase, phosphatase and steroidogenic enzymes, except in 2.6 mg dosage group and 10 days duration treatment with 5.2 mg phosphamidon treatment decreased the activity of 17β-HSD significantly when compared with those of the control mice (Table 26.6).

Study on the activity of dehydrogenase and phosphatase enzymes in epididymis revealed that, treatment with 3.9 and 5.2 mg dosage for 20 days duration with 5.2 mg phosphamidon, caused significant decrease in the activity of SDH, Na^{+}-K^{+}ATPase, Mg^{++}ATPase, Ca^{++}ATPase and ACP enzymes, whereas the activity of LDH and AKP were increased significantly. However the mice treated with 1.3 and 2.6 mg dosage and 5 and 10 days duration treatment with 5.2 mg phosphamidon caused no significant change in the phosphatase and dehydrogenase enzymes activity whereas SDH and Mg^{++}ATPase activity were decreased significantly when compared with those of the control mice (Table 26.7).

Table 26.7: Dose and Durational Effect of Phosphamidon on Serum Testosterone Level in Male Mice

Groups	*Treatment*		*Testosterone Level (ng/ml serum)*
	Dose (mg)	*Duration (days)*	
I	Control	30	53.99±2.15
II	1.3	30	34.54±7.33
III	2.6	30	23.63±5.10*
IV	3.9	30	12.78±6.30*
V	5.2	30	0.672±8.53*
VI	5.2	20	16.75±11.12*
VII	5.2	10	30.83±10.05*
VIII	5.2	5	46.12±6.31

Values are mean±SE M of 10 animals.

* Significant $P \leq 0.05$ compared to control.

Testosterone Concentration

Phosphamidon decreased the serum testosterone levels in 2.6, 3.9 and 5.2 mg dosage and 10 and 20 days duration exposure with 5.2 mg phosphamidon treatment. However, treatment with 1.3 mg dose and 5 days duration treatment with 5.2 mg phosphamidon showed no significant change in the serum testosterone levels when compared with those of control mice.

Discussion

Body Weight, Testis and Accessory Reproductive Organs Weight

This study is first to show that sub-acute oral exposure of a known environmental toxicant phosphamidon with broad spectrum of activity, to a male adult mice altered reproductive functions by decreasing the paired testicular mass and accessory reproductive organs weight inhibiting testicular androgenesis by decreasing concentrations of testosterone. Paired testicular mass, a valuable index of reproductive toxicity in male animals (Aman, 1982) decreased in phosphamidon treated animals and this decrease in testicular mass was consistent with elimination of germ cells (Nakai *et al.*, 1992; Narayan *et al.*, 2000). Similarly, rats exposed to different pesticides showed reduced weight of the accessory reproductive organs due to decrease in the testosterone levels which might be due to lowered sperm density and degenerative changes in the testis and decreased protein quantity (Stocker, *et al.*, 2000; Pant and Srivastava, 2003; Yu *et al.*, 2004; Poon *et al.*, 2004; Narayan *et al.*, 2005). Many of the reproductive toxicants have primary effects on the testis, which potentially overshadow effects downstream on the efferent ductus and epididymis. Reports on male reproductive effects by earlier investigators reveal the testicular toxicity reflected by different ways such as biochemical effects, histopathology, impaired neurotransmitter metabolism and hormonal imbalance (Krause and Homola, 1974; Goldman *et al.*, 1994; Usha *et al.*, 2003; Ryan *et al.*, 2004; Sharpe and Irvine, 2004; Narayana *et al.*, 2006; Amina *et al.*, 2007).

Histological and Histometric Changes

In the present histometric study with high dosage and prolonged duration of exposure to phosphamidon resulted significant decrease in the number and diameter of spermatogenic and Leydig cells, whereas, phosphamidon exposure for lower dose and shorter duration treatment groups showed no significant change in the spermatogenic and Leydig cells. The histopathological changes in the testis were observed in the higher dose and prolonged duration of exposure to phosphamidon. The major defects were marked reduction in spermatogenic cells, formation of giant cells, vacuoles, lumen with tissue debris and clumped sperms and Leydig cells with deformed condition, cellular degeneration and finally lead to tubular atrophy (Hess and Nakai, 2000). Cellular degeneration was evident in terms of halo appearance, nuclear pyknosis, and giant cell formation. However, such effects were not seen in lower dosage and shorter duration treatment groups. The mechanism was akin to the one seen with other chemicals (Sarkar *et al.*, 2000; Okazaki *et al.*, 2001; Narayana *et al.*, 2005; 2006; Prashanti *et al.*, 2006).

The observed results indicated that high dose and prolong exposure of phosphamidon affect the spermatogenesis showing antispermatogenic and anti-androgenic property as reflected by the effect on number and diameter of spermatogenic and Leydig cells, testes and accessory reproductive organs weights as testicular steroids influences spermatogenesis and function of accessory reproductive organs (Stocker *et al.*, 2000; Okazaki *et al.*, 2001; Latchoumycandane *et al.*, 2002; Poon *et al.*, 2004; Narayana *et al.*, 2005; 2006; Prashanti *et al.*, 2006). The other possibility might be due to germ cells apoptosis and chromosomal damage resulting into decreased number of germ cells and formation of giant cells (Akbarsha and Shivaswamy, 1997; Debnath and Mandal, 2000; Usha *et al.*, 2003; Prashanti *et al.*, 2006) or due to deprived level of androgens mediated through the gonadotropins of the pituitary due to the effect on hypothalamus (Goldman *et al.*, 1990; Sarkar *et al.*, 2000). Maitra and Sarkar, (1996) and Sarkar *et al.*(2000) have observed decreased acetyl cholinesterase activity along with severe testicular damage in methyl parathion treated munias and quinolphos treated rats. Thus it is presumable that the testicular damage induced in the present study might have relation with decreased acetyl cholinesterase activity. It is unknown at present whether or not any effects of phosphamidon on the nerve supplying these reproductive organs mediates the structural changes observed in the present study. However, further investigation is needed in this regard.

Testosterone Levels

A dose and duration dependent decrease in serum concentrations of testosterone in phosphamidon treated mice may be due to the inhibition of 3β-HSD and 17β-HSD, a testicular androgenic enzymes activities, because these enzymes are responsible for the regulation of testosterone biosynthesis (Ghosh *et al.*, 1980; Jana *et al.*, 2005). The decrease in the levels of serum testosterone concentration may be explained by the fact that the phosphamidon treatment might have caused decrease in the release of gonadotropins, LH and FSH by its direct effect on the anterior pituitary or hypothalamus which might have lead to decreased secretion of testosterone (Ellis *et al.*, 1982) or due inhibition of acetyl cholinesterase activity as phosphamidon a well known inhibitor of AChE resulting in alternations in the pituitary gonadotropins which could have influenced the gonadal function directly through the effect on the pituitary AChE by affecting androgen levels.

Biochemical Studies

DNA, RNA, Proteins and Glycogen and Cholesterol

The response of the Testis and epididymis for nucleic acid (DNA and RNA), Total proteins, and glycogen contents of mice were significantly decreased and cholesterol content significantly increased when compared to control mice. In the present study, it has been found that with increasing dose and prolong exposure to phosphamidon caused decrease in the levels of DNA, RNA, proteins and glycogen. Thus, the observed results in the levels of DNA, RNA, proteins and glycogen in testis and epididymis may be due to decrease in mitotic index and disturbed cell division (Flessel *et al.*, 1993; Topaktas *et al.*, 1996) or due to oxidative stress as a result of which synthesis of macro molecules are reduced as different types of DNA lesions (Debnath and Mandal, 2000; Ranjbar *et al.*, 2002; Sharma *et al.*, 2005; Pina-Guzman *et al.*, 2006) and altered normal metabolic process and caused oxidative damage to important macromolecules such as DNA, RNA, proteins and lipids and lead to physiological attritions (Eun-Sun Hwang and gun-Heekin, 2007) or due to its effect on CNS (Central Nervous System) which might have suppressed the brain's release of gonadotropic hormones, FSH and LH through their effect on acetyl cholinesterase activity (Lyons, 1969; Flessel *et al.*, 1993) or due to inhibition of DβH (Dopamine-β-hydroxylase) and release of GnRH gonadotropin (Maj and Vetulani, 1969; Prezewlocka *et al.*, 1975) thereby affecting the production of gonadal steroids. The synthesis of RNA is potentially influenced by testosterone or dehydrotestosterone (Coffey *et al.*, 1968). Therefore, the observed results in the levels of DNA, RNA, proteins and glycogen in testis and epididymis under the influence of phosphamidon treatment in mice may be due to genotoxic action of phosphamidon as reported by Behera and Bhunya (1987) and Saxena *et al.* (1997) or effect on hormones which are essential for the regulation of DNA and RNA synthesis which in turn influence protein synthesis (Hamilton, 1968) or lower content of protein and glycogen content was possibly due to direct effect of pesticide phosphamidon on protein and glycogen metabolism or due to enhanced proteolytic activity as a consequence of increased metabolic demands following exposure to the toxic stress or increased catabolism to meet the energy demand of the animal under stress induced by pesticide (Ivanova-Chemishanka *et al.*, 1971). Thus the effect was reflected by testicular damage and biochemical synthesis in *albino* mice. In the present study the increase in testicular concentration of total cholesterol level is almost an index of reduced steroidogenesis (Dorfman, 1963). The increase in cholesterol in testis and high sudanophilic lipid accumulation in the interstitial tissue of drug treated mice supports the inhibition of androgen production which may be due to inadequate availability of LH which is essential to carry out steroidogenesis in the Leydig cells (Hansson *et al.*, 1973).

Enzyme Activities

3β-HSD and 17β-HSD plays a key role, as these are the prime enzymes in testicular androgenesis (Jana and Samanta, 2006). The reduced activities of the important enzymes of steroidogenesis observed in the present study employ that toxic action of phosphamidon was exerted by impaired steroidogenesis in the Leydig cells and Sertoli cells of the seminiferous tubules (Shivanadappa and Krishnakumari, 1983) or due to inhibition of certain pathways of steroidogenesis (Sujata and Chaitra, 2001; Sarkar *et al.*, 2003) or may be a result of low levels of FSH and LH as these two gonadotropins are prime regulator of testicular androgenic enzymes activities (Shaw *et al.*, 1979) or might be due to high affinity of binding of pesticide (Zarh *et al.*, 2002) causing testicular toxicity either by its direct action on reproductive system or by indirect action through neuroendorine system as reflected by impaired spermatogenic disorder by the diminution in the number of spermatogenic cells.

Testicular hormones are known to regulate the activity of LDH and SDH in testis (Srivastava and Vijayan, 1996). The rise in the activity of LDH and reduced activity of SDH in the present study may be due to the effect of pesticide on carbohydrate metabolism in the tissue as indicated by decrease in SDH activity (Preidkalns and Weber, 1968) as this enzyme is related with high metabolic activity such as absorption and secretion (Padykula, 1952). The rise in LDH activity in tissue suggested high turnover of pyruvate to lactate and vice-versa to yield required energy to overcome pesticide induced metabolic stress (Pant and Srivastava, 2003) or due to impaired steroidogenesis as these enzymes are regulated by testosterone (Kuladeep *et al.*, 2006) or due to deterioration of germinal epithelium of seminiferous tubule and depletion in the number of spermatogenic cells (Srivastava and Vijayan, 1996; Pant *et al.*, 2004).

It has been found that increase in dose and prolong exposure to phosphamidon caused significant decrease in the activity of ATPases in the testis and epididymis. Similar findings have been reported in rats treated with organophosphate pesticides such as methyl parathion and parathion (Basha and Nayeemmunnisa, 1993; Blasiak, 1995). It has been reported that carbamate pesticide thiram inhibited Na^+-K^+ATPase in rat testis (Mishra *et al.*, 1998). Similarly, it has been reported in earlier studies in gonads and other tissues of pesticide treated rats (Seeth *et al.*, 1976; Srivastava *et al.*, 1978). Therefore, the reduced activity of Na^+-K^+ATPase, Mg^{++}ATPase and Ca^{++}ATPase in testis and epidiymis may be due to mitochondrial disorganization as reported during pesticidal toxicosis (Pardini *et al.*, 1980) which might have caused inhibition of ATPases in testis and epididymis or due to Pesticide induced effect on cell membrane because of their strong affinity for interaction with number of lipids (Antuner-Mudeira and Madeir, 1978) causing inhibition of membrane bound ATPase enzymes by affecting enzyme complex (Kinter *et al.*, 1972; Rauchova *et al.*, 1995) or due to inhibition of testicular androgenesis as testosterone is known to stimulate the Na^+-K^+ATPase pump in many tissues including brain (Fraser and Swanson, 1994) which lead to impaired gonadal function.

ACP and AKP are the sensitive functional indicators of the reproductive status of animal and AKP is associated with transport of metabolites across the cell membranes (Rackallio, 1970). Both AKP and ACP serve as markers for androgen action in the target organs (Neima and Karmano, 1963; Mann, 1964; Mann *et al.*, 1981). In the present study, with increase in dose and prolong exposure to phosphamidon caused decrease in the activity of ACP and increase in the activity of AKP in testis and epididymis. Similarly, decreased and increased activity of ACP and AKP in liver, testis and serum were also reported by methyl parathion, diazinon, monocrotophos treated rats and mice (Narayana *et al.*, 2006; Prashanti *et al.*, 2006). ACP activity is androgen dependent and its activity has been shown to rise and fall with the levels of androgen (Stafford *et al.*, 1949). Alkaline phosphatase (AKP) is the characteristic enzyme of male accessory sex organs as a whole and its distribution differs from that of acid phosphatase (ACP) and both are sensitive functional indicators of the reproductive status of the animal (Ghosh *et al.*, 1980). Both the enzymes serve as markers for androgen action in the target organ (Mann and Lutwak, 1982).

Thus, the observed results in the present study may be due to absorptive or secretory surface of the cell membrane causing cell damage hence by reducing the activity of ACP and elevated AKP activity as an adaptive rise to the persistent stress (Abraham and Wilfred, 2000; Pant and Srivastava, 2003) or due to testicular degeneration, which may likely be a consequence of suppressed testosterone and indicative of lytic activity (Kaur *et al.*, 1999) or due to inhibition of testicular androgenesis as these enzymes are androgen dependent (Ghosh *et al.*, 1980; Mann and Lutwak, 1982).

In conclusion, the present results suggests that phosphamidon is a reproductive toxicant in males affecting the synthesis of protein, DNA, RNA, glycogen, cholesterol, steroid enzymes, cell membrane and lysosomal enzyme activities resulting in the testicular degeneration by causing cell death in the testis. Further, the loss of gonadal macromolecular constituents DNA, RNA and protein may be due to increased lysosomal activity (catabolism) or independent of the reduced mitochondrial and microsomal activities or increased catabolism of the biomolecules to meet the enhanced energy demand of the animals under stress on their reduced function of various biochemical enzymes. It appears, therefore, that phosphamidon treatment produces degenerative changes in spermatogenic cells by inhibiting the production of androgen acting primarily at the level of pituitary by inhibiting the release of FSH and LH causing testicular toxicity either by its direct action on reproductive system or by its indirect action through neuroendocrine system as reflected by impaired spermatogenic disorder by the diminution in the concentrations of testosterone, number of spermatogenic cells and physiological observations.

Acknowledgements

We thank UGC for financial assistance for sanction of (COSIST-II No. F-4-5/2000/2564/dated October 30, 2000) for the DRS Department of Zoology, Karnatak University, and Dharwad. The authors express their sincere thanks to Chairman, Post-Graduate Department of Studies in Zoology, Karnatak University, Dharwad for providing research facilities.

References

Abdollahi M, Ranjabar A, Shadina S, Mikfar S, Razaice A. 2004. Pesticide and oxidative stress: a Review. *Med. Sci. Moni.*, 10: RA 144-147.

Abell LL, Levy BB, Brodie, BP, and Kendal FE, 1952. Simplified method for estimation of total cholesterol in serum and demonstration of its specificity. *J. Biol. Chem.*, 195: 357-361.

Abraham P, Wilfred G. 2000. Lysosomal enzymes in the pathogenesis of carbon tetrachloride induced injury to the kidney and testis in the rat. *Indian J. Pharmacol.*, 32: 250-251.

Akbarsha MA, and Sivaswamy P. 1997. Apoptosis in male germinal line cells of rat *in vivo* caused by phosphamidon. *Cytobios.*, 91: 33-34.

Aman RP. A 1982. Critical review of methods for evaluation of spermatogenesis from seminal vesicle characteristics. *J. Androl.*, 2: 37-58.

Amina T, Faraq Ahmed F, El-Aswad and Nasra A Shaban 2007. Assessment of reproductive toxicity of orally administered technical dimethoate in male mice. *Reproductive Toxicology*, 23(2): 232-238.

Antuner-Mudeira M, Madeir VMC. 1987. Partition of malathion in synthetic and native membranes. *Biochem. Biophys Acta.*, 901: 61.

Averal HI, Stanley a, Akbarsha MA. 1996. Apoptic death of male germinal line cells of rat caused by vincristinc light and electron microscopic study. *Biochem. Letter*, 52: 171-180.

Basha PM, and Nayeemunnisa1993. Effect of methyl parathion on Na^{+}-K^{+} and Mg^{2+} adenosine triphosphatase activity in developing central nervous system in rats. *Ind. J. Exptl. Biol.,* 31: 785.

Behera BC, Bhunya SP. 1987. Genotoxic-potential of an organophosphate insecticide phosphamidon (dimecron): an *in vivo* study in mice. *Toxicol. Letter,* 37(3): 269-77.

Bergermeyer HV and Bernt E. 1963. In: Methods of enzymatic analysis (Ed) Bergermeyer HV. Academic Press. Weinheim NV and London. 837.

Blasiak J. 1995. Inhibition of erythrocyte membrane (Ca^{2+}+Mg^{2+}) ATPase by organophosphorus insecticides parathion and methyl parathion. *Comp. Biochem. Physiol Pharmacol Toxicol. Endocrinol.,* 110-119.

Bockelheide K, Neely D, Sioussat TM. 1989. The sertoli cell cytoskeleton: A target for toxicant induced germ cell loss. *Toxicol. Appl. Pharmacol.,* 101: 373-389.

Carrol NV, Lungchy RW, Raw RH. 1956. Glycogen determination in the liver and mode by use of anthrone reagent. *J.Biol.Chem.,* 25: 583-593.

Coffey DS, Shimaraki T, and Williams-Ashman HG. 1968. Polymerization of deoxyribonucleotides in relation to androgen induced prostatic growth. *Arch. Biochem., Biophys.,* 124: 184

David Plummer, 2001. An introduction to practical biochemistry, 3rd Edn. Tata McGraw Hill Publishing Company, New Delhi.

De Krester DM, Kerr JB. 1994. The cytology of the testis, In: Knobil E, Neil JD. (Eds). The physiology of reproductive. Vol. 2, Raven Press, New York, pp: 1177-1240.

Debnath D, Mandal TK. 2000. Study of quinolphos (an environmental oestrogenic insecticide) formulation (Ekalux 25 E.C) induced damage of the testicular tissues and antioxidant defense system in Sprague-Dawley albino rats. *J. App. Toxociol.,* 20: 197-204.

Dikshith TSS, Datta KK. 1972. Pathologic changes induced by pesticides in the testes and liver of rats. *Exp. Pathol.,* 7: 309-316.

Dorfman RI. 1963. Anti-androgen in a castrated mouse testis. *Steroid,* 2: 185.

Dym M, Fawcett DW. 1971. Further observations of the number of spermatogonia, spermatocytes, and spermatids connected by intercellular bridges in the mammalian testis. *Biol. Reprod.*, 4: 195-215.

Ellis GB, Desjardins C. 1982. Male rats secrete luteinizing hormone and testosterone episodically. *Endocrinol.*, 110: 1637-1648.

Ellis LC. 1970. Radiation effects. In: Johnson AB, Gomes WR, and Vandermark NL Ed.The testis; Vol. III, New York: Academic Press, 333-76.

Eun-Sun Hwang, and Gun-Heekim. 2007. Biomarkers for oxidative stress status of DNA, lipids and proteins *in vitro* and *in vivo* cancer research. *Toxicology*, 229: 1-10.

Fawcett DW, Neaves WR, Flores MN. 1973. Comparative observations on inter tubular lymphatics and the organization of interstitial tissue of mammalian testis. *Biol. Reprod.*, 9: 500-532.

Flessel P, Quintana PJE. and Hooper K. 1993. Genetic toxicity of malathion: a review. *Environ. Mol. Mutagen*, 22: 7-17.

Foster PMP, Sheard CM, and Lloyd SC. 1986. Dinitrobenzene: a sertoli cell toxicant? Excerpta. *Med. Int.Cong. Ser.*,716: 281-8.

Fraser CL, and Swanson RA. 1994. Female sex hormones inhibit volume regulation in rat brain astrocyte culture. *Am. J. Physiol Cell Physiol.*, 267: 909-914.

Georgellis A, Purvinen M, and Rydstrom J. 1989. Inhibition of stage-specific DNA, synthesis in rat spermatogenic cells by polycyclic aromatic hydrocarbons. *Chem. Biol. Interact.*, 72: 79-92.

Ghosh PK, Sarkar M, Ghosh P, Ghosh AG, Ghosh P. 1980. Effect of lithium chloride on spermatogenesis, testicular D^5,3β and 17β-hydroxysteroid, dehydrogenase activities in food (Bufo malanogastices). *Andrologia*, 21: 199-203.

Giwercman A, Skakkebaek NE. 1992. The human testis-an organ at risk. *Int. J. Androl.*, 15(5):373-5.

Goldman JM, Stocker JE, Cooper RL, McElory WK, and Hein JE. 1994. Blockade of ovulation in the rat by fungicide sodium N-methyl dithiocarbamate relationship between effects on the luteinizing hormone surge and alterations in hypothalamic catecholamines. *Neurotoxicology and Teratology*, 16: 257-268.

Goldman, J.M., Cooper, R.L., Laws, S.C. 1990. Chlordimeform induced alterations in endocrine regulation within the male rat reproductive system. *Toxicol. Appl. Pharmacol.*, 22: 467-72.

Hamilton TH. 1968. Control by estrogen of genetic transcription and translation. *Science*, 161: 649.

Hansson V, Ronesech E, Trygstod O, Ritzen EM, and French FS. 1973. Nature, *New Biol.*,246: 56.

Hess RA, and Nakai M. 2000. Histopathology of the male reproductive system induced by the fungicide benomyl. *Histol. Histopathol.*,15: 207-224.

Ivanova- Chemishanka L, Markov DV, Deshev G. 1971. Light and electron microscopic observation on rat thyroid after administration of some dithiocarbamates. *Environ. Res.*, 4: 201-202.

Jacques R and Bein HJ. 1960. Toxicology and pharmacology of new systemically acting insecticides of the phosphoric acid ester group phosphamidon. *Arch. Toxicol.*, 18: 316-330.

Jana K, Ghosh D, Samanta PK. 2005. Evaluation of single intratesticular injection of calcium chlorine on non-surgical sterilization of male goats (*Capra hiraus*): a dose-dependent study. *Anim. Reprod. Sci.*, 86: 89-108.

Jana K, Samanta PK. 2006. Evaluation of single intratesticular injection of calcium chloride for non-surgical sterilization in adult albino rats. *Contraception*, 73: 289-300.

Jarabak J, Adams JA, Williams-Ashman HG, Talalay P. 1962. Purification of a 17β-hydroxysteroid dehydrogenase of human placenta and studies on its transhydrogenase function. *J. Biol. Chem.*, 237:345-357.

Jinna RR, Uzodinma JE, and Desaiah, D. 1989. Age related changes in rat brain ATPases during treatment with chlordecone. J. Toxicol. *Environ. Hlth.*, 27: 199-208.

Kaur R, Dhanuja CK, Kaur K. 1999. Effect of dietary selenium on biochemical composition in rat testis. *Ind. J. Expt. Biol.*, 37: 509-511.

King J, 1965. In: Practical clinical enzymology. Edn. Van Z, Nor strand D. Co. London. p 83.

Kinter WB, Merkens LS, Janiki RH, Guarino AM. 1972. Studies on the mechanism of toxicity of DDT and PCBS; Disruption of osmoregulation in marine fish. *Environ. Hlth. Persps.*, 8: 169-173.

Krause W, and Homola S. 1974. Alterations of the seminiferous epithelium and the Leydig cells of the mouse testis after the application of dichlorovos (DDVP). *Bull. Env. Contam and Toxicol.*, 11: 429-433.

Kuladip Jana, Subanna Jana, and Prabhat Kumar Samanta, 2006. Effect of chronic exposure to sodium arsenite on hypothalamo-pituitary testicular activities in adult rats. Possible an estrogenic mode of action. *Reproductive Biol. and Endocrinol.*, 4: 9.

Latchoumycandane C, Chitra KC, Mathur PP. 2002. The effect of methoxychlor on the epididymal antioxidant system of adult rats. *Reprod. Toxicol.*, 16: 161-172.

Lowry OH, Rosenberg NJ, Far AL, Ronald RJ. 1951. Protein measurement with Folin phenol reagent. *J. Biol. Chem.*, 193: 265-378.

Lyons G. 1999. Endocrine disrupting pesticides. Pesticide News. 46: 16-19.

Maitra and Sarkar. 1996. Morphological study of the testes in relation to the brain and testicular acetyl cholinesterase activity in an organophosphate pesticide ingested wild passerine bird (*Lonchura malabarica*). *Folia Biologica*, 43: 143-149.

Maitra SK, and Sarkar R. 1996. Influence of methyl parathion on gametogenic and acetyl cholinesterase activity in the testis of white throated munia (*Bonchura malabarica*). *Arch. Environ. Contam. Toxicol.*, 30: 384-389.

Maj J, Vetulani J. 1969. Effect of some N, N-disubstituted dithiocarbamates on catecholamines level in rat brain. *Biochem. Pharmacol.*, 18: 2045-2047.

Mann T, and Lutwak C Mann. 1981. Epididymis and epididymal semen. In: Male reproductive function and semen. Springer-Verlag, Berlin. pp. 139.

Mann T, Lutwak-Mann C. 1982. Passage of chemicals into human and animal semen: Mechanisms and significance. *Crt. Rev. Toxicol.*, 11: 1-14.

Mann T, Lutwak-Mann C.. 1982. Passage of chemicals into human and animal semen: Mechanisms and significance. *Crt. Rev. Toxicol.*, 11: 1-14.

Mann T. 1964.Biochemistry of semen and the male reproductive tract. Mathuen and Ltd., London.

Mathew G, Vijayalaxmi KK, and Rahiman MA. 1992. Methyl parathion induced sperm shape abnormalities in mouse. *Mutation Res.*, 280: 169-173.

Meeker JD, Ryan L, Barr DB, Herrick RF, *et al.*, 2004 b. The relationship of urinary metabolites of carbonyl naphthalene and chlopropyrifos with human semen quality. *Environ. Health. Perspect.*,112: 1665-1670.

Meeker JD, Singh NP, Ryan L, Duffy SM. *et al.* (2004a). Urinary levels of insecticides metabolites and DNA damage in human sperm. *Hum Reprod.*, 19: 2573-2580.

Meistrich ML. 1986. Critical components of testicular function and sensitivity to disruption. *Biol. Reprod.*,34:17-28.

Mishra VK, Srivastava MK, Raizada RB. 1998. Testicular toxicity in rat to repeated oral administration of tetramethyl thiuram disulfide (Thiram). *Ind. J. Exptl. Biol.*, 36(4): 390-394.

Nachles MM, Morgulius SI, Sellirgman AM. 1960. Site of electron transfer to tetrazolium salts in succinoxidase. *J. Biol. Chem.*, 235: 2739.

Nakai M, Hess RA, Moore BJ, Guttriff RF, Strader LF and Linder RE. 1992. Acute and long-term effects of a single dose of the fungicide carbendazim (methyl 2-benzimididazole carbamate) on the male reproductive system in the rat. *J. Androl.*, 13: 507-518.

Narayan K, D'Souza VJA, Sanyal AK and Rao KPS. 2000. 5-Fluorouracil (5-FV) induces the formation of giant cells and sloughing of seminiferous epithelium in the rat testis. *Indian J. Physiol. Pharmacol.*, 44: 317-322.

Narayana K, Prashanthi N, Nayanatara A, Ganesh Kumar S, Harish H, *et al.*, 2006. A broad-spectrum organophosphate pesticide *O,O*-dimethyl *O*-4-nitrophenyl phosphorothioate (methyl parathion) adversely affects the structure and function of male accessory reproductive organs in the rat. *Environ. Toxicol. Pharmacol.*, 22 (3): 315-324.

Narayana K, Prashanti N, Narayanatara A, Kumar HH, Abhilash K, Bairy KL. 2005. Effect of methyl parathion (*O, O*-dimethyl *O*-4-nitrophenyl phosphorothioate) on rat sperm morphology and sperm count, but no fertility, are associated with decreased ascorbic acid level in the testis. *Mutation Res.*, 588: 28-34.

Niemi M, and Kormano, 1963. Cyclical changes and significance of lipids and acid phosphatase activity in the seminiferous tubules of the rat testis. *Anat. Rec.*, 151:156.

Okazaki K, Okazaki S, Nishimura A, Tsudo T, Katamura Matayama K, Nakamura A *et al.*, 2001. A repeated 28-day oral dose toxicity study of methoxychlore in rats, based on the enhanced OECD test guideline 407 for screening endocrine disrupting chemicals. *Arch. Toxicol.*, 75: 513-521.

Padungtod C, Savitz DA, Oversheat JW, Christiani DC, Ryan LM, Xu X. 2000. Occupational pesticide exposure and semen quality among Chinese workers. *J. Occup. Environ. Med.*, 42: 982-992.

Padykula HA. 1952. The localization of succinic dehydrogenase in tissue sections of the rat. *Amer.J.Anat.*, 91: 107.

Pant N, Murty RC, and Srivastava SP. 2004. Male reproductive toxicity of sodium arsenate in mice. *Human Exp. Toxicol.*, 23: 399-403.

Pant N, Srivastava SP. 2003. Testicular and spermatotoxic effects of quinolphos in rats. *J. Appl. Toxicol.*, 23: 271-274.

Pardini RS, Heidken JC, Baker TA, Payme B. 1980. Toxicology of various pesticides and their decomposition products on mitochondrial electron transport. *Arch. Environ. Contam. Toxicol.*, 9: 87-97.

Pina-Guzman B, Solis-Heredia MJ, Quintarilla-Vega B. 2005. Diazinon alters sperm chromatin structure in mice by phospharylating nuclear protamines. *Toxicol. Appl. Pharmacol.*, 202: 189-198.

Pina-Guzman B, Solis-Heredia MJ, Rojas-Garcia AE, Vriostegui-Acosta M, Quintanilla-Vega B. 2006. Genetic damage caused by methyl parathion in mouse spermatozoa is related to oxidative stress. *Toxicology and Applied Pharmacology*. 216(2): 216-224.

Poon R, Rigden R, Chu I, Valli VE. 2004. Short-term oral toxicity of plantyl ether, 1,4-diethoxybutane, and 1,6-dimethoxyhexane in male rats. *Toxicol. Sci.*, 77: 142-150.

Prashanti M, Narayana K, Narayanatara A, Chandrakumar HH, Bairly KL, D'Souza JA. 2006. The reproductive toxicity of the organophosphate pesticide *O, O*-dimethyl-*O*-4-nitrophenyl phosphorothioate (methyl parathion) in the male rat. *Folia Morphology*, *65*: 309-321.

Prezewlocka B, Sarnek J, Szmielski A, Niewiakomsha A. 1975. The effect of some dithiocarbamic acids on dopamine-β-hydroxylase and catecholamines level in rats brain. *Pol. J. Pharmacol. Pharm.*, 27: 555-559.

Rackallio J. 1970. Enzyme histochemistry of wound healing. *Prog. Histochem. Cytochem.*, 1: 1.

Ranjbar A, Pasalar P, Abdollahi M. 2002. Induction of oxidative stress and AchE inhibition in organophosphorus pesticide manufacturing workers. *Hum. Exp. Toxicol.*, 21: 79-182.

Rauchova H, Ledvinkova J, Kalous M, Drahota Z. 1995. The effect of lipid peroxidation on the activity of various membrane-bound ATPases in rat kidney. *Int. J. Biochem. Cell Biol.*, 27: 251.

Ryan T, Goad John T, Good Bassam H, Atich Ramesh C Gupta, 2004. Carbofuran-induced endocrine disruption in adult male rats. *Toxicology Mechanisms and Methods*, 14: 233-239.

Sanchez-Pena LC, Reyes BE, Lopez-Carrillo L, Reciol R, *et al.*, 2004. Organophosphates pesticide exposure alters sperm chromatin structure in Mexican agricultural workers. *Toxicol. Appl. Pharmacol.*, 196: 108-113.

Sarkar R, Mohankumar KP and Chowdhury M. 2000. Effects of organophosphate pesticide quinolphos on the hypothalamo-pituitary gonadal axis in adult male rats. *J. Reprod. Fertility.*,118: 29-38.

Sarkar R, Ray Choudhari G, Chattopadhyay A, Biswas NM. 2003. Effects of sodium arsenite on spermatogenesis, plasma gonadotropins and testosterone in rats. *Asian J. Androl.*, 1: 27-31.

Saxena S, Ashok BT, Musarraf J. 1997. Mutagenic and genotoxic activities of four pesticides, captan, foltaf, phosphomidon and furdan. *Biochem. Mol. Biol. Int.*, 41(6): 1125-1136.

Schneider WC. 1957. Determination of nucleic acids in tissues by pentose analysis, In: Sowick SP, Kaplan ND, (Eds), methods in enzymology, Academic Press, New York, pp 680-684.

Sciffer S, Dayton S, Novic B, Myntiyer E. 1950. The estimation of glycogen with anthrone reagent. *Arch. Biochem.*, 25: 191.

Seth PK, Srivastava SP, Agarwal DK and Chandra SV. 1976. Effects of di (2-ethylhexyl) phthalate (DEHP) on rat gonads. *Environ. Res.*, 12: 131-138.

Sharma Y, Bashir S, Irshad M, Gupta SD, Dogra TD. 2005. Effects of acute dimethoate administration on antioxidant status of liver and brain of experimental rats. *Toxicology*, 209: 49-57.

Sharpe RM, and Irvine DS. 2004. How strong is the evidence of a link between environmental chemicals and adverse effects on human reproductive health? *Brit. Med. J.*, 328: 447-451.

Shaw MJ, Georgopopouls LE, Payne AH. 1979. Synergistic effect of FSH and LH and testicular D^5-3β-hydroxysteroid dehydrogenase isomerase. Application of a new method for the separation of testicular compartments. *Endocrinology*,104: 912-918.

Shivanandappa T, and Venkatesh S. A. 1997. Colorimetric assay method for 3βHydroxy, Δ^5-steroid dehydrogenase. *Anal. Biochem.*, 254: 57-61.

Shivanandappa T, Krishnakumari MK. 1983. Hexachlorecyclohaxane induced testicular dysfunction in rats. *Acta. Pharmacol. Toxicol.*, 52: 12-17.

Srivastava P, Vijayan E. 1996. Testicular lactate dehydrogenase and sorbitol dehydrogenease activity after intratesticular injection of dymorphin and morphin in male rats. *Ind. J. Expt. Biol.*, 34: 363-365.

Srivastava SP, Agarwal DK, Mustaq M, and Seth PK. 1978. Effect of di (2-ethylhexyl) phthalate (DEHP) on chemical constituents and enzymatic activity of enzymatic activity of rat lifer. *Toxicology*, 11: 271-275.

Srivastava TG. 2002. Enzyme linked immunosorbent assay for steroid hormones. In Orientation training course on research methodology in reproductive biomedicine, New Delhi: *National Institute of Health and Family Welfare*, 55-58.

Stafford RO, Rubinsten JN, and Meyer RKJ. 1949. Effects of testosterone propionate on phosphates in the seminal vesicle and prostate of the rat. *Proc. Soc. Exp. Biol. Med.*, 71:353.

Stoker TE, Laws SC, Guidici DI, Cooper RL. 2000. The effect of atrazine on puberty and thyroid function in male wistar rats. An evaluation in the protocol for the assessment of pubertal development and thyroid function. *Toxicol. Sci.*, 58: 50-59.

Sujatha R., Chitra KC, Latchoumycandave C, Mathur PP. 2001. Effect of lindane on testicular anti oxidant system and stroidogenic enzymes in adult rats. *Asian J. Androl.*, 3: 135-138.

Topaktas M, Rencizogullari E, Ila HB. 1996. *In vivo* chromosomal aberrations in bone marrow cells of rats with marshal. *Mutation. Res.*, 371: 259-264.

Usha RA, Raja M Rathore, Monalisa Mishra, 2003. Role of vitamin C and lead acetate induced spermatogenesis in Swiss albino mice. Environ. *Toxicol. Pharmacology*, 13: 9-14.

Waulkes MP, Ward JM, Liu J, Diwan BA. 2003. Transplacental carcinogenicity of inorganic arsenic in the drinking water: induction of hepatic, ovarian pulmonary and adrenal tumors in mice. *Toxicol Appl. Pharmacol.*, 186: 7-17.

Yu WJ, Lee BJ, Nan SY, Ahn B, Hong JT, Do JK, Kim YC, Lee YS and Yun YW. 2004. Reproductive disorders in pubertal and adults phase of male rats exposed to vinlozalin during puberty. *J. Vet. Med. Sci.*, 66: 847-853.

Zarh JA, Bruschweiler BJ, Schlatler JR. 2002. Azole fungicides affect mammalian steroidogenesis by inhibiting sterol 14α-demethylase and aromatase. *Environ. Health. Perspect.*, 111: 255-261.

Chapter 27

Active Immunization Against Riboflavin Carrier Protein Prevents from Pregnancy Establishment in Sub-Human Primate (*M. radiata*)

Mukesh Kumar*

Department of Zoology, M.S.J. (Govt. P.G.) College,
Bharatpur, Rajasthan, India

ABSTRACT

Estrogen–inducible riboflavin carrier protein (RCP) obligatorily involved in yolk deposition of the water-soluble vitamin in the developing oocyte in the egg – laying hen is evolutionarily conserved in mammals including primate in terms of physicochemical, immunological and functional characteristics and serves as primary mediator of transplacental vitamin transport to support uninterrupted embryonic vitamin nutrition and hence its development. In order to understand the role of RCP in female fertility, proven fertile female bonnet monkeys were recruited and actively immunized against SDS–treated carboxymethylated RCP (300 µg/sc in alugel for priming and subsequent booster in dose regimen of 200 µg/sc in alugel). Blood samples were collected for detecting circulating

* E-mail: prof.mukeshkumar@rediffmail.com

antibody titers by ELISA and level of progesterone and estrogen by RIA.

Immunized animals elicited good circulating antibody titres. When these immunized females were mated (between day 9-16) with normal proven fertile males to assess their fertility performance, none of the female became pregnant. So far, 101 ovulatory cycles of these actively immunized female monkeys have been protected from the pregnancy establishment without affecting the general well being and normal cycle length. In conclusion, active immunization against SDS-RCM-RCP provides good protection from pregnancy development.

Keywords: *Riboflavin carrier protein, Active immunization, Female bonnet monkey, Fertility suppression.*

Introduction

Riboflavin carrier protein (RCP) an estrogen inducible protein obligatorily involved in oocyte deposition of vitamin in egg laying hen, is an evolutionarily highly conserved reproductive entity which, in mammals, mediates transplacental Flavin transport of this during gestation (Adiga *et al.*, 1988). The functional importance of this vitamin carrier in fetal development and hence pregnancy progression has been unequivocally demonstrated in rat and mice (Adiga and Murthy, 1983; Natraj *et al.*, 1987), immunoneutralization of endogenous maternal RCP leads to fetal wastage resulting into pregnancy termination without concomitant deleterious effect on vitamin status and general well being of the maternal system. The above phenomenon of carrier protein mediated vitamin delivery to subserve pregnancy establishment and subsequent fetal development appears to be conserved during further evolution of the species since such protein were subsequently detected by immunological and biochemical means and extensively characterized after purification in a variety of animals including sub human primates and human (Visweswariah and Adiga, 1987a,b; 1988).

Extensive evolutionary conservation of such a vital protein provided for the first time, a handle to use heterologous RCP as an antigen from cheap and abundant source like chicken egg to actively immunize the primate so that the effect of such a process on fetal growth and development can be monitored. It is anticipated that provided that high titre antibody to RCP capable of effective bioneutralization of endogenous vitamin carrier are produced, such a process should curtail vitamin supply to the primate embryo and should interfere with pregnancy establishment without any deterimental consequence on maternal heath well being. Therefore, present study was planned choosing female bonnet monkey as sub-human primate because of the close proximity to human (Srinath, 1979: Rao *et al.*, 1984) with the possibility to develop newer approach (Immuno) to regulate female fertility.

Materials and Methods

Purification of RCP from Chicken White

Riboflavin carrier protein (RCP) was purified to homogeneity from chicken egg white (Murthy and Adiga, 1978).

Preparation of Reduced Carboxy-methylated RCP

Reduced Carboxy-methylated RCP (RCM-RCP) was prepared (Konigsberg, 1972).

SDS – Treatment to RCP

RCM-RCP was dissolved in water to obtain 1 mg/ml stock solution. The solution was boiled with 1 per cent SDS for 10 min and then extensively dialysed against water. The protein contents were estimated (Lowry *et al.*, 1951).

Animals

The Sexually mature South Indian female bonnet monkey (*M. radiata*) were chosen for animal experimentation as a sub-human primate model during study being carried out. The animals were exclusively husbanded for laboratory investigation for fertility studies and guide lines set out by NIH, USA were generally followed with regard to husbandry practices in Primate Research Laboratory (PRL) at Indian Institute of Science [IISc], Bangalore, India.

Active Immunization with SDS – RCM – RCP in Bonnet Monkey

Each recruited female monkey was administered 300 µg of SDS – treated reduced Carboxy-methylated riboflavin carrier protein (in

Alugel) subcutaneously at 3 different sites on the dorsal surface and first three booster doses (200 µg/sc in Alugel) were given at monthly interval and subsequent boosters were given at 4 monthly intervals.

Determination of Circulating Antibody Titre by ELISA

To determine the immune response; ELISA was done with serum from immunized animal. Each well of a microtitre plate (Greiner) was coated with 1 µg of antigen (RCM – RCP) in 100 µl of phosphate buffer saline (PBS) pH 7.2 for 3 h at room temperature. The unoccupied sites were then blocked by filling the well with 0.5 per cent gelatin in PBS for 2 h at room temperature. The plates were washed thrice with PBS for 2 min each. The plates were incubated with 100 µl monkey serum from 1:100 to 1:51000 for 3 h at room temperature. Pre immune serum of corresponding dilution was used as a negative control. The plates were washed again as described earlier with PBS-Tween and PBS and then incubated with 100 µl (1:100) α-goat human IgG alkaline phosphates conjugate (Bangalore Genei) for 1 h. After washing the plates, the binding of antibody was quantitated using 200 µl of para–nitrophenyl phosphate (PNP) as a substrate. After incubation 20 min in the dark the reaction was stopped by adding 1 M NaOH. The antibody titres were expressed as the dilution of antiserum capable of binding 50 per cent of the immobilized antigen (1 µg/well).

Fertility Performance Experiment

When the circulating antibody titres were high, fertility performance of immunized female bonnet monkey was assessed by mating (between day 9-16) with normal fertile male monkey, the brief description of reproductive physiology of female bonnet monkey is as follow:

This primate model has got 28 days menstrual cycle, the initial 9-10 days constitute the follicular phase and rest the luteal phase. The pre-ovulatory surge of estrogen (E) occurs on day 9-11, the circulatory concentration of the estrogen vary from 0.2-1 µg/ml. This is followed by gonadotropin surge (FSH and LH) on day 10-12 and ovulation the luteal phase elevation of progesterone (P) reaches its maximum (4-8 ng/ml) on day 10 and falls to basal level on day 25 of the cycle. The fertile period of mating is between days 9-14 of the cycle. Implantation occurs on day 9 after fertilization (*i.e.* between day 22-24 cycle) following pregnancy establishment, both E an P

Table 27.1: Protection from Pregnancy in Bonnet Monkeys Immunized with SDS–Treated RCM-RCP

Monkeys No.	*Parameter*	*Sequential Ovulatory Cycle During Sept. 93 – May 94*									*Total No. of Cycles Protected so far*
		I	*II*	*III*	*IV*	*V*	*VI*	*VII*	*VIII*		
909	Ovulation*	+	+							To continue	2
	Progesterone@	+	+								
	Cycle (Days)	33	43								
945	Ovulation*	+	+	+	+	+	+	+		To continue	7
	Progesterone @		+	+	+	+	+	+	+		
	Cycle (Days)	35	30	30	32	34	36	32			
946	Ovulation*	+	+	+	+	+	+	+	+	To continue	8
	Progesterone@	+	+	+	+	+	+	+	+		
	Cycle (Days)	24	27	23	29	28	25	32	35		
956	Ovulation*	+	+	+	+	+	+	+	+	To continue	8
	Progesterone@	+	+	+	+	+	+	+	+		
	Cycle (Days)	26	27	27	26	31	27	31	48		
920	Ovulation*	+	+	+	+	+	+			To continue	6
	Progesterone@	+	+	+	+	+	+				
	Cycle (Days)	21	25	27	29	22	34				
941	Ovulation*	+	+	+	+	+	+	Died due to Diarrhoea		To continue	6
	Progesterone@	+	+	+	+	+	+				
	Cycle (Days)	21	27	25	53	26	36				

Contd...

Table 27.1–Contd...

Monkeys No.	*Parameter*	*Sequential Ovulatory Cycle During Sept. 93 – May 94*									*Total No. of Cycles Protected so far*
		I	*II*	*III*	*IV*	*V*	*VI*	*VII*	*VIII*		
4001	Ovulation*	+	+	+						To continue	3
	Progesterone@	+	+	+							
	Cycle (Days)	23	25	36							
4003	Ovulation*	+	+							To continue	2
	Progesterone@	+	+								
	Cycle (Days)	28	38								

* Normal Ovulation is assessed when (Day 8 – 10) E2 200 pg/ml Progesterone 2.5 ng/ml on days 18-20.

++ indicates persistent P after 26 (early pregnancy?).

Table 27.2: Protection from Pregnancy in Bonnet Monkeys Immunized with SDS–Treated RCM-RCP

Monkeys No.	Parameter	Sequential Ovulatory Cycle During Sept. 93 – Apr. 96																Total No. of Cycles Protected so Far
		I	*II*	*III*	*IV*	*V*	*VI*	*VII*	*VIII*	*IX*	*X*	*XI*	*XII*	*XII*	*XIV*	*XV*	*XVI*	
909	Ovulation*	+	+	+	+	+	+	+	+	+	+	+						11
	Progesterone@	+	+	+	+	+	+	+	+	+	+	+						
	Cycle (Days)	44	20	37	28	25	28	36	25	30	27	29						
956	Ovulation*	+	+	+	+	+	+	+	+	+	+	+	+	+	+	+	+	16
	Progesterone@	+	+	+	+	+	+	+	+	+	+	+	+	+	+	+	+	
	Cycle (Days)	32	24	28	34	34	32	26	36	29	33	29	28	27	27	31	36	
2	Ovulation*	+	+	+	+	+	+	+	+	+	+	+	+	+	+	+	+	16
	Progesterone@	+	+	+	+	+	+	+	+	+	+	+	+	+	+	+	+	
	Cycle (Days)	28	27	29	35	33	36	29	24	27	29	32	36	29	27	31	30	
4	Ovulation*	+	+	+	+	+	+	+	+	+	+	+	+	+	+	+	+	16
	Progesterone@	+	+	+	+	+	+	+	+	+	+	+	+	+	+	+	+	
	Cycle (Days)	37	20	28	29	33	26	28	31	34	32	30	31	36	29	28	29	

* Normal Ovulation is assessed when (Day 8 – 10) E2: 200 pg/ml.

@ Progesterone: 2.5 ng/ml on days 18-20.

++ indicates persistent P after 26 (early pregnancy?).

rise from day 20 onwards. The level of E increases further and are maintained high during early stage of pregnancy. Similarly P level during day 20-26 are clearly higher if pregnancy ensues compared to non-gravid cycle. The duration of pregnancy is 166±4.8 days and the outcome is a single baby. The colony pregnancy index is 80-90 per cent for the proven fertile and regularly cycling animals. After parturition, lactational amenorrhoea lasts 6-7 months. During summer 40 per cent of the females show summer amenorrhoea at which time fertility experiments are not conducted, the reproductive span of this primate is 5-17 years (Srinath 1979; Rao *et al.*, 1984).

Estimation of Progesterone and Estrogen by RIA

Estrogen and progesterone levels in maternal sera (collected from day 7-10 daily and from day 16 to 31 alternate days) were measured by specific radio immunoassay for the steroid. Briefly, aliquot of plasma (100 µl) were deprotinized with methanol (20 µl), steroid extracted with dimethyl ether (5 µl) and ether layers evaporated to dryness. Samples were reconstituted in PBS, pH 7.4 (1 µl) containing 0.1 per cent W/V gelatin (G-PBS). Unknown samples and various concentration (30-100 pg) of progesterone (1.9 – 1000 pg) of estrogen were taken in duplicate. After making up the volume to 300 µl with G-PBS to all standard and samples tubes, 100 µl or [^{3}H] – progesterone and estrogen (approximate 10,000 cpm) and 100 µl of 1:30,000 antiserum of progesterone and 1:10,000 antiserum of estrogen were added and incubated at 4°C for 3 h for P and E assay. Free and bound steroid were separated by adding 300 µl of dextran coated charcoal 11 per cent W/V charcoal and 0.1 per cent (W/V) dextran in PBS and kept on ice for 10 min. The samples were centrifuged at 4°C at 1600 g for 10 min and the supernatant for using toluene: methanol (1:1 v/v) containing 0.5 (w/v) PPO and 0.05 per cent POPOP as the scintillation fluid. A standard graph was drawn by using various known amounts of unlabelled progesterone and estrogen to inhibit labeled progesterone and estrogen and amount of progesterone and estrogen in unknown samples were calculated from standard curve.

Results and Discussion

Our study shows that active immunization against SDS-RCM-RCP elicits good circulating antibody titres. Our data also show that when these female monkeys were mated with normal proven fertile male to assess their fertility, these females were protected from pregnancy establishment. This reflects that riboflavin carrier protein (RCP) plays importance role in delivery of vitamin riboflavin from maternal system to developing embryo. The requirement of riboflavin carrier protein mediated delivery to embryo from maternal system is so essential that if carrier protein is immunologically neutralized endogenously it hampers the delivery of vitamin to developing embryo which ultimately results into failure of pregnancy establishment. As we know that in mammals fetoplacental unit obtains a continuous supply or nutrients from mother and accumulates against concentration gradients (Miller *et al.*, 1976). Riboflavin is one of the such nutrients, which is indispensable for normal embryonic growth and survival and its transport-requires a specific carrier protein from maternal system to the developing embryo across the placental barrier (Adiga *et al.*, 1986). Experimental evidences have shown that active and passive immunoneutralization of maternal RCP in rodents (Murthy and Adiga, 1982, Adiga and Ramanamurthy, 1983) results in early embryonic loss which emphasizes the importance of RCP in the transplacential for sustained delivery to support-embryonic growth but is dispensable for general well being of the mother (Adiga *et al.*, 1986; Adiga and Ramanmurthy, 1983).

Early studies done in our lab have shown that RCP is conserved from egg laying birds to mammals including primates despite marked differences in their pattern of embryonic development. It has been found that protein exhibiting immunological cross reactively with chicken RCP capable of binding riboflavin is detectable by radio-immunoassay in pregnancy sera of bonnet monkey (Visweswariah, 1987). The protein purified from primate exhibits physicochemical mobility, molecular weight and ligand binding very similar to chicken RCP (Visweswariah and Adiga 1988). Furthermore, the circulatory levels of bonnet monkey's RCP appears to be physiological modulated by estrogen status of the animals as in the case with avian and rodent systems, thus, for example, increase in the concentration of estradiol in adult female during the menstrual cycle and early pregnancy has very good correlation with enhanced serum RCP level, additionally, estradiol 17-β administered to both immature female and male monkey (with undetectable for RCP in blood) could bring about elevated level of RCP in both the sexes, however, male seems to respond relatively less efficiently than females in terms of the magnitude of the response (Viswesariah and Adiga, 1988).

In present study, SDS-denatured, unfolded (by reduction, carboxymethylation of disulphide bond) chicken RCP as the antigen for immunization has been used instead of RCP because earlier studies done in Prof. Adiga's lab have shown that immunogenic efficiency of chicken RCP in terms of eliciting bioneutralizing antibody in monkey and rat can substantially improve and pregnancy suppression can be achieved upto 100 per cent by active immunization. This shows that many of the native epitopic conformation at the surface chicken RCP may be rundant in terms of bioneutralization which can be accomplished by a restricted population of antibodies recognizing segmental epitopes (Adiga *et al.*, 1991). This was again substantiated by the present study in terms of response of high level of bioneutralizing antibody and protection from pregnancy development. It appears that interference with pregnancy may perhaps operates prior to the establishment of placental barrier, pregnancy, interference could even at the peri-implantation stage itself. The above hypothesis is supported by elegant *in vitro* experiment with zona denuded hamster eggs coincubated with sperm in the presence of mono-specific anti-cRCP-anti serum raised in rabbit and examined after 24 h and 48 h later, out of 20 eggs examined after 24 h. 15 eggs had disrupted plasma membrane, one eggs had intact plasma membrane but no pronuclei and 4 eggs had pronuclei (R.P. Das personal Communication). It appears that in the presence of adequate level of sequence specific anti-chicken RCP antibodies in the uterine lumen, the viability of the zygote may be jeopardized. Recent studies done in our lab shows that RCP is a component of sperm, oocyte and trophoblast which indicates new possibility that anti RCP antibodies may be effective in interfering with pregnancy even at the fertilization, post-fertilization or blastocyst stage itself.

From the above discussion it is clear that the RCP which is utilized by avian system as a remarkably conserved in primate and serves as primary mechanism of vitamin transport to developing embryo hence active immunization against RCP provides good protection from pregnancy development and this approach may be developed also as a female immunocontraceptive.

Acknowledgements

This study was done under the supervision of Late Professor P.R. Adiga, CRBME (Now renamed as MRDG), at IISc, Bangalore. Dr. Mukesh Kumar was awarded Research Associateship under DBT sponsored project. Cooperation of Primate Research Laboratory (PRL), IISc, Bangalore is acknowledged in conducting monkey's experiments.

References

Adiga, P.R., Karande, A.A., Visweswariah and Seshagiri, P.B., 1988. Estrogen-induced riboflavin carrier protein and its role in fetal development in progress in Endocrinology 11 pp. 343 – 348 (Eds. H. Imura, K Shizume and S. Voshida, Amsterdam Medica Excerpts).

Adiga, P.R. and Murthy, C.V.R., 1983. Vitamin carrier protein during embryonic development in birds and mammals in Molecular Biology of Egg. Maturation 98 pp. 111 – 136. (Eds. R. Porter and J. Whelan, Ciba Foundation Symposium, Landon Pitaman Books).

Adiga, P.R., Seshagiri, P.B. Malathy, P.V. and Sandhya, S.V., 1986. Reproduction specific vitamin carrier protein involved in transplacental vitamin transport primates. In Pregnancy Proteins in Animals (Edited by Han., J.) pp 317 – 329, Walter de Gruyter, Berlin.

Adiga P.R. Karande, A.A., Visweswariah, S.S. and Seshagiri, P.B., 1991. Carrier protein mediated transplacental riboflavin transport in the primates. In: Perspectives in Primate Reproductive Biology (Ed. N.R. Moudgal, K. Voushinaga. A. J. Rao and P.R. Adida) pp. 129 – 140. Wiley Eastern, New Delhi.

Jagannada Rao. A., Kotagi, S.G. and Moudgal, N.R., 1984. Serum concentration of chorionic gonadotropin, estradiol–17β and progesterone during early pregnancy in South Indian Bonnet Monkey (*Macaca radiata*). *J. Reprod. Fert.*, 70: 449 – 455.

Konigsberg, W., 1977. Reducing of disulphide bond in protein with dithiothereitol. *Meth. Enzymol.*, 25B: 185 – 188.

Lowry, O.H., Rosebrough, N.J. Farr. A.L. and Randall, R.Y., 1951. *J. Bio. Chem.*, 193: 265 – 275.

Murthy, U.S. and Adiga, P.R., 1978. Estrogen induced synthesis of riboflavin binding protein in immature chicks. *Biochem. Biophys. Acta.*, 558: 364 – 378.

Miller, R.K., Koszalk, T.R. and Brent, R.L., 1976. The transport of molecules across placental membranes. In: Poste, G. and Nicolson, G.L. eds. Cell Surt Rev. 1. Amsterdam: Elsvier Publication 145 – 207.

Murthy, C.V.R. and Adiga, P.R., 1982. Pregnancy suppression by active immunization against gestation specific riboflavin carrier protein. *Science,* 216: 191.

Natraj, U., Kumar, R. A. and Kadam, P., 1987. Termination of pregnancy in mice with antiserum to chicken riboflavin carrier protein. *Bio. Reprod.*, 36: 677-685.

R.P. Das Personal Communication.

Srinath, B.R., 1979. Husbandary and breeding of bonnet monkeys (*Macaca radiata*). In non-human primate models for study of human reproduction (Edited by T.C. Anand Kumar) S. Kargar Basel, pp. 17 – 22.

Visweswariah, S.S., 1987. Studies on riboflavin carrier protein: Physiochemical, biosynthetic and immunological aspects, Ph.D. Thesis, Indian Institute of Science, Bangalore, India.

Visweswariah, S.S. and Adiga P.R., 1987a. Purification of circulatory riboflavin carrier protein from pregnant bonnet monkey (M. Radiata. Comparison with chicken egg vitamin carrier. *Biochim. Biophys. Acta.*, 615.

Visweswariah, S.S. and Adiga P.R., 1987b. Isolation of riboflavin carrier protein from pregnant human and umblical cord serum. Similarities with chicken egg riboflavin carrier protein. *Biosci. Rep.*, 7: 563-571.

Visweswariah, S.S. and Adiga P.R., 1988. Estrogen modulation of riboflavin carrier protein in the bonnet monkey (*M. radiata*). *J. Steroid Biochem.*, 31: 916.

Previous Volumes

— Volume 1 —

2002, xiii+308p., figs., tabls., ind., 23 cm Rs. 650

ISBN 81-7035-272-X

Section I: Animal Ecology

Section II: Animal Reproduction

— Volume 2 —

2004, xi+387p., figs., tabls., ind., 23 cm Rs. 850

ISBN 81-7035-322-X

Section I: Animal Ecology

Section II: Animal Reproduction

— Volume 3 —

2006, xvii+443p., figs., tabls., ind., 23 cm **Rs. 1150**

ISBN 81-7035-424-2

Section I: Animal Ecology

Section II: Animal Reproduction

— Volume 4 —

2007, xiv+550p., figs., tabls., ind., 23 cm Rs. 1400

ISBN 81-7035-459-5

Section I: Animal Ecology

Section II: Animal Reproduction

— Volume 5 —

2008, xiv+476p., col. plts., figs., tabls., ind., 23 cm Rs. 1600

ISBN 81-7035-563-X; 978-81-7035-563-2

Section I: Animal Ecology

Section II: Animal Reproduction

— Volume 6 —

2010, xiv+497p., col. plts., figs., tabls., ind., 23 cm Rs. 1500

ISBN 81-7035-635-0; 978-81-7035-635-6

Section I: Animal Ecology

Section II: Animal Reproduction

Index

B

C

H